Nature Translated

Edinburgh Critical Studies in Literary Translation
Series Editors: Stuart Gillespie and Emily Wilson

The series reflects the current vitality of the subject, and will be a magnet for future work. Its remit is not only the phenomenon of translation in itself, but the impact of translation too. It also draws on the increasingly lively fields of reception studies and cultural history. Volumes will focus on Anglophone literary traditions in their foreign relations.

Published Titles

The English Aeneid: *Translations of Virgil, 1555–1646*
Sheldon Brammall

The Many Voices of Lydia Davis: Translation, Rewriting, Intertextuality
Jonathan Evans

Nature Translated: Alexander von Humboldt's Works in Nineteenth-Century Britain
Alison E. Martin

www.edinburghuniversitypress.com/series/ECSLT

Nature Translated

Alexander von Humboldt's Works in Nineteenth-Century Britain

Alison E. Martin

EDINBURGH
University Press

Edinburgh University Press is one of the leading
university presses in the UK. We publish
academic books and journals in our selected
subject areas across the humanities and social
sciences, combining cutting-edge scholarship
with high editorial and production values to
produce academic works of lasting importance.
For more information visit our website:
edinburghuniversitypress.com

© Alison E. Martin, 2018, 2020

Edinburgh University Press Ltd
The Tun – Holyrood Road
12(2f) Jackson's Entry
Edinburgh EH8 8PJ

First published in hardback by Edinburgh University Press 2018

Typeset in 10.5/13 Sabon by
Servis Filmsetting Ltd, Stockport, Cheshire

Croydon, CR0 4YY

A CIP record for this book is available from the
British Library

ISBN 978 1 4744 3932 9 (hardback)
ISBN 978 1 4744 3933 6 (paperback)
ISBN 978 1 4744 3934 3 (webready PDF)
ISBN 978 1 4744 3935 0 (epub)

The right of Alison E. Martin to be identified
as the author of this work has been asserted in
accordance with the Copyright, Designs and
Patents Act 1988, and the Copyright and Related
Rights Regulations 2003 (SI No. 2498).

Contents

List of Figures	vi
Acknowledgements	viii
Series Editors' Preface	x
List of Abbreviations	xi
Introduction	1
1. Styling Science	22
2. Dispute and Dissociation: John Black's *Political Essay on the Kingdom of New Spain* (1811)	40
3. 'A Colossal Literary and Scientific Task': Helen Maria Williams and the *Personal Narrative of Travels to the Equinoctial Regions of the New Continent* (1814–1829)	75
4. 'A Plain and Unassuming Style': Thomasina Ross and Humboldt's *Travels* (1852–1853)	117
5. The Poetry of Geography: The *Ansichten der Natur* in English Translation	150
6. *Cosmos*: The Universe Translated	187
Conclusions	233
Bibliography	243
Index	259

List of Figures

2.1 Alexander von Humboldt, *Political Essay on the Kingdom of New Spain*, trans. John Black, 4 vols (London: Longman, Hurst, Rees, Orme and Brown, 1811), II, p. 96. (Reproduced by permission of the author.)   67

3.1 Stipple engraving of Helen Maria Williams (London: Dean and Munday, 1816), The Carl H. Pforzheimer Collection of Shelley and His Circle. (Reproduced from Wikimedia Commons <https://commons.wikimedia.org/wiki/File:HelenMariaWilliams.jpg) (last accessed 7 September 2017).   77

3.2 Humboldt's handwritten feedback on Williams's draft translation of the *Personal Narrative*. Top: 'de grace faites disparaitre <u>godets</u> – n'est-ce pas <u>little cups</u>'. (Reproduced by permission of The Royal Archives, The Hague, G016-A439, nr. 95.)   97

3.3 'Plate 69: *Fucus vitifolius*', Alexander von Humboldt and Aimé Bonpland, *Plantes équinoxiales*, 2 vols (Paris: Schoell, 1808–9) II, unpaginated. (Reproduced by permission of The Bodleian Libraries, The University of Oxford, CR.L.35/1-2 (V.2), plate 69.)   98

3.4 Humboldt's feedback on Williams's draft translation of the *Personal Narrative*. Middle: 'c'est presque grappes'. (Reproduced by permission of The Royal Archives, The Hague, G016-A439, nr. 95.)   100

4.1 Front matter to Alexander von Humboldt, *Personal Narrative of Travels to the Equinoctial Regions of America, During the Years 1799–1804*, trans. Thomasina Ross, 3 vols (London: Bohn, 1852–3). (Reproduced by permission of the author.)   123

List of Figures   vii

5.1 Oil colour printing of Chimborazo, facing the title page of the *Views of Nature* (London: Bohn, 1850). (Reproduced by permission of the author.) 152

5.2 Photograph of Elise C. Otté, AR1438A, Hester Pengelly Collection, Torquay Museum. (Courtesy and Copyright of Torquay Museum.) 169

5.3 Facsimile of a letter of 20 October 1848 from Alexander von Humboldt to Henry George Bohn, printed at the end of the front matter, *Views of Nature* (London: Bohn, 1850). (Reproduced by permission of the author.) 171

5.4 Title page of Alexander von Humboldt, *Views of Nature: Or, Contemplations on the Sublime Phenomena of Creation; With Scientific Illustrations*, trans. Elise C. Otté and Henry G. Bohn (London: Bohn, 1850). (Reproduced by permission of the author.) 172

6.1 *Literary Gazette*, 10 March 1849, 1677 (1849), p. 162. (Reproduced by permission of The Bodleian Libraries, The University of Oxford, N. 269 d.7, page 162.) 188

Acknowledgements

This book has evolved from a post-doctoral research project (*Habilitation*) completed at the Martin-Luther-Universität (MLU) Halle-Wittenberg in 2013, and I particularly thank Sabine Volk-Birke for giving me the freedom to tackle such a broad, interdisciplinary subject. Daniel Göske, Vera Kutzinski, Jürgen Meyer, Werner Nell, Charlotte Sleigh and Joachim Whaley also gave unstintingly of their time to comment on individual chapters or even the whole manuscript. Susan Pickford read substantial parts of this study and offered characteristically swift, cheerful and expert advice. I am indebted to her for this, and for almost two decades of stalwart intellectual companionship.

I am grateful to colleagues at the Modern Languages Library of the MLU Halle-Wittenberg – particularly Ulrike Dorn, Angelika Hagenbruch and Susanne Wißmann – who carried several hundredweight of works to and fro in the course of this project. Staff at the following libraries helped considerably in numerous ways: the Leopoldina in Halle, the Universitätsbibliothek Leipzig, the Leibniz-Institut für Länderkunde in Leipzig, the University of St Andrews Library (particularly Liz Smith and Moira Mackenzie in Manuscripts), the British Library, the Rare Books Room of Cambridge University Library, Leeds University Library, University of Reading Special Collections (especially Verity C. Andrews), the Massachusetts Historical Society, the Royal Society Centre for History of Science, the John Murray Archive at the National Library of Scotland and the Geheimes Staatsarchiv Preußischer Kulturbesitz in Berlin. I particularly thank Barry Chandler, Curator of Collections at Torquay Museum, for enabling me to access archive material on Elise Otté. Much of my travel to foreign archives and libraries was financed by the *Frauenförderung* of the MLU Halle-Wittenberg, for which I am most grateful.

I also particularly wish to thank Ulrich Päßler for helping me to navigate the archives at the Alexander-von-Humboldt-Forschungsstelle in Berlin and enabling me to present my work at the conference for young Humboldt researchers, held at the Berlin-Brandenburgische Akademie der Wissenschaften in September 2010, under the aegis of Eberhard Knobloch. Peter Hulme and Marina Warner kindly invited me to present a paper at the British Academy conference on 'Alexander von Humboldt and America' in November 2009, subsequently published as '"These Changes and Accessions of Knowledge": Translation, Scientific Travel Writing and Modernity – Alexander von Humboldt's Personal Narrative', in *Studies in Travel Writing*, 15.1 (2011), pp. 39–51, which constitutes a much abridged version of Chapter 4. Feedback from a guest lecture I gave in October 2011 at the Institute of Germanic and Romance Studies, University of London, organised by Elinor Shaffer, greatly helped me sharpen points made in Chapter 3.

Stuart Gillespie's tenacity and encouragement when I first approached Edinburgh University Press (EUP) were essential in injecting a new dynamism into this project, and careful criticism of the manuscript from him and from Emily R. Wilson has been invaluable in honing it further. The two anonymous readers for EUP also gave me stimulating food for thought. Michelle Houston, Ersev Ersoy and James Dale at EUP, together with my vigilant copy-editor Wendy Lee, have all smoothed the passage from manuscript to book immeasurably.

Keeping me of good cheer during this long-haul voyage were Heike Heklau, Regine Klimpfinger, Ursula Mack, Ruth and Ernst-Heinrich Prinz, Annegret Rakers(†) and Julia Semmer.

My parents, Patricia and Ian Martin, have woven science firmly into my life. This study is testament to their own spirit of intellectual enquiry, rigour and determination: it draws together the threads that give our worlds meaning and purpose.

I could never have embarked upon this undertaking, still less completed it, without the generous, patient and unwavering support of my fellow traveller, Björn Zehner, over so many years. His constant appeals to relevance and readability were crucial in shaping the work I hope it has become.

Series Editors' Preface

Translators, Pushkin's 'post-horses of enlightenment', play a central role in every society's reception of other cultures. The study of translation – in theory, in practice and in relation to broader narratives in literary and cultural history – is now a vibrant scholarly field. It is key to current debates on literary canons in an increasingly global world, and on the possibility of World Literature. Edinburgh Critical Studies in Literary Translation addresses translation as a literary and historical phenomenon and is the first monograph series to do so.

Some of these studies engage with the approaches individual authors have taken to translation. Some deal with the impact of particular source texts or of particular translations on the societies in which they were produced. A central concern of the series is with interactions between translation and other forms of creative work, and with the part translation can play in forging the identity of individual authors. We are no less interested in the way translation can set directions for literary cultures at large.

There are no constraints on historical period. The emphasis of the series is in the first instance on translations involving the English language, whether in the context of ancient or modern literature. Our scholarly territory straddles the disciplines of English Literature, Classical Studies, Comparative Literature and Modern Languages. Contributors necessarily work at their frontiers, using innovative tools on interdisciplinary topics.

Stuart Gillespie and Emily Wilson

List of Abbreviations

AdN Alexander von Humboldt (2004), *Ansichten der Natur, mit wissenschaftlichen Erläuterungen*, ed. Anette Selg, Frankfurt am Main: Eichborn.

AN Alexander von Humboldt (1849), *Aspects of Nature, in Different Lands and Different Climates; with Scientific Elucidations*, trans. Mrs Sabine, 2 vols, London: Longman, Brown, Green and Longmans; and John Murray.

CO Alexander von Humboldt (1849–58), *Cosmos: A Sketch of a Physical Description of the Universe*, trans. E. C. Otté (and B. H. Paul [vol. 4] and W. S. Dallas [vol. 5]), 5 vols, London: Bohn.

CS Alexander von Humboldt (1846–58), *Cosmos: Sketch of a Physical Description of the Universe*, trans. under the Superintendence of Lieut.-Col. [from 1858 'Major-General'] Edward Sabine, London: Longman, Brown, Green and Longmans; and John Murray.

EP Alexandre de Humboldt (1811), *Essai politique sur le royaume de la Nouvelle-Espagne*, 2 vols, Paris: Schoell.

K Alexander von Humboldt (2004), *Kosmos: Entwurf einer physischen Weltbeschreibung*, ed. Ottmar Ette and Oliver Lubrich, Frankfurt am Main: Eichborn.

KP Alexander von Humboldt (1845–8), *ΚΟΣΜΟΣ: A General Survey of the Physical Phenomena of the Universe*, trans. by Augustin Prichard, M.D., M.R.C.S., 2 vols, London: Hippolyte Baillière.

PE Alexander de Humboldt (1811), *Political Essay on the Kingdom of New Spain*, trans. John Black, 4 vols, London: Longman, Hurst, Rees, Orme and Brown.

PNR Alexander von Humboldt (1852–3), *Personal Narrative of*

Travels to the Equinoctial Regions of America, during the Years 1799–1804, trans. Thomasina Ross, 3 vols, London: Bohn.

PNW   Alexander von Humboldt (1814–29), *Personal Narrative of Travels to the Equinoctial Regions of the New Continent, During the Years 1799–1804*, trans. Helen Maria Williams, 7 vols, London: Longman, Hurst, Rees, Orme, and Brown, J. Murray and H. Colburn.

R   Alexander von Humboldt (1814), *Researches Concerning the Institutions and Ancient Monuments of the Ancient Inhabitants of America, with Descriptions and Views of Some of the Most Striking Scenes in the Cordilleras!*, trans. Helen Maria Williams, 2 vols, London: Longman, Hurst, Rees, Orme and Brown, J. Murray and H. Colburn.

RH   Alexander von Humboldt (1970), *Relation historique du voyage aux régions équinoxiales du nouveau continent*, ed. Hanno Beck, 3 vols, Stuttgart: Brockhaus.

VN   Alexander von Humboldt (1850), *Views of Nature: Or, Contemplations on the Sublime Phenomena of Creation; With Scientific Illustrations*, trans. Elise C. Otté and Henry G. Bohn, London: Bohn.

Introduction

'All the while I am writing now my head is running about the Tropics,' wrote Charles Darwin to his sister Caroline from his student rooms in Cambridge in April 1831. 'I go and gaze at Palm trees in the hot-house and come home and read Humboldt: my enthusiasm is so great that I cannot hardly sit still on my chair.'[1] Early the next year, Darwin sailed into the harbour of Santa Cruz aboard the *Beagle*, picturing to himself 'all the delights of fresh fruit growing in beautiful valleys, & reading Humboldt's descriptions of the Island's glorious views'.[2] Shaping the young Darwin's scientific imagination was the work of the Prussian scientist and explorer Alexander von Humboldt (1769–1859), whose bold new vision of the natural world understood the forces of nature as a series of dynamic, interconnected systems. Humboldt was not interested in collecting isolated facts but in representing nature 'as one great whole, moved and animated by internal forces' (CO I, ix). Not only did his narratives revolutionise how people thought about their world. Their accompanying illustrations also brought scientific findings into visual dialogue with each other in ways that continue to inform our understanding of ecosystems today. Humboldt's account of his voyage to the Americas with the French botanist Aimé Bonpland between 1799 and 1804 inspired many others besides Darwin. The poet Robert Southey considered Humboldt so eminent that he was 'among travellers what Wordsworth is among poets' (Southey 1965: II, 231). To fellow scientists like the British botanist Joseph Dalton Hooker, Humboldt was, quite simply, a 'God' (Hooker 1918: II, 127).

But what had fascinated the young Darwin was not the rich prose of Humboldt's travel account in its French original, the *Relation historique du voyage aux régions équinoxiales du nouveau continent* (1814–25). It was that of the English translation, the *Personal Narrative of Travels to the Equinoctial Regions of the New Continent* (1814–29). The 'glorious views' of Santa Cruz were therefore not entirely of Humboldt's making.

They were also indebted to the linguistic creativity of this work's translator, the sentimental poet and radical writer Helen Maria Williams. The role played by translators – particularly women – in the development of science still remains a neglected field of study. Yet translation was central in extending the reach of those scientific disciplines swiftly emerging in (pre-)Victorian Britain. Humboldt's works began to appear in English editions in the first half of the nineteenth century, a period in which Latin, the *lingua franca* of the European male scholarly elite, was on the wane as the means of circulating scientific ideas beyond national borders. Translation rapidly gained momentum as men of science – and, increasingly, an ever-expanding audience of more general readers – awaited impatiently the transformation of a foreign work into a language they could read with ease.

Nature Translated contributes to our evolving understanding of this 'golden age' of translation by focusing on the cultural dimension of scientific knowledge transfer. It is well known that in this period scientists like Darwin travelled to the furthest corners of the globe, returning home with specimens, sketches and measurements that would transform how their scientific contemporaries comprehended the diversity of wildlife, habitats and climates on the earth. But we still know relatively little about how this knowledge 'travelled', whether on a linguistic level to reach new audiences speaking other languages, or on a material level as objects of print culture. Humboldt's works appeared in Britain in an astonishing range of rival translations, editions and formats, sold at markedly divergent prices and with rather different audiences in mind. This book sets out to demonstrate that translators did not just ensure that scientific knowledge generated on the Continent circulated in Britain and the Anglophone world. They were crucial in evaluating and explaining, criticising and censoring this knowledge in ways that influenced how both the material itself and its author were received by British readers. These translators, working together with illustrators, editors and publishers, set their stamp on these editions in ways that have influenced our understanding of Humboldt well beyond his lifetime. *Nature Translated* tells the story of these forgotten mediators who ensured that 'reading Humboldt' was a pastime enjoyed by the many, not the few, in nineteenth-century Britain.

Humboldt's major works appeared in their first English editions between 1811 and 1858, a period in which science 'transformed European understanding of the natural and human worlds and simultaneously transformed European modes of acting in those worlds' (Fulford, Lee and Kitson 2004: 5). The new 'scientists' – the word was coined by William Whewell in 1833 – considered themselves an innovatory and rapidly evolving species of knowledge-makers. As John Holmes and Sharon Ruston have emphasised, it was not only the scale of scientific achievement but also its significance as a distinctive discourse that established it 'as the hallmark of modern civilisation, with a cultural authority to rival the Church, the Classics and the arts' (Holmes and Ruston 2017: 2). Scholarship in the 1980s and 1990s – particularly the work of Gillian Beer (1983; 1996), Sally Shuttleworth (1981) and George Levine (1987) – highlighted the centrality of literary structures in Victorian scientific writing. More recently, numerous anthologies and critical introductions have amply demonstrated that in juxtaposing texts by literary authors with those by scientific writers we can productively investigate common and diverging patterns in their narrative techniques. Charlotte Sleigh's *Literature and Science* (2011) argues persuasively that 'science cannot be conducted without language, and language is not a neutral tool' (Sleigh 2011: 6). The distinction between 'objective' science and 'subjective' literature becomes particularly questionable if we reflect that science and literature 'remain human activities undertaken for very human motives' (Sleigh 2011: 3) and that the presentation of scientific endeavour is shaped by the passion, curiosity and wonder that motivate its practitioners.

Scholars of Humboldt's writing have not been slow to explore the literary dimension of his scientific narrative. Ottmar Ette's study *Literatur in Bewegung* [*Literature in Motion*] (2001) was one of the first works to investigate in more detail the literary qualities of scientific travel literature and to understand this as a genre that tests boundaries not only geographically and culturally, but also stylistically. In his first monograph devoted to Humboldt, *Weltbewußtsein* [*World-Consciousness*] (2002), Ette drew attention to the global dimension of the Prussian intellectual's writings, their intertexts and notional hyperlinks, and illustrated how they articulated a perpetually evolving, unfinished engagement with modernity. Appearing that same year, Nigel Leask's *Curiosity and the Aesthetics of Travel Writing, 1770–1840* (2002) likewise focused on the literary dimension of Humboldt's writing, while also locating it within the complex relationship between the natural sciences and Romanticism. Laura Dassow Walls's multidimensional account of the influence of Humboldt's work on the American Renaissance – notably Emerson, Thoreau, Poe and Whitman – examines how Humboldt's writing stresses the importance of process over product and that, growing 'organically from an inner impulse that arises ultimately from nature', it deploys language to create a vision in which science is combined with aesthetics, using the imagination to fuse these elements into a whole (Dassow Walls 2009: 223). More recently, Johannes Görbert's comparative study (2014) of the narrative techniques used by the German naturalist Georg Forster, the poet and botanist Adelbert von Chamisso, and Humboldt has distinguished between the presentation of scientific fieldwork 'outwards' to an international scientific audience, and the autobiographical dimension of these works that carry a strong 'inward' slant.

While literary scholars and historians of science have therefore eagerly embraced the notion of scientific writing as narrative, scholars of translation studies have been surprisingly reluctant to turn their attention to non-fictional writing in general, and scientific writing in particular. Yet Scott Montgomery's seminal work *Science in Translation* (2000) emphasises that knowledge transfer

> involves the transfer of certain powers: powers of expression in the case of literary or artistic knowledge; powers over the patterns and organization of life in the case of political, legal, or religious ideas; and, in the case of science, powers of imagination and practice with regard to the material world and uses of it. (Montgomery 2000: 2)

Scientific writing is a language that 'contains many words and expressions with multiple and potential meanings': scientific translators make important decisions about how to formulate ideas in another language and are 'potent actors in the globalisation of knowledge' (Montgomery 2010: 303–4).

Nicolaas Rupke rightly observes that the existence of translations of a given scientific text has generally been understood simply as a barometer of its wider success or of a scientist's standing on the international intellectual stage (Rupke 2000: 209). But as Rupke's examination of the Dutch and German versions of Robert Chambers's best-seller *Vestiges of the Natural History of Creation* (1844) demonstrates, the local circumstances underpinning the activity of translation are intimately connected to the meanings that scientific texts acquire in a different cultural, political and religious community. Understanding translation not as a derivative process, which generates a poor 'copy' of an authoritative 'original', but rather as a creative 'rewriting' in a different language has been essential in teasing out the complexities of translation as a social practice. As André Lefevere (1992) has indicated, such rewritings are produced in the service, or under the constraints, of certain ideological motivations and poetological currents, and different interpretative communities influence the production of such rewritings in different ways. Brian Nelson and Brigid Maher cogently explain that translation is an extremely complex activity 'involving a multiplicity of exact choices about voice, tone, register, rhythm, syntax, echoes, connotations and denotations, the colour and texture of words' (Nelson and Maher 2013: 1, 3). Translators remould a text with a different set of interests and norms in mind. Together with printers and publishers, they determine which texts circulate internationally and which do not. As Marwa Elshakry and Carla Nappi point out, translation has, over centuries, influenced in very practical ways 'the historical transmission, preservation, and even innovation of ideas and practices of science' (Elshakry and Nappi 2016: 372). Translation therefore marks what Walter Benjamin sees as 'the stage of continued life' for any given work, lending it the potential to endure rather than vanish, to develop its own history and to engage with the new (Benjamin 1992: 73).

For nineteenth-century naturalists in Britain, translation and the problems raised by language lay at the heart of concerns to reform science. Bernard Lightman argues that by reading continental scientific publications they would encounter different visions of science, as well as the ground-breaking ideas of French and German scientists whose work represented some of 'the most advanced scientific research' of the period (Lightman 2015: 411). Moreover, by showing greater interest in seeing their own work in translation, they could engage actively in strategies to export their own ideas across the globe. In his *Reflections on the Decline of Science in England* (1830), Charles Babbage, Lucasian Professor of Mathematics at Cambridge, attacked Britain's apparent indifference to Continental science. He blamed monoglot complacency, nurtured by British insularity, for the decline in scientific achievement in Britain. While 'everything, even moderately valuable in the scientific publications of this country [...] is instantly repeated, verified and commented upon, in Germany, and we may add too, in Italy', he claimed, such practices were not found in Britain (Babbage 1830: viii). Indeed, he continued, 'whole branches of continental discovery are unstudied, and indeed almost unknown' in Britain, a nation that was 'fast dropping behind' (Babbage 1830: viii).

Humboldt – highly proficient in French, English and Spanish, besides his native German – put many British scientists to shame. They were notoriously weak at foreign languages, having been schooled rigorously in Latin and Greek but rarely in any modern tongues, with the exception of French. While Darwin did reconcile himself to ordering some of Humboldt's more technical works in the original, since no translation seemed forthcoming – the *Fragmens de géologie et de climatologie asiatiques* [*Fragments of Asiatic Geology and Climatology*] (1831) would be a case in point – he preferred to procure his works in English translation, where possible.[3] Letters written while on board the *Beagle* show him pestering his family back home to acquire volumes of the

Personal Narrative (in the translation by Williams) to update his set of Humboldt's works, the first two volumes of which had been presented by the British botanist John Stevens Henslow 'to his friend C. Darwin on his departure from England upon a voyage around the World 21 Septr 1831'.[4] While French was a language of which gentleman scientists had at least a passive reading knowledge, German, an emerging scientific language in nineteenth-century Europe, was more problematic. 'I would give £100 if I knew German decently,' admitted the geologist Charles Lyell (K Lyell 1881: I, 249). Darwin, writing to the German physician Ludwig Büchner in 1862, noted with characteristic self-effacement, 'I am, like so many Englishmen a very poor proficient at Languages, & German is to me, excepting the simplest descriptions, extremely difficult.'[5]

By casting scientific communication as 'knowledge in transit', James Secord (2004) has drawn attention to the power of translation to intervene in the knowledge-sharing process at the heart of scientific endeavour. If we focus on the conduits through which scientific theories and findings gained broader significance, we can also disentangle a network of people, places and institutions that were, to a greater or lesser degree, gatekeepers to that knowledge. More recently, Kapil Raj has suggested that by making circulation our main line of enquiry, we can move beyond essentialist notions of science as a 'system of formal propositions or discoveries' towards understanding better the 'construction, maintenance, extension, and reconfiguration of knowledge' (Raj 2013: 341). Mobility was central to Humboldt's world as a scientific traveller, and he was himself a nodal figure in myriad conduits of knowledge, whether scientific, geopolitical or economic. As Nicholas Jardine, James A. Secord and Emma Spary emphasise, scientific knowledge was never generated by isolated individuals 'working wholly within the domain of the mind', but was always 'the product of conglomerates of people [. . .] all linked by a range of practices of different kinds' (Jardine, Secord and Spary 1996: 8). Humboldt was the ultimate 'go-between': a mediator who operated in spaces of encounter and negotiation, and enabled complex and contested knowledge flows to take place (Schaffer, Roberts, Raj and Delbourgo 2009: xiv–xix). Indeed, the transdisciplinary, relational nature of Humboldt's thought has been productively visualised as a *perpetuum mobile* in which knowledge was generated as much through his own peripatetic movements around the globe as through the intersecting, interlinking and realigning of ideas (Ette 2009: 27).

Many of Humboldt's ground-breaking ideas were, of course, to be found in scientific papers translated in specialist British journals, but it is on the best-selling book-length works that this study focuses. His first major publication to appear in English was the two-volume *Essai politique sur le royaume de la Nouvelle-Espagne* (Paris: Schoell, 1811), translated by John Black as the *Political Essay on the Kingdom of New Spain* (London: Longman, 1811). While it was avidly read by economists, political scientists and anthropologists fascinated by the insights it gave into Spain's South American colonies, the *Essai* also articulated Humboldt's democratic convictions and his impassioned criticism of colonial exploitation.

Humboldt's next work to appear in English was the extensive (if essentially unfinished) account of his voyage to the Americas, the *Relation historique du voyage aux régions équinoxiales du nouveau continent*, put into English by Helen Maria Williams as the seven-volume *Personal Narrative*, that earned him wider recognition among the international scientific community as he worked from his adopted home of Paris. Some thirty years later, Thomasina Ross would revive and repackage Williams's translation in an abridged version (1852–3) for the publisher Henry Bohn in his recently launched 'Scientific Library' series.

Humboldt's short essay collection, the *Ansichten der Natur* – initially published in 1808, revised before Humboldt left Paris for Berlin in 1826 and, finally, modernised in 1849 – appeared in English in two different editions that openly competed with each other. The first was by Elizabeth Sabine, wife of the Irish explorer and geophysicist Colonel Edward Sabine, and appeared as *Aspects of Nature* (1849) with Longman and Murray; the second, by the linguist and marine biologist Elise C. Otté, was published one year later as *Views of Nature* with Bohn.

The fierce competition between the rival publishing houses of Bohn and of Longman and Murray was sustained as the Anglophone editions of *Kosmos*, Humboldt's best-known work today, gradually emerged volume by volume on to the British market. Essentially conceived as a work in progress, it remained incomplete on Humboldt's death in 1859, when he was midway through the fifth volume. A challenging book to read, it was just as challenging to translate. Three versions nevertheless appeared in English. The first, by the Bristol eye-surgeon Augustin Prichard, working for the London publisher Hippolyte Baillière (1845–8), covered only the first two volumes and quickly disappeared from public memory. It was followed by two more comprehensive editions, notably Elizabeth Sabine's translation for Longman and Murray (1846–58) and Otté's version for Bohn (with the aid of Benjamin H. Paul and William S. Dallas in volumes four and five, respectively), which appeared between 1849 and 1858.

Nicolaas A. Rupke's highly readable *Alexander von Humboldt: A*

Metabiography (2007) explores the different 'lives' that Humboldt and his work have acquired over time as the Prussian savant has been clothed in different ideological garbs. By the end of the nineteenth century Alexander and his brother, the statesman Wilhelm, had become central to Germany's self-portrayal as a *Kulturnation*, a nation of intellectual accomplishment on a par with classical Greece (Rupke 2007: 31, 69–72). Some three decades letter, the National Socialists used the universality of *Kosmos* to reinforce German claims to supremacy in Europe. In post-war East Germany, Marxists made of the young scientist a man of the people concerned to improve conditions for miners with his designs for a safety lamp, thus carefully downplaying his aristocratic title of Alexander 'von' Humboldt (Rupke 2007: 86–91, 121). Meanwhile, queer theorists have avidly taken up Humboldt's early close friendships with men and later his long-standing relationship to his valet Johann Seifert as examples of his homosexuality, thus neatly explaining the enigma of his lifelong bachelorhood (Rupke 2007: 199). Most recently, Humboldt has gone global as his 'cosmopolitics' have seen him increasingly drawn into environmental and ecological debates.

Rupke suggests that if all these various Humboldts were brought together, seated round a table, and the 'real Humboldt' asked to stand up, all would rise to their feet, or none (Rupke 2007: 210). This image is helpful in visualising how the various English translations cast Humboldt in different ways according to the agenda of their individual translators, editors and publishers. This study does not argue that the avatars of Humboldt created by his British translators are figures as radically different as those whom Rupke sketches across a 150-year span of reception history. Nor did the divergences present nineteenth-century British readers with irreconcilably refracted images of Humboldt. But it is none the less essential to realise that the changes of tone and register and the omissions and additions that restyled his writing in English translation cast Humboldt as a scientist, traveller and author in subtly different lights.

Over the past two decades, important work has been undertaken into the international reception of Humboldt's writing that is fundamental to the present study. The bibliography painstakingly compiled by Horst Fiedler and Ulrike Leitner (2000) of Humboldt's writing – not just in its original forms, but also its later editions and, above all, its translations into French, German, English, Spanish and a host of other languages besides – provides the authoritative account of who translated Humboldt's writing, when and for which publisher. As Humboldt scholars have started to ask more searching questions about the nature of these translations, the circumstances surrounding their production and their (lack of) faithfulness to the original, this has generally heightened awareness that these translations are mediated texts. While Mary Louise Pratt, in her study *Imperial Eyes: Travel Writing and Transculturation* (1992), discusses Humboldt's use of language in the *Relation historique* by quoting from the first English translation (Pratt 1992: 111–43), with astonishing disregard for Williams's own intervention in the text, other researchers, notably Nigel Leask (2002) and Jason Lindquist (2004) have shown greater caution.

More recently, research by Vera Kutzinski and Ottmar Ette has cast fresh light on how Humboldt was translated for nineteenth-century American and German audiences. Focusing on John Sidney Thrasher's translation of Humboldt's *Essai politique sur l'île de Cuba* (1826), which was published as *The Island of Cuba* with the New York publishers Derby and Jackson thirty years later, Kutzinski (2009) has examined how Thrasher manipulated and restructured the text, primarily by omitting Humboldt's highly critical observations on slavery, drawn first-hand from his visits to a sugar plantation near Havana, but also by changing the title. A translation that Humboldt famously 'disowned', this is an arresting example of how his work was radically repositioned by a translator who did not even work from the original French, but produced a relay translation based on José de Bustamanté's Spanish edition. Thrasher's 'infamous defacement', to borrow Kutzinski's succinct formulation (Kutzinski 2009: 303) – Fernando Ortiz (2009) speaks more caustically of 'mutilation' and 'falsification' – is an example of abridgement that fundamentally rethought the source text on a political and stylistic level. While Thrasher's reorientation of Humboldt's writing to address a rather different agenda serves as a stark example of the translator's transformational powers, these did not always have a negative effect. Ette, investigating the abridged four-volume German translation (1859–60) of Humboldt's narrative of his travels to the Americas by Herman Hauff, has explored how such 'surrogates' can be vital in enhancing the popularity of a given account (Ette 1996: 111).[6]

The British editions of Humboldt's writing offer rich fodder for scholars of book history and translation studies precisely because of the double translation of three of Humboldt's main works – the *Relation historique*, the *Ansichten der Natur* and *Kosmos*. Jeremy Munday proposes that a close comparison of different translations of the same work can be extremely productive in revealing repeated patterns indicative of an individual translator's style and ideology (Munday 2008: 8). These stylistic differences help us to gain a purchase on why these competing editions could exist side by side on the nineteenth-century British book market and ensure that Humboldt's writings percolated through to different sectors of society that engaged with his works in different ways. The recent sociological turn in translation studies (Wolf and Fukari 2007; Milton and Bandia 2009) has focused attention on understanding translation as a social practice carried out by translators who are themselves 'agents' in the translation process. By contextualising the situations in which translations are produced and received, and analysing the working environment of these translators and their role and status in society, we can investigate more fully how these realities influence the translation process and the implications they have for transfer situations. While those involved in the technologies of publishing and information dissemination – Jonathan Topham's neatly termed 'technicians of print' (2004) – are beginning to be granted more attention by historians of science, there is still much to be done if we are to understand the role of translators and translation in this process. Maeve Olohan's discussion (2014) of the emerging methodological parallels between the history of science and of translation highlights the need to turn away from a sole focus on canonical figures – in essence the figures credited with making great scientific breakthroughs – towards studying those far more marginalised people, including their translators, who played a significant part in bringing these scientific endeavours to prominence.

Gender, Science, Translation

The role played by women – still highly marginal figures in historical accounts of science – is a case in point. 'It is curious to observe the native eloquence of Humboldt struggling with the encombrance [sic] of all the sciences,' wrote Lucy Aikin to her mineralogist brother Edmund in 1818, on reading the *Personal Narrative* in Williams's translation. 'Did ever mortal man study so many *ologies*, or travel with so many *ometers*!' she exclaimed, before praising Humboldt for his 'magnificent passages of description' (Aikin 1864: 114). While women like Aikin were increasingly becoming consumers of science through their reading, they were also establishing a role as energetic, unashamedly public, producers of scientific literature. The most striking example of women's engagement with science in the nineteenth century is surely Mary Shelley's *Frankenstein* (1818/1831). But Patricia Fara reminds us that women had been engaged less visibly in science over centuries as wives, sisters and patrons of men of science, acting as their translators, collectors, editors and illustrators, and thus 'playing indispensable parts in achieving the results for which their men became renowned' (Fara 2004: 10). Well into the Victorian era, these women were still struggling against the pervasive cultural categories of women/nature and men/mind, which curbed their ability to speak with authority on scientific matters. Nevertheless, introductory works like Jane Haldimand Marcet's *Conversations on Vegetable Physiology* (1829), textbooks like Margaret Bryan's *Comprehensive Astronomical and Geographical Class Book* (1815) and broad-ranging studies such as Isabella Gifford's *The Marine Botanist* (1848) indicate that, by mid-century, women had fashioned a visible role for themselves in science. And as Rebekah Higgitt and Charles W. J. Withers (2008) demonstrate, the very success of organisations such as the British Association for the Advancement of Science (BAAS) was dependent on the presence of ladies to create a 'sociable' milieu within the emerging domain of public science.

Translations of scientific texts still offer rich, largely untapped, resources for exploring a different dimension of women's engagement with the making of scientific knowledge. Ground-breaking work in the 1980s and 1990s by Pnina Abir-Am and Dorinda Outram (1987), Patricia Phillips (1990), Ann B. Shteir (1996; also 2003) and Barbara Gates (1998) identified and subtly teased out the contribution of women, many previously unknown, to Victorian science as writers of elementary scientific textbooks and educators, but their role as translators was largely overlooked. The title of Michèle Healy's dissertation, *The Cachet of the 'Invisible' Translator: Englishwomen Translating Science (1650–1850)*, speaks volumes about why even prodigious female scientific translators such as Elizabeth Sabine, Ada Lovelace and Mary Somerville long went unnoticed.[7] An extensive amount of close textual analysis and archival digging is often required to unearth the significant scientific contribution that such women made.

This book explores how women gradually removed or circumvented barriers to deploy, as Londa Schiebinger has termed it, 'their culturally defined difference as a platform for critique' (Schiebinger 1987: 307). Joy Harvey's account of Darwin's notorious French translator, Clémence Royer, who appended to the French edition of Darwin's *Origin of Species* extensive footnotes and 'a startling anticlerical preface' (Harvey 1997: 64), is an unusual example of female visibility in a translation. Kathryn A. Neeley's study of the mathematician and astronomer Mary Somerville has shown that other women were more deferential, diffident or diplomatic. Somerville's 1831 translation of Pierre Simon Laplace's *Traité de mécanique céleste* (1798–1825), which described the laws of nature operating in the heavens, saw her approach the text under a series of different guises, sometimes giving a 'transparent' translation of Laplace's account, sometimes guiding the reader by resolving apparent contradictions or simply explaining the strategy Laplace was using (Neeley 2001: 97). More recently, Mary Orr has emphasised, through her example of the English explorer, taxidermist and translator Sarah Bowdich Lee, how the 'exchange, circulation and advancement of new ideas [...] always depended on multi-lingual, and intra-lingual, re-articulation' (Orr 2015: 28): historical research has hitherto ignored how women's multilingual proficiency enabled them to make significant independent contributions to the transmission of scientific findings.

This study is wary of presenting Humboldt's female translators as proto-feminists blazing a trail through the domain of 'male' science. Formative statements of feminist translation theory – notably Lori Chamberlain's 1988 essay on 'Gender and the Metaphorics of Translation' – tended to operate with binary categories of dominant versus submissive behaviour, reflecting male versus female authorial practices. These ill reflect the fluid nature of these categories and the ambivalent relationship that women had towards translation. While some women were constrained by their spouses, editors and publishers to adopt a submissive attitude, Susanne Stark has shown that others, like Sarah Austin, were self-reflective intellectuals that actively chose to secure for themselves a place 'behind the welcome defence of inverted commas' from which they could develop successful careers as literary professionals (Stark 1999: 39).

Almost all of the texts by Humboldt discussed here are travel accounts or engage in some way with the genre of travel writing. The recent burgeoning of academic interest in travel literature has seen new investigations of an extremely inclusive genre that lies somewhere 'between aesthetic and functional forms of writing: the "literary" and the "non-literary"' (Thompson 2011: 3). Precisely because it is 'not hermetically sealed off from other forms of writing' (Korte 2000: 8), this generic indeterminacy has allowed it to enquire into myriad aspects of human experience and convey these experiences using a range of narrative techniques. This has also made it a non-fictional genre particularly accessible by women writers and translators. Judith Johnston's account of Victorian women active both as travellers and as translators is illuminating in highlighting the significant contributions figures like Jane Percy Sinnett, Anna Jameson and Sarah Austin made to contemporary debates about political reform, nationhood and identity in Britain, acting as 'engineers of culture and cultural exchange' (Johnston 2013: 8).

Other accounts of women as linguistic mediators of travel literature have offered more fine-grained analyses of how translation was key to their own self-positioning. Susan Pickford's 2012 discussion of the Norwich translator Anne Plumptre's English rendering of François Pouqueville's *Voyage en Morée, à Constantinople, en Albanie et dans plusieurs autres parties de l'empire Othoman* [*Travels through the Morea, Albania, and Several Other Parts of the Ottoman Empire*] (1805; trans. 1813) amply demonstrates that women translators were not automatically low in social capital, but could be agents in significant intellectual networks who benefited greatly from the stimulating surroundings in which they lived and worked. In line with Mirella Agorni's call to examine the 'specificity of women's contribution to literary production' and their webs of connection and communication (Agorni 2005: 818–19), this study investigates other ways of thinking about women, power and translation. It explores how the relationship of these women translators to their editors, publishers and, indirectly, to their readers enhanced or limited their ability to be significant agents in furthering scientific knowledge. Equally importantly, it examines how they deployed scientific translation for their own intellectual ends and, with varying levels of success, developed polymathic profiles to mirror the encyclopaedic range of Humboldt's own writing.

Popularising Knowledge

'Two or three years my senior, Sam, like myself, was acquiring a taste for books,' recalled the Northumberland miner Thomas Burt of his life in the 1850s. 'Sam's more solid reading was in science, especially in astronomy and geology [. . .]. Sam's favourite books at this time were Alison's *History of Europe* and Humboldt's *Cosmos*' (Burt 1924: 124). That Samuel Bailey, a coalminer like Burt, should have got hold of a copy of Humboldt's *Cosmos* (in whichever English translation), let alone declared its contents to be among his preferred reading, might initially seem surprising. It is evidence, though, of the extent to which science had become more accessible in Britain by mid-century and of Humboldt's success (due in no small part to his translators) in cultivating a readable, engaging writing style that made this work attractive as much to the lay person as to the scientific specialist.

As Alan Rauch notes, 'direct access to knowledge, through popular, cheap, and readable texts, became a central factor in both the production of knowledge and the structuring of social order' (Rauch 2001: 14). The gradual emergence of the public library in mid-nineteenth-century Britain was, for many, a godsend. Rapidly expanding literacy and access to print culture fuelled public demand for books. Statistics from 1838 on holdings for parish libraries in London indicate that 'Voyages, Travels, History and Biography' represented the largest category of books borrowed, after 'Fashionable Novels, well known', 'Novels of the lowest character' and 'Novels by Walter Scott' (Wyatt-Edgell 1838: 485).[8] The *Daily News*, which reported in 1854 on the frequency of books borrowed from the newly opened Marylebone Free Library over the previous six months, noted that Humboldt's *Cosmos* (edition unclear) was on a par with Swift's *Gulliver's Travels* and Chaucer's *Canterbury Tales*. It could therefore hold its own against classics of English literature and, at thirty-nine borrowings over half a year, drew more interest than Benjamin Franklin's *Memoirs* or Francis Bacon's collected works. It could not compete with Dickens's latest blockbuster *Bleak House* (1852) or other best-sellers like Scott's *Ivanhoe* (1820). But the fact that Humboldt's *Cosmos* deserved mention in a line-up of otherwise homegrown titles – predominantly fiction, rather than science – highlights the avid curiosity of British readers about Humboldt's writing at mid-century.

Among the most public of Humboldt's readers were his critics. The founding of the quarterlies – notably the *Edinburgh Review* and the *Quarterly Review*, in 1802 and 1809, respectively – ushered in an era in which these solid, authoritative and ultra-respectable publications became highly influential (Shattock 1989: 5). With large print runs and multiple readership in clubs, libraries and reading rooms, the quarterlies enjoyed great popularity among the wealthy middle classes and those wielding cultural power. In his study of the reception of Darwin's work in the British periodical press, Alvar Ellegård warns against assuming that the press necessarily voices the opinions of its readers or tells them what to think (Ellegård 1990: 19). Nevertheless, the comments made about Humboldt's writing in British journals must, necessarily, have supplied material for discussion, if not agreement. Moreover, as Jon Klancher (1987) has stressed, reading audiences were not a static entity, merely fed with information by periodicals: rather, the relationship between reader and periodical was a symbiotic one, in which readers become aware of themselves as members of a particular audience.

Institutions such as the Society for the Diffusion of Useful Knowledge (1826–48) and the BAAS – established in 1831, the same year that Darwin 'discovered' Humboldt – responded to the desire for self-education shown by the rapidly expanding working and middle-class readership. They made cheap scientific works presenting 'general concepts and simple laws' accessible to a wider audience (Secord 2000: 55). The sixpenny biweekly *Library of Useful Knowledge*, which sustained middle-class interest in popular science, was one of several publications that sought to bring scientists' work into the realm of public discourse and argued for the significance of their discoveries in as compelling a way as possible. But their aim was not purely to inform. By integrating science into society at large, they attempted to transform the image of the Man of Science as the lonely individual named in the preface to Wordsworth's *Lyrical Ballads*, who 'seeks truth' and 'cherishes and loves it in his solitude' (Wordsworth 2007: 76), into one seeking integration and sociability by making his work more easily accessible to a wider public.

The popularisation of scientific knowledge had its critics. Southey voiced concern both at the sudden increase in educated people from the 'lower orders', as a result of which 'philosophical words and phrases [...] now form part of the language of ordinary conversation', and at the accessibility of works 'of a subordinate character [...] which present science or literature in a ready and familiar form' (Southey 1829: 495–6). The result, according to Southey, was the spread of superficial knowledge and the loss of respect for profounder works. Everyone was expected to be so well rounded and well informed, in a period in which science was rapidly branching out in all directions, to include the disciplines of agriculture, natural history, mechanics, chemistry, mineralogy and geology (Southey 1829: 496). Coleridge worried more at the folly of trying to make everyone into a natural philosopher: 'you may wish for *general* illumination', he carped, but 'with the attempt to *popularize* science [...] you will only effect its *plebification*' (Coleridge 1830: 70–1).

Humboldt himself warned unequivocally in *Kosmos* against using the 'inappropriate term of *popular knowledge*' for those texts that presented scientific knowledge in a greatly generalised, imperfect manner (CO I, 32).

> I take pleasure in persuading myself [Humboldt declared] 'that scientific subjects may be treated of in language at once dignified, grave and animated, and that those who are restricted within the circumscribed limits of ordinary life [...] may thus have opened to them one of the richest sources of enjoyment by which the mind is invigorated by the acquisition of new ideas. (CO I, 32)

The English term 'popular', and its German equivalents 'populär' and 'populärwissenschaftlich', were problematic. As Andreas Daum (2002) and Jonathan Topham (2009a) have shown, popularisation tended to be allied either with simplification and dumbing-down, or the production of texts that 'exclude non-specialists from knowledge production and render them passive consumers' (Topham 2009a: 12). In reality, though, this model is inadequate because it fails to emphasise the extent to which such traffic was actually two-way: the regard in which such seemingly passive consumers held the scientific author they were reading was central to the processes of narrative self-fashioning deployed by emergent scientists in early nineteenth-century Britain (Topham 2000: 560). Moreover 'popular science', with its implicit appeal to ordinary people, was actually used, initially at least, to engage new readers in scientific enterprise and maximise market share. Indeed, the term 'popular', Topham observes, is intimately related to commercial developments in print culture in the 1820s and 1830s, since those who first used the phrase 'popular science' largely did so to increase profitability, deploying it as a way of 'signalling affordability, accessibility, utility and attractiveness' (Topham 2009a: 12).

Affordability was high on Henry Bohn's agenda as he boldly advertised his *Cosmos* at three shillings and sixpence a volume (to Humboldt's blank astonishment), thus undercutting Longman and partners' price of twelve shillings a volume by almost two-thirds and forcing them to lower the price of their fourth edition to two shillings and sixpence (Humboldt 2009: 359). Simon Eliot has shown that book pricing changed significantly in the late 1840s and 1850s – the period in which the bulk of the English translations of Humboldt's works appeared – so that cheap books, namely those with a cover price of three shillings and sixpence or below, occupied an important market share (Eliot 1995: 30). But in deliberately entering into competition with Longman's by publishing rival editions of the *Ansichten der Natur* and *Kosmos*, Bohn had the unenviable task of trying to make his own product sufficiently different from the version already in existence. Style played a key part in pointing up the individuality of the various English renderings of Humboldt's works as they came on to the market. Indeed, it was fundamental to the way in which Bohn carved out a niche for his translations as distinctive, authoritative (because more comprehensive) and, arguably, superior.

Aileen Fyfe and Bernard Lightman stress that the marketplace for nineteenth-century science lay well beyond the confines of the bookstore, to include soirées, lectures, museums and zoological gardens. As the century progressed, the audience for science extended beyond 'polite' London society to those who attended working men's lectures and Mechanics' Institutes, read penny magazines or visited public gardens and civic museums, thus enabling a broader range of social groups cast across a wider area of Britain to come into contact with science (Fyfe and Lightman 2007: 5–6). Considerable growth in consumer culture, particularly from mid-century on, also made a wide range of commodities and experiences available.

As the audience for scientific knowledge expanded to include the working and the middle class, as well as more women, practices of reading and writing necessarily changed. The Edinburgh publisher John

Murray warned Lyell that there were 'very few authors, or have ever been, who c^d write profound science & make a book readable', adding, morosely, 'depend on it, if you try to abridge & condense more & still be *popular* you may fail' (quoted in Wilson 1972: 344); yet, increasingly, scientific specialists in the second quarter of the nineteenth century were being asked to contribute to a top-down notion of 'popular' science. These new forms of scientific narrative often included 'sweeping narratives of natural progress and majestic visions of all-encompassing natural laws, rather than merely the sober technicalities of specialist science' (Topham 2009b: 832). Humboldt and his translators had found their niche.

Humboldt among the Leopards

To date, the bulk of research on Humboldt's reception outside Germany has focused on America. Ulrich-Dieter Oppitz's impressively detailed account (1969) of the dissemination of the name 'Humboldt' across the globe in anything from towns to parks, hotels, pharmacies and ships, convincingly argued that the Americas adopted Humboldt in a way that Britain apparently did not.[9] As Aaron Sachs's extremely readable study *The Humboldt Current: Nineteenth-Century Exploration and the Roots of American Environmentalism* (2006) illustrates, Humboldtian thought had an immense impact on nineteenth-century explorers who ventured to places that Humboldt never managed to see himself – the US western frontier, the Antarctic and the Northern Polar regions. Neil Safier's *Measuring the New World: Enlightenment Science and South America* (2008) offers a rather different narrative of Humboldt's engagement with science by setting his account against those of other European scientific travellers to the Americas, including Charles-Marie de la Condamine, Pierre Bouguer and Jean-Baptiste Bourguignon d'Anville. Since then, Kutzinski's thought-provoking volume *Alexander von Humboldt's Transatlantic Personae* (2012) has made considerable inroads into restoring a 'circumatlantic, and indeed global, flow to the forefront of our scholarly consciousness' (Kutzinski 2012: 7) by demonstrating the centrality of large-scale intellectual networks to Humboldt's legacy.

This study primarily seeks to understand better how Humboldt became a prominent figure in British scientific and cultural history through the translations of his work into English. Brief accounts by Wolf-Dieter Grün (1983), Peter Alter (1984) and William H. Brock (1993) give some sense of how Humboldt fared in what he termed the *Leopardenland*, the

'Land of the Leopards': a jest at the lion on the British coat of arms as much as the rapacious mercantilism he encountered there. One of the great achievements of Andrea Wulf's energetic Humboldt biography (2015) has been to reveal his connections with the intellectual and cultural elite of Victorian Britain, dispelling the lingering misconception of Humboldt as a lone genius, sustained by Daniel Kehlmann's *Die Vermessung der Welt* (2005) [*Measuring the World* (2006)]. Humboldt was a consummate networker and (honorary) member of eighteen British societies, which ranged from the Royal Society and Royal Horticultural Society to the Royal College of Surgeons, the Royal Society of Literature and the British Foreign Aborigines Protection Society (Théodoridès 1966: 54). His visits in the 1820s to London, where his brother Wilhelm was working as Ambassador of Prussia, helped him to forge or renew acquaintance with scientific figures such as Somerville and the botanist Robert Brown, as well as with politicians including the future Prime Minister Robert Peel and the Foreign Secretary Lord Dudley, both of whom he met in 1827 (Théodoridès 1966: 45–6).

Humboldt exploited these English connections to market his publications and bolster his international profile. Writing to his Swiss friend, the natural philosopher Marc-Auguste Pictet, in 1806, Humboldt described the warm reception offered by his English friends 'Lord Harrowby, Mr. Hammond, and Lord Gower, Mr. Pierpoint and the whole diplomatic tribe' (Humboldt 1869: 48). They had expected to see him 'completely frenchified' and were 'astounded that I can make myself understood in English and that I hold the fork in my left hand', a connection, he added, which had been 'very useful to me for the sale of my books' (Humboldt 1869: 48). Only much later, once he was well established within the scientific community, would he self-deprecatingly style himself 'a savage of the Orinoco' to the talented Norwich translator Sarah Austin (Ross 1893: 206).

But Humboldt was also a prominent figure in more 'common' environments. 'Alexander von Humboldt grasps universal nature; Alexis Soyer exhausts cookery,' ran an 1852 advertisement in the *Daily News* for the French celebrity chef's best-seller *The Modern Housewife* (1849).[10] Humboldt was clearly a household name (and a by-word for universality) by mid-century. Wide coverage of his publications in the British press enhanced this prestige. The *Leisure Hour* declared him 'The Patriarch of Modern Science', heading nothing less than a 'dynasty of genius', while *The Times* later equated him with the British greats Robert Peel and the Duke of Wellington.[11] Although Humboldt was involved in several cutting-edge scientific projects – notably the 'magnetic crusade' of the 1830s, which involved the coordination of magnetic observatories across Britain and the Empire – he also contributed to more 'popular' undertakings, and sat on the committee of patrons supporting the project to turn the geographer and map-seller James Wyld's 'Great Globe' at London's Leicester Square into a more permanent edifice for scientific education (Altick 1978: 489–91). By the 1850s, he had been awarded the highest of accolades, a street named after him in London – even if, with characteristic English disregard for the minutiae of foreign spelling, it ended up as 'Humbolt' Road, now in London W6.[12]

Humboldt also did not evade the beady eye of the British satirical journal *Punch*. He did not receive the same acerbic treatment meted out to the chemist Carl Ludwig von Reichenbach for a lumbering 1850 treatise on the animal magnetism generated by a kiss.[13] Nor was he exposed to the direct satire levelled at other Germans, notably Queen Victoria's consort Prince Albert, who presented his 'gomblimends' to 'Herrn Herrn Punsch' in a spoof article in 1844.[14] Revered for his scientific achievements, Humboldt was installed in the *Punch* 'House of Fame' alongside François Arago, Galileo Galilei, René Descartes and Michael Faraday, and was only gently satirised, in his capacity as a statesman, for his utopian hopes of an international peace settlement at the Peace Congress of 1850.[15] Humboldt's works were mined most frequently for their satirical application to everyday life. In 1858, for example, 'Mr Punch, M.A., F.R.S., F.L.S., F.R.S.L.' gave a mock address to the Natural History Section of the British Association in which he compared the numbing shocks generated by the *Gymnotus electricus*, the electric eel vividly described in Humboldt's *Personal Narrative*, to the torpor triggered by dull public orators.[16] The lasting impression Humboldt made on the British as a meticulous, methodical scientist is best demonstrated by an article from 1908, almost half a century after his death, in which *Punch* was quipping, 'Don't exceed the speed limit. Think of Baron Humboldt. Good solid work is as necessary to peace of mind as it is for the health of the body.'[17]

The premise of this study is that without the significant input of his British translators, Humboldt's success in the Anglophone world would have been slower and less spectacular. Initially, the intention was to trace the emergence of Humboldt's writings on the British book market in chronological order, addressing step by step how they interconnect with changes in readership, the prevailing discourse on science in the public arena and the increasingly aggressive strategies deployed by Humboldt's British publishers. However, this is complicated by the position of Ross's retranslation of the *Personal Narrative* (1852–3), which appeared while the competing translations by Sabine and Otté of *Kosmos* (1846–58 and 1849–58, respectively) were in progress, and just after the two competing translations of the *Ansichten der Natur* issued by Longman and Murray and by Bohn (1849 and 1850, respectively) had appeared. Working chronologically, we would discuss the first volumes of the English translations of *Kosmos*, then move on to the British renderings of the *Ansichten der Natur*, shift on to the next volumes of *Kosmos*, interrupt this with our discussion of Ross's *Personal Narrative* and then conclude the study with an analysis of the last volumes of *Cosmos*.

This would certainly emphasise the fragmentary, fast and furious approach to translating Humboldt into English in the 1840s and 1850s. But this study adopts a different methodological perspective by making productive comparisons between translations that centre on their textual genealogies and the layers of meaning that translators added to them. Ross's translation of the *Relation historique* is therefore discussed in the chapter immediately following the account of Williams's first rendering of this work. Then the fate of the *Ansichten der Natur* is examined in an account of its competing editions for Longman and for Bohn. Finally, the analysis turns to the translations of *Kosmos*, which – more than all his other works – have defined Humboldt's legacy in the Anglophone world. By proceeding in this way, we are able, for the first time, to gain a truer understanding of the role played by translation in shaping Humboldt's impact on the scientific imagination of nineteenth-century Britain.

Notes

1. Darwin Correspondence Project, available at <https://www.darwinproject.ac.uk/letter/DCP-LETT-98.xml>, letter of 28 April 1831 from Charles Darwin to Caroline Darwin (last accessed 24 September 2017).
2. Darwin Correspondence Project, available at <https://www.darwinproject.ac.uk/letter/DCP-LETT-158.xml>, letter of 8 February–1 March 1832 from Charles Darwin to Robert Waring Darwin (last accessed 13 June 2017).
3. Darwin Correspondence Project, available at <https://www.darwinproject.ac.uk/letter/DCP-LETT-182.xml>, letter of 18 August 1832 from Charles Darwin to Erasmus Alvey Darwin (last accessed 16 June 2017).
4. Darwin Correspondence Project, available at <https://www.darwinproject.ac.uk/letter/DCP-LETT-206.xml>, letter of 22 May–14 July 1833 from Charles Darwin to Caroline Darwin (last accessed 16 June 2017); Darwin Correspondence Project, available at <https://www.darwinproject.ac.uk/letter/DCP-LETT-96.xml>; letter of 7 April 1831 from Charles Darwin to William Darwin Fox (last accessed 16 June 2017), see footnote 2.
5. Darwin Correspondence Project, available at <https://www.darwinproject.

ac.uk/letter/DCP-LETT-3810.xml>, letter of 17 November 1862 from Charles Darwin to Ludwig Büchner (last accessed 16 June 2017).
6. On 'surrogates' and 'proxies', see also Kutzinski 2010.
7. Submitted to the University of Ottawa in 2004. This dissertation is still, regrettably, unpublished. Thanks to Luise von Flotow, Healy's supervisor, for making a copy of this work available to me.
8. See also the figures compiled by Alexis Weedon for the quantity of books printed, organised according to different genres. The category of 'Geography and History' represented a good quarter of books published in 1836 and in 1846, but dropped progressively over the next thirty years (Weedon 2003: 94).
9. On Humboldt's South American legacy, see also Ette 1999; for more on the translation of his works on Cuba and South America, see Prüfer Leske 2001 and Schuchard 1998.
10. *Daily News*, 1751 (2 January 1852), p. 8.
11. *Leisure Hour*, 62 (3 March 1853), p. 156; *The Times*, 9 May 1859, p. 12; *The Times*, 16 May 1859, p. 10.
12. *Daily News*, 3301 (15 December 1856), p. 3.
13. *Punch*, 19 (1850), p. 254.
14. *Punch*, 6 (1844), p. 32.
15. *Punch*, 25 (1853), p. 106; *Punch*, 19 (1850), p. 121.
16. *Punch*, 34 (1858), p. 131.
17. *Punch*, 134 (1908), p. 86.

Chapter 1

Styling Science

'The mad fancy has seized me', wrote Humboldt to the German diplomat Karl Varnhagen von Ense in autumn 1834, 'of representing in a single work the whole material world' (Humboldt 1860: 15). It would contain

> all that is known to us of the phenomena of heavenly space and terrestrial life, from the nebulæ of the stars to the geographical distribution of mosses on granite rocks, and this in a work in which a lively style shall at once interest and charm. (Humboldt 1860: 15–16)

Humboldt's final, magisterial, work, *Kosmos*, was arguably the greatest holistic narrative of the natural world to be written in the nineteenth century. It was not just its content that was so ambitiously all-embracing. By employing a 'lively' style, Humboldt deliberately aimed his work at scientific specialists and general readers alike. Knowledge, he recognised, could never be divorced from the techniques of its presentation. Yet even for an experienced author like Humboldt, writing a work such as *Kosmos* was fraught with difficulty. 'The besetting sins of my style are', he confessed to Varnhagen von Ense, 'an unfortunate propensity to poetical expressions, a long participial construction, and too great concentration of various opinions and sentiments in the same sentence'; these were evils, he reflected, that only a strict agenda of simplicity could remedy (Humboldt 1860: 19).

Like Humboldt, Darwin was obsessed with style. 'Some are born with a power of good writing,' he wrote to the entomologist Henry Walter Bates in December 1861. 'Others like myself', he continued:

> have to labour very hard and slowly at every sentence. I find it a very good plan, when I cannot get a difficult discussion to please me, to fancy that some one [*sic*] comes into the room and asks me what I am doing; and then try at once and explain to the imaginary person what it is all about. I have done this for one paragraph to myself several times, and sometimes to Mrs. Darwin, till I see how the subject ought to go. [. . .] But style to me is a great difficulty.[1]

Darwin's concern to be direct and clear underpinned the conviction that language should facilitate the communication of scientific ideas, making them comprehensible to a range of different audiences. As Richard Yeo has stressed, in the first half of the nineteenth century science was still in the process of acquiring definition (Yeo 2003: 31–2). There was no real consensus about the place of science in British culture and therefore an urgent need to discuss and defend its position in the public sphere.

Style and rhetoric are central to this study of Humboldt's writing in English translation, which investigates how language was used to communicate scientific knowledge in the nineteenth century, in a largely book- and journal-based information world. We tend to think of scientific narrative as stylistically neutral, but even the 'driest experimental account' can be subjected to an analysis of the kind we might more commonly apply to literature, and any scientific text is 'meaningful only within a structure of other meanings' (Sleigh 2011: 9). As Lawrence J. Prelli (1989) has shown, language reflects the logic and traditions of scientific method as well as the norms to which scientific narrative practices adhere. Rhetorical decisions are closely linked to the authorisation, justification and authentication of the assertions made in the text: indeed, they are constitutive of the scientific knowledge they seek to convey (Gross 2006: 7).

Scientific narrative of the kind we are investigating here was never intended to be read solely by a specialist readership. Scientists like Darwin recognised the importance of locating their writing within broader perceptions and experiences of the natural world that informed Victorian literary and visual culture. The cultural dimension of scientific knowledge-making in the Victorian period has been widely explored in a number of studies (see Allen 2001; Dawson 2007; Jenkins 2007a; Buckland 2013; Secord 2014) that investigate how science influenced other cultural fields and vice versa, and how it permeated people's understanding of the world around them. As Maria Freddi, Barbara Korte and Josef Schmied elaborate, 'much of the cultural relevance of science is effected through, and made manifest in, discourse', and the rhetoric of scientific communication is closely bound up with concerns to represent science effectively to a particular target audience (Freddi, Korte and Schmied 2013: 221).

Such rhetoric is now frequently thought of as hermetic – intended for a small group of readers with specialised knowledge – yet given the expanding popular interest in science in the period, there was an urgent need for scientific discourses to be publicly understood and other communicative strategies were required. The first half of the nineteenth century represented an interlude of popular access to science, located between two periods of elite practice – the earlier one characterised by low levels of general literacy and the later by swift advances towards specialisation and professionalisation. Capturing an audience's interest depended, as Darwin and Humboldt well knew, on generating scientific prose that was factually accurate, yet well paced and readable. Lightman has noted that experimentation with narrative form was one of the key ways in which more 'popular' writers sought to attract their public, and the storytelling dimension remained an essential aspect of scientific writing well into the Victorian period (Lightman 1997: 191–2). This placed considerable demands upon translators, who needed the expertise to translate scientific passages accurately, yet also the creative flair to craft engaging prose. As Patrice Bret (2012) has shown in his detailed survey of science and translation in the nineteenth century, they sometimes produced work less in keeping with the standards of the scientific community than with the desires of the market.

By focusing on the individual stylistic choices made by the British translators of Humboldt's works, we can also analyse at close quarters how exactly his texts were reconfigured for an Anglophone audience. This enables us to give prominence to their role as mediating agents and investigate how their complex cultural and institutional allegiances, coupled with their individual linguistic repertoires, governed the stylistic preferences that shaped their translations. The lexical and rhetorical decisions translators make are, to some degree, decided by the norms and models established by other practitioners in their field (Milton and Bandia 2009: 8), and this is certainly true of the environment in which Humboldt's nineteenth-century translators worked. But Daniel Simeoni's discussion (1998) of 'translational habitus' – translators' learnt behaviours, which are the outcome of their individual personal, professional and cultural history – is also key to uncovering how they respond to a piece of scientific prose as they are translating it. Understanding the process of textual transfer in terms of habitus reflects disciplinary developments in translation studies over the past couple of decades towards making 'finer-grained analyses of the socio-cognitive emergence of translating skills and their outcome, in particular at the micro-level of stylistic variation' (Simeoni 1998: 33).

This chapter starts by investigating developments in thinking about rhetoric and style in relation to both scientific and more general communicative practices before and during the (pre-)Victorian period. It then focuses on just some of the multiple relationships between literature and science in nineteenth-century Britain to illustrate how literary texts could provide important structuring and representational impulses for the diffusion of scientific knowledge and ideas. As we explore the style of Humboldt's writing in the original and in translation, we consider how well his narrative strategies managed to transcend the specialist language in which scientific practitioners communicate with each other and be accessible by a wider audience. In a final section we turn to investigating what is at stake stylistically when a text is translated from one language into another, how we can approach issues of stylistic variance and how this will inform our comparison of the competing translations into English of Humboldt's works in the chapters that follow.

Language, Style and Knowledge

Darwin and Humboldt were by no means the first scientists to reflect on how scientific matters should be presented and discussed in writing. Thomas Sprat, writing his history of the Royal Society in 1667, a mere four years after its foundation, had warned that the language of science appeared to be under threat from a 'vicious abundance of *Phrase*', 'trick of *Metaphors*' and 'volubility of *Tongue*, which makes so great a noise in the World' (Sprat 1667: 112). Rather, it should actively seek 'to reject all the amplifications, digressions, and swellings of style' and return to 'the primitive purity, and shortness, when men deliver'd so many *things*, almost in an equal number of *words*' (Sprat 1667: 113). He was therefore advocating the precise, unpretentious employment of language, which itself implied that those adopting such a writing style were themselves men of rigour and precision. By using 'a close, naked, natural way of speaking; positive expressions; clear senses; a native easiness' (Sprat 1667: 113), they would be making discoveries and results more accessible by a wider audience. In 'preferring the language of Artisans, Countrymen, and Merchants, before that of Wits and Scholars', their work would benefit not a lettered elite but those who contributed concretely to the common good through mastery of a craft or trade.

Sprat's association of good style with 'a natural way of speaking' resonates well into the nineteenth century. The astronomer John Herschel, in his *Preliminary Discourse on the Study of Natural Philosophy* (1831), emphasised the importance of stripping scientific writing 'of all such technicalities as tend to place it in the light of a craft and a mystery, inaccessible without a kind of apprenticeship' (Herschel 1831: 70). He acknowledged that, like any other area of learning, science had its own peculiar terms and idioms that it would be difficult to relinquish. But everything that tended 'to clothe it in a strange and repulsive garb' and lend it 'an unnecessary guise of profundity and obscurity' should be eschewed (Herschel 1831: 70). Knowledge – whether it was generated by practitioners of the arts or the sciences – was not to be cultivated or enjoyed only by the few; it could only truly gain authority by 'passing through the minds of millions' and being 'made accessible to all' (Herschel 1831: 69).

Elsewhere, others were reflecting on how knowledge and ideas could be put down on paper effectively. William Godwin, journalist, political philosopher and British radical, published a revised version in 1823 of his essay 'Of English Style' (1797), in which he proclaimed that the English now being written was 'closer and neater, more free from laxity of structure, and less subject to occasional incongruities, superfluities, unnaturalness and affectation' than ever before (Godwin 1993: 340). The modern era was one in which the 'spirit of philosophy has infused itself into the structure of our sentences', such that 'we express our thoughts in precise words, directly flowing out of the subject to be treated' (Godwin 1993: 286). Perfect style was, according to Godwin, 'the transparent envelop [sic] of our thoughts; and, like a covering of glass, is defective if, by any knots and ruggedness of surface, it introduces an irregularity and obliquity into the appearances of an object' (Godwin 1993: 239). The aim of language was to express ideas with clarity; any blemishes that drew attention to the medium would distort the thoughts being imparted and distract the reader.

Writing at mid-century, the journalist, critic and translator Thomas De Quincey was less convinced that the British were writing more elegantly than ever before. 'In England', he observed, 'it is strange enough that, whilst the finest models of style exist, and sub-consciously operate effectively as sources of delight, the *conscious* valuation of style is less perfectly developed' (De Quincey 2001: XVII, 57). De Quincey ascribed the decline of 'good' English prose to the rise of a certain type of writer and a certain class of reader. Style had become the slave of journalism and it was to newspapers, the 'main reading of this generation', that he attributed its demise. The natural, spontaneous 'language of life', hitherto used in general communication, was now subordinate to the 'whole artificial dialect of books' that 'encrusted and stiffened' natural expression into erudite idiom (De Quincey 2001: XII, 14–15). Moreover, journalism had deformed the very sentence structure of the English language by increasing sentence length and generating 'clause within clause *ad infinitum*'; gone was the graceful succession of long and short sentences, and in its place a bloated 'monster model' (De Quincey 2001: XII, 15).

De Quincey was also well placed to comment on stylistic differences between languages, having translated the work of several German writers and thinkers in the mid-1820s. In his essay 'On the Present State of the English Language' (1850), he asked which of the three nations 'that walk at the head of civilisation' – the French, Germans or the English – took greatest pains to write language accurately and elegantly (De Quincey 2001: XVII, 63). For him, French sustained republican ideals by using an economical sentence structure that was easily accessible by its audience (De Quincey 2001: XVII, 64). English, by contrast, was now characterised by sentences of near 'hyperbolical length', even if they could at least boast a continuous movement of 'flux and reflux, swell and cadence' (De Quincey 2001: XII, 20). German fared worse. 'Whatever is bad in the English ideal of prose style, [. . .], we see there carried to the most outrageous excess' in a 'zealotry of extravagance that really seems like wilful burlesque', De Quincey groused (De Quincey 2001: XII, 20–1). The sheer elasticity of the German sentence, and the structural complexity it could sustain, distanced readers so much from the main idea that they lost the thread of what they were reading. 'We doubt, indeed,' De Quincey asserted elsewhere, 'whether any German has written prose with grace, unless he had lived abroad' (De Quincey 2000: VI, 182).

Style obviously governed whether a text was considered to be readable and accessible, or laboured, affected and falsely erudite. It also characterised the identity of its writer – nationally, professionally and culturally. By prescribing a 'plain' style for the Royal Society, Sprat was on a patriotic mission to perfect how knowledge was constituted by the most prestigious of British scientific institutions. Godwin and De Quincey likewise suggested that English had the potential to convey ideas with clarity, where other European languages might not. Style not only was related to national identity: it was the very index of the man of science. The way in which natural philosophers communicated scientific knowledge, whether specialised or more popular, was intimately related to their presentation of themselves as gentlemen of authority and integrity. The linguistic choices that they made were necessarily encoded within their particular world view and a sense of their place in the communication of knowledge. As Sprat and Herschel emphasised, stylistic choices made by the writer also revealed assumptions about the reader's identity. To ensure that a text was meaningful for diverse audiences, scientific practitioners needed to cultivate an awareness of the cultural, political and social significance of their work.

Models of Emulation

To win the approval of an educated readership, nineteenth-century scientists frequently drew on literary texts with which readers would have been familiar to present new ideas, theories and findings. Laura Otis has emphasised how demonstration of a close acquaintance with the British literary canon could play a crucial role in ensuring that a scientist's work was favourably received (Otis 2002: xix). Intertextual references to literary sources were, however, important indicators of a knowledge that was more than just 'literary'. Rather, they functioned as what John Guillory (following Pierre Bourdieu) terms '*symbolic* capital, a kind of knowledge-capital whose possession can be displayed upon request and which thereby entitles its possessor to the cultural and material rewards of the well-educated person' (Guillory 1993: ix). Lyell's first main publication, the *Principles of Geology* (1830–3), is a striking example of how scientific authors used literature to appeal to British cultural traditions. The *Principles* quickly became recognised as a leading work in the field, and would remain a standard geological text for the next fifty years. Its very subject matter was embedded in controversies about Creation, the Fall and the Flood. How was Lyell to engage with issues concerning the age of the earth, the powerful forces that shape the natural world or the progressive development of life? The answer, Secord has demonstrated, was to deploy a rhetoric of respectability (Lyell 1997: xiv–xv). Lyell did this by drawing heavily on classical sources from Horace, Ovid, Pindar and Virgil, but also canonical works of British literature: Shakespeare's *Henry V* and *As You Like It*, Milton's *Paradise Lost*, Byron's poem *Childe Harold's Pilgrimage* and Scott's *Lady of the Lake*.

Lyell used some literary sources to help him develop his argument: controversial issues were presented in measured prose and supported by quotations from canonical works to lessen their potentially divisive impact. Others enhanced the visual dimension of his narrative. For example, as he was discussing the ability of marine plants to move far beyond their native shores on ocean currents, Lyell wove in a couple of lines from Byron's *Childe Harold* (Canto iii, 17–18). 'All are familiar with the sight of the floating sea-weed,' he observed, '*Flung from the rock on ocean's foam to sail, Where'er the surge may sweep, the tempest's breath prevail*' (Lyell 1997: 243). He therefore harnessed the energy of Byron's surging waves to lend his own narrative added dynamism and enhance the imaginative appeal of a discussion of the global distribution of seaweed, which general readers might otherwise not have found riveting. That he did not reference the quotation is an indication that Lyell presumed his readers would be sufficiently well versed to recognise that it was borrowed from one of the literary best-sellers of his day.

Not all early British writers of science used literary quotation so explicitly. Alice Jenkins (2008) has analysed how Michael Faraday, definitely not a 'gentleman scientist' by birth, used literary sources to consider how to improve his scientific writing. Faraday had already begun to reflect on how best to address a public audience, organise material and deliver arguments when he became Sir Humphry Davy's chemical assistant in 1813. He quickly realised that vocabulary, grammar, register and expression were all key components in the successful communication of scientific knowledge. He worried that his workmanlike approach to writing up scientific findings generated a heavily structured prose that:

> introduces a dryness and a stiffness into the style of the piece composed by it for the parts come together like bricks one flat on the other [. . .]. I would if possible imitate a tree in its progression from roots to a trunk to branches twigs & leaves, where every alteration is made with so much ease & yet effect that though the manner is constantly varied the effect is precise and determined. (James 1991: I, 149)

The self-improvement circle he set up to cultivate a good writing style turned, inevitably, to literature. The range of texts on which they drew in the nearly fifty essays they produced between August 1818 and July 1819 has been neatly tabulated by Jenkins in her detailed study of Faraday's 'Mental Exercises' (Jenkins 2008: 32–3). The Bible was a key source of quotations, direct and indirect. Other references were made to Alexander Pope, Samuel Johnson, Milton's *Paradise Lost*, James Thomson (writer of the eighteenth-century best-selling poem 'The Seasons'), Shakespeare's *As You Like It*, *Hamlet*, *Julius Caesar* and *Othello*, the Scottish poet Robert Burns and the English lyricist William Cowper. Drawing on canonical literary figures as well as more contemporary writers, Faraday's fellow writers were therefore using intertextuality to connect their own writing with the British literary tradition, while also claiming its relevance to contemporary culture.

Humboldt as Reader and Author

Humboldt, like Faraday, was committed to encouraging the popularisation of science as much through public lectures as through the written word. The lectures he gave in Berlin in 1827 and 1828, which undergirded much of *Kosmos*, were certainly not aimed at an audience of specialists. Free entry meant that they drew a wide social spectrum of listeners – from stonemasons to King Friedrich Wilhelm III – and the hall was packed each time. Jürgen Hamel and Klaus-Harro Tiemann estimate that around 13,000 listeners attended – evidence of the immense success that Humboldt enjoyed as he sought to democratise access to scientific knowledge (Hamel and Tiemann 1993: 11). His lectures were described by the *Spenersche Zeitung* [*Spener's Gazette*], one of the city's leading newspapers, as examples of 'calm clarity', in which detail was neatly placed within the wider context of Humboldt's argument, and the interrelatedness of emerging scientific disciplines continually emphasised.[2]

Orality was also vital to Humboldt's narrative projects. It had the potential to lend his writing an immediacy and vivacity that could help to 'transport' his readers imaginatively to the scenes described. As Ette reminds us, Humboldt was both a sparkling conversationalist and a highly 'oral' writer (Ette 1991: 1576). Goethe noted to the musician and pedagogue Carl Friedrich Zelter in October 1831 that he had just finished perusing Humboldt's *Fragments de géologie et de climatologie asiatiques* and could only marvel at Humboldt's extraordinary powers of expression. The treatises in it read like speeches (Goethe 1993: 473). Indeed, Goethe enthused, this conqueror of the natural world was arguably the greatest rhetorician of all time: 'Unserer Welteroberer ist vielleicht der größte Redekünstler' (Goethe 1993: 474).

Yet the eloquence that characterised Humboldt's Berlin lectures is not always reflected in his published prose. Indeed, throughout his life and, successively, in the *Relation historique*, the *Ansichten der Natur* and *Kosmos*, Humboldt would repeatedly raise the problem of how to meld precise, fact-laden scientific narrative with more impressionistic, imaginative prose. Michael Dettelbach's exploration of the tensions in Humboldt's writing – between a devotion to measurement and quantification, and a concern with emotional and aesthetic responses – dismantles the apparent polarity between Enlightenment empiricism and Romantic idealism by reassessing the nature of Humboldt's commitment to measurement and observation. Dettelbach argues that Humboldt was 'engaging in an Enlightenment redefinition of the authority of the philosopher' by understanding experimental philosophy not as an attempt to build theories but 'to observe the co-variation of phenomena through more or less precise instruments and languages' (Dettelbach 2001: 17). Rooted in the eighteenth century (he was thirty-one when the century turned), Humboldt remained heavily indebted throughout his life to an Enlightenment view of knowledge construction that, as George Rousseau has emphasised, was underpinned by both rational and imaginative components (Rousseau 2003: 763).

From the outset, Humboldt struggled to weave these 'rational' and 'imaginative' elements into a seamless narrative. Christian Helmreich (2009) contends that Humboldt's account of travel to the Americas is an uneasily hybrid piece, a 'tableau synthétique' of styles that pulled his work in different directions. As we have seen, the *Relation historique* fascinated Darwin because of its aesthetic evocation of tropical fecundity and its picturesquely sinuous valleys and sublime *cordilleras*. Writing home to his father from the Brazilian city of São Salvador after a couple of months on board the *Beagle*, he noted, 'If you really want to have a [notion] of tropical countries, *study* Humboldt,' but added, 'Skip th[e] scientific parts & commence after leaving Teneriffe.'[3]

The verdant fecundity and profusion of the natural world described by Humboldt could therefore exert a destabilising effect on the aesthetic sensibility, disrupting the author's intentions to represent the Americas with scientific coherence. In whole swathes of the main body of the *Relation historique*, Humboldt and Bonpland set about measuring air temperature, pressure and humidity, taking astronomical readings, calculating their position according to the sun or stars, describing the rock formations in the terrain they were crossing, and recording the flora and fauna they observed around them. This information, presented soberly, gave an account its authority. Helmreich suggests that the more literary passages with which the narration is interspersed are a valuable way of conveying the totality of the experience, and reorganising the multitude of observations into a meaningful whole (Helmreich 2009: 306).

Humboldt's *Personal Narrative* had a significant impact on Darwin's *Beagle* account. Originally published as the third volume of the *Narrative of the Surveying Voyages of His Majesty's Ships Adventure and Beagle between the Years 1826 and 1836* (1839), and then in a substantially revised and extended edition as the *Journal of Researches into the Geology and Natural History of the Various Countries Visited by H.M.S. Beagle* (1845), it was an important apprentice piece that demonstrated Darwin's proficiency as a scientist and writer. It reveals Darwin's sophisticated skills of 'meticulous observation, combined with imagination, logical interference, and analogical thinking' (Levine 2011: 41). Key to Darwin's narrative style is a way of seeing the world, in which the foreign is evoked through descriptions that reveal careful attention to detail, 'opening the unfamiliar to the familiar' through simile and comparison (Levine 2011: 65). Darwin not only teaches his audience to explore the natural world around them with new eyes. He also enables the reader to visualise phenomena that are tangible and intangible, present but not visibly present, and neatly forges links between observation and reasoning, advancing from detailed description to a wider understanding of the world (Levine 2011: 58).

Darwin's contemporaries noted how the *Journal* bore the imprint of Humboldt's narrative style. As the orchidist Hermann Kindt enthused to Darwin in 1864, those

> who have been happy enough to pursue your graphic Zoological descriptions and annotations in your 'Zoology of the voyage of H. M. Ship Beagle' in the original language, are delighted with the vivid, Humboldt-like pictures you bring before their mental eyes.[4]

Indeed, Darwin internalised the style of Humboldt's (translated) prose so thoroughly that it threatened to colour his own prose. His sister Caroline, on reading the first parts of Darwin's journal in October 1833, cautiously observed:

> I have been reading with the greatest interest your journal & found it <u>very</u> entertaining [. . .]. I am very doubtful whether it is not *pert* in me to criticize, [. . .] but I *will* say just what I think – I mean as to your style. I thought in the first part (of this last journal) that you had, probably from reading so much of Humboldt, got his phraseology & occasionly [sic] made use of the kind of flowery french [sic] expressions which he uses, instead of your own simple straight forward & far more agreeable style.[5]

Her criticism of the 'flowery' French expressions proliferating in Williams's translation of Humboldt's *Relation historique* implicitly affirm what Caroline valued most in her brother's writing: that 'straight forward' and 'agreeable' style that Herschel, and others well before him, had equated with the plain style of English science.

Structurally, too, there were similarities between Humboldt's *Personal Narrative* and Darwin's *Beagle* account. As Leask notes, Humboldt's preference for synchronic over diachronic description and for narrative organisation according to 'place' rather than 'time' were echoed in Darwin's *Journal* (Leask 2003: 32). It shares with the *Personal Narrative* full tables of contents heading each chapter, which well convey the comprehensiveness of each section, and it carries the same extensive footnoting that locates the narrative at the centre of a web of related research. Darwin, like Humboldt, was not afraid to be anecdotal, deploy direct speech to reconstruct conversations, use 'I' to refer to his own thoughts or suppositions, or ask rhetorical questions. These devices, aimed variously at changing the pace of the narrative, lending it more immediacy and intimacy, or constructing a direct rapport with the reader, helped to make Darwin's account of the *Beagle* voyage lively, varied and engaging.

The difficulties that Humboldt had already encountered in negotiating literary and scientific discourses in the *Relation historique* would recur in the *Ansichten der Natur* and be something he addressed in a preface to the later editions of this work.

> The combination of a literary and a purely scientific aim [he observed (in Otté's translation)], the desire to engage the imagination, and at the same time to enrich life with new ideas by the increase of knowledge, render the due arrangement of the separate parts, and what is required as unity of composition, difficult of attainment. (*VN* xi)

Humboldt was acutely aware that the aesthetic dimensions of scientific writing had an essential role to play in appearing to lend the text a vibrancy and immediacy, which made it engaging reading. Humboldt's deployment of a range of different literary strategies, such as the insertion of anecdotal or autobiographical passages, constant shifts of narrative perspective or the drawing of striking parallels, gave his work the energy and dynamism that would enhance its appeal to a 'popular' audience (Ette 2004: 42).

But it was really in the immense narrative project *Kosmos* that Humboldt would founder in his attempts to keep his material under control. Ette has highlighted the restlessness, incompleteness and fragmentariness central to Humboldtian narrative (Ette 2001a: 35). Casting Humboldt's œuvre as a vast corpus of work 'in progress', Ette suggests that he was deeply conscious of the constantly evolving nature of his research, energised at different intervals by new impulses from a range of different disciplines. In *Kosmos* in particular, Humboldt always aimed to be as comprehensive as possible. Yet he was unable to bring any section to completion because science was always marching on in the time that he was trying to bring sections of this work to press. In reality, he was working away at a series of immense fragments, themselves broken down fractally into a myriad of subsections, in a process that Ette has characterised as 'nomadic writing' – 'écriture nomade' – itself perpetually on the move, roaming across different texts, subjects and disciplines (Ette 2001a: 35).

As Humboldt emphasised in the preface to the third volume of *Cosmos*, he sought to 'avoid the accumulation of isolated facts', since '[o]ur impressions of nature will [...] be essentially weakened, if the picture fail [*sic*] in warmth of colour by the too great accumulation of minor details' (*CO* III, 1, 3). This recollects the narrative aims of Forster, Humboldt's great mentor, in whose company he first encountered Britain in 1790. Forster had been particularly influential in rethinking how the foreign experience could be represented. In his preface to the

Voyage Round the World, he described his attempts to break with the empirical recording of fact:

> The learned, at last grown tired of being deceived by the powers of rhetoric, and by sophistical arguments, raised a general cry after a simple collection of facts. They had their wish; facts were collected in all parts of the world, and yet knowledge was not increased. They received a confused heap of disjointed limbs, which no art could reunite into a whole. (Forster 1777: I, xii–xiii)

The art of 'reuniting into a whole' was one that Humboldt, almost three-quarters of a century later, was still seeking to perfect by conveying the totality of the foreign experience through a wealth of data coupled with sensory experience. As he explicitly stated with regard to *Kosmos*, the object of this book was 'to give a scientific and at the same time an animated description of nature' (CO III, 5). Nevertheless, he was conscious that the literary component of his text could potentially outlive the scientific findings of his work. Science focused on the new: new facts, new findings, and new gadgets that would allow scientists to produce still more new knowledge. These quickly grew dated and made science antiquated, in contrast to a literary delineation of nature 'in all its vivid animation and exalted grandeur', which generated images that would not fade (CO I, xiv).

Kosmos well illustrates Humboldt's awareness of literature's power to evoke the natural world by appealing to the imagination. This is precisely what Humboldt illuminated in his second volume of *Kosmos*, which was concerned with 'the impressions reflected by the external senses on the feelings, and on the poetic imagination of mankind' (CO II, 370). Its first section, entitled 'Incitements to the Study of Nature', explored how the contemplation of nature had changed over time and how it differed from race to race. For the modern period, Humboldt drew (in Otté's translation) on a number of writers, including Shakespeare:

> who, in the hurry of his animated action, has hardly ever time or opportunity for entering deliberately into the descriptions of natural scenery, yet paints them by accidental reference, and in allusion to the feelings of the principal characters, in such a manner that we seem to see them, and live in them. [. . .]
>
> 'A true description of nature occurs, however, in *King Lear*, where the seemingly mad Edgar represents to his blind father, Gloucester, while on the plain, that they are ascending Dover Cliff. The description of the view, on looking into the depths below, actually excites a feeling of giddiness.' (CO II, 430)

Rather than directly citing his sources, Humboldt analysed the rhetorical means by which the dramatist drew the observer into these scenes such that 'we seem to see them, and live in them'. The quoted extract at the end, taken from a letter by the great nineteenth-century German translator of Shakespeare, Ludwig Tieck, is revealing of both the literary circles within which Humboldt moved and his acknowledgement of the ability of literature to transport its audience sensually to a different time and place. Humboldt concluded his assessment of the contribution that 'literary' landscape description could make to scientific writing by noting that '[d]escriptions of nature [. . .] may be defined with sufficient sharpness and scientific accuracy, without on that account being deprived of the vivifying breath of imagination' (CO II, 438).

Humboldt therefore shared with British counterparts like Faraday and Lyell an understanding of the imaginative potential of literature to create an image in the mind's eye of the reader. The reference to Shakespeare in *Kosmos* exemplifies how some of his discussions would be culturally translatable for a British audience and could usefully employ a literary intertextuality that would give this work a familiar feel in its English editions. The sheer complexity of his narrative structure and writing style would, however, pose greater challenges for his translators. They would need to convey the content of his original text accurately, while perhaps downplaying the fragmented nature of parts of his prose and above all negotiating the tensions inherent in the scientific and imaginative components of his narrative.

Translating Style

The concept of style was clearly central to nineteenth-century scientists' understanding of how texts could construct arguments, convey ideas and persuade their readers of the importance of the knowledge they were communicating. We have seen that scientific narrative of the type that Humboldt and others were producing cannot comfortably be pigeonholed as 'non-literary' prose. Rather, it shares with literary narrative the capacity to use language in a way that exploits the creative possibilities of lexis and grammar, and can employ particular techniques to extend the imaginative reach of their texts, while at the same time enabling their work to appeal to a wider audience than just scientific specialists. But how do translators take into account issues of style as they set about putting a text into a foreign language, what kinds of challenges do they face and how much stylistic freedom can they have to articulate their own agenda through translation?

Jean Boase-Beier's *Stylistic Approaches to Translation* (2006) is a particularly illuminating study because it considers stylistic issues in relation to literary and non-literary texts. She argues that any analysis of prose in translation is complicated by the fact that the styles of two texts have to be taken into account: that produced by the original author and that by the translator. There are therefore at least four potential viewpoints to examine in terms of style, as we compare the 'original' or source text and the translation or 'target' text. The style of the source text is an expression of the author's choices, but it can also be analysed in terms of its effects on the reader (and on the translator as reader). Likewise, the style of the target text is clearly an expression of the choices made by its author (who is the translator), but also of its effects on the reader (Boase-Beier 2006: 5). Reading, Boase-Beier contends, depends on what the reader perceives as the author's manner of expression and is therefore only rarely an uncontroversial practice. More often it is a 'dynamic, active, participatory, open-ended process', unlikely to be clearly separated in the translator's mind from the act of textual recreation in another language (Boase-Beier 2006: 32–3).

By thinking about the ideological motivations behind translation, Munday's *Style and Ideology in Translation* (2008) also makes an extremely valuable contribution to the analysis of style in translated texts. As Munday rightly argues, the translated text is always an amalgam of the stylistic choices made by the author of the original text and the translator who reconfigures this text: variations between the stylistic make-up of the original and the translation are the result of the translator's conscious and unconscious decision-making (Munday 2008: 13). The translator, or rather, the 'implied translator' (Munday 2008: 13) – if we consider those other individuals alongside the translator who are involved in the production of the translation, such as the publisher, editor and copy-editor – is in a powerful position. He or she can opt quietly or more provocatively to reshape the translation so that it, in turn, alters the voice of the author in the foreign language. Much in the same way that the narrators can be 'unreliable', so too can translators (Munday 2008: 14).

Mona Baker's assertion that all narratives are 'dynamic entities' and can 'change in subtle or radical ways' (Baker 2006: 3) as they are exposed to a new contextual framework is fundamental to our understanding of the transformative possibilities of translation. Some of the impulses for change may come from external forces acting upon the translator – prevailing norms and expectations about how a translation should be done – which determine the end result. Christina Schäffner suggests that an understanding of how norms have operated at a given time enables us to compare general concepts of translation among a certain community and enquire after the authorities regulating such norms or catalysing changes in them (Schäffner 1999: 6). The style that translators choose is also subject, consciously or otherwise, to a range of different constraints and influences imposed upon them by the target language and culture. Such norms, Theo Hermans contends, are constituted not just by 'regularities in behaviour' but by 'prevailing normative and cognitive expectations', while the selective aspect of the individual translator's voice must be viewed within the context of a 'limited range of realistically available alternatives' (Hermans 1999: 51). If norms are about preference and exclusion, it is just as important to explore the alternatives emphatically not chosen (Hermans 1999: 134).

Humboldt's translators were working in a period before the professionalisation of the European translation industry and, as such, had few ethical and legal obligations towards their clients. Unlike the global language industry today, they did not need to conform to any issues concerning regulation and accreditation. However, discerning readers (and particularly reviewers) would have had a clear idea of what they felt constituted good practice in terms of fidelity to the style and meaning of the original, and in terms of the fluency and readability of the translation. Translation theory in the first two decades of the nineteenth century was a vibrant, swiftly evolving field, dominated by the works of the German Romantics. Friedrich Schleiermacher's landmark Berlin lecture 'Ueber die verschiedenen Methoden des Uebersezens' [*On the Various Methods of Translation*] (1813) reflected on the difficulty inherent in the translator's task of bringing an author and a reader together. He concluded that if readers were to understand the foreign text in the same way the translator had done, they had to comprehend the spirit of the language, the '*Geist der Sprache*'. Two strategies open to translators – paraphrase and imitation – failed to convey the reading experience of the original: the first because it succeeded in reproducing the content but not the impact of the original, the second because it did not render the spirit of the original language but tried to reproduce the same reaction as that enjoyed by the source text readers (Schleiermacher 1838: 214). The solution, Schleiermacher proposed, was either for the translator to emphasise the linguistic and cultural differences of the source text and 'foreignise' it for the reader ('Verfremdung'), or to 'domesticate' the text by giving the impression that the target language text was a product of their own culture ('Einbürgerung'). Foreignising strategies lacked cultural capital at this time, Venuti asserts (Venuti 2008: 84–6), but they were also potentially a form of experimentalism, in which the formation of a national culture could be controlled by an educated elite with innovative ideas about how to employ the registers and styles available in their language.

Munday correctly observes that we tend all too often to read translated texts 'in isolation', as if they were wholly detached from their source text (Munday 2008: 14). Only by making a close comparison of the language of the original text and of its translated counterpart is it possible to identify the style of the author and that of the translator, and hence the voices present in the translation (Munday 2007: 19). To understand the contribution that the translator makes, Hermans contends that we need to 'confront' the authorial voice with that of the translator to illuminate what kinds of shift the text has undergone (Hermans 1996: 27). By determining the nature of the translator's discursive presence in the text, we can better understand 'the way in which the translator's voice insinuates itself into the discourse and adjusts to the displacement which translation brings about' (Hermans 1996: 43). If we listen carefully for the voices of Humboldt's English translators – sometimes concealed, sometimes more conspicuous – in their translations, and investigate these moments of intrusion or their sustained 'insinuation', we are then better placed to appreciate what they tell us about the challenges and rewards of mediating his work to an Anglophone audience.

Conclusions

Rather than simply being a stylistic inventory of the changes Humboldt's texts underwent in translation and retranslation, the analyses in the chapters that follow will be concerned with identifying the variations between the source text and the translation, and mapping these divergences to the cultural, social and ideological environment in which these texts circulated. The stylistic interventions discussed introduce broader issues pertaining to the overall presentation of a book, the publisher with whom it appears and the series within which it might be marketed. In the book itself, paratextual material such as a translator's preface or postscript, footnotes or endnotes, together with organisational features such as indexes, all indicate how those individuals subsumed under the notion of the 'imagined translator' work to reframe a text for a different reading public and direct how a reader accesses and responds to the knowledge a book contains. A close textual analysis of the English editions of Humboldt's French and German source texts will be able to reveal other forms of stylistic intervention, including what the translation foregrounds and whether it omits source text material, how it employs narrative perspective, point of view and (in)direct speech, and how it negotiates metaphorical and allusive forms of language. By comparing the different contexts of cultural production, we aim to uncover what motivated a translator's choices and understand how competing translations of his works were lent an individual stylistic identity that enabled them to circulate side by side on the Anglophone book market.

Research on style in translation has tended to focus on relatively large text corpora. Mona Baker's study of the œuvre of two British literary translators has, for example, shown how certain 'patterns of choice' point to individual stylistic tendencies (Baker 2000: 246). These patterns enable us to uncover the particular 'thumb-print' of an individual translator by highlighting repeated translation strategies, typical manners of expression and characteristic uses of language. In the chapters that follow we will, for obvious reasons, be focusing primarily on how Black, Williams, Otté and the Sabines put key texts by Humboldt into English. But by contextualising this translation work within their wider creative enterprise, we gain a clearer sense of how their stylistic choices are encoded in assumptions about the text and its author, their own role as mediators of scientific knowledge and the translation's intended readership. An investigation of the experimental energy that these translators threw into 'Englishing' Humboldt brings into sharper relief the difficulties they encountered – and successfully circumvented – in negotiating different national styles of writing, cultural traditions and audience expectations.

Notes

1. Darwin Correspondence Project, available at <https://www.darwinproject.ac.uk/letter/DCP-LETT-3338.xml>, letter of 3 December 1861 from Charles Darwin to H. W. Bates (last accessed 6 October 2017).
2. *Berlinische Nachrichten von Staats- und Gelehrten Sachen* (*Spenersche Zeitung*), 8 December 1827.
3. Darwin Correspondence Project, available at <https://www.darwinproject.ac.uk/DCP-LETT-158.xml>, letter of 8 February–1 March 1832 from Charles Darwin to Robert Waring Darwin (last accessed 6 October 2017).
4. Darwin Correspondence Project, available at <https://www.darwinproject.ac.uk/DCP-LETT-4615.xml>, letter of 16 September 1864 from Hermann Kindt to Charles Darwin (last accessed 6 October 2017).
5. Darwin Correspondence Project, available at <https://www.darwinproject.ac.uk/DCP-LETT-224.xml>, letter of 28 October 1833 from Caroline Darwin to Charles Darwin (last accessed 6 October 2017).

Chapter 2

Dispute and Dissociation: John Black's *Political Essay on the Kingdom of New Spain* (1811)

In 1811 the first of Humboldt's major works appeared in English translation. An account of the population, climate, industry and agriculture of Mexico, the *Essai politique sur le royaume de la Nouvelle-Espagne* (1808–11), translated by the young Scottish journalist John Black (1783–1855) as the *Political Essay on the Kingdom of New Spain*, attracted immediate interest in Britain. It gave important new insights into mining, manufacturing, defence and revenue in the Spanish colonies. Hailed by the *New Universal Magazine* as 'the most laudable travels ever undertaken by individuals for the progress of science', Black's translation of the *Essai politique* was declared by the *Critical Review* to throw 'more light on the state of New Spain than any which has been hitherto published', and offer a store of useful information to the 'philosopher, the merchant and the statesman' alike.[1] The appearance of the *Political Essay* in Britain was timely. In September 1810, the liberal priest and revolutionary Miguel Hidalgo y Costilla had declared Mexico's independence from the Spanish crown. The Peninsular War, which Spain had been waging against France from 1808 following Joseph Bonaparte's seizure of the Spanish throne, made the volatile future of Spain and its colonies a subject closely followed in the British press. And the fact that the *Essai politique* was by Humboldt heightened its market appeal: 'The attraction excited by the subject receives also much addition from the name of a traveller, who adds to the activity of an inquisitive mind the stores of extensive erudition,' enthused the *Monthly Review*.[2]

But if British critics were fascinated by the author of the *Political Essay*, they were less enthusiastic about its translator. The *Literary Panorama* remarked sourly that 'his labour could have been more honourable to his abilities had he carefully re-inspected it, before it was committed to the press'.[3] It was not just the text that bore signs of haste. The plates annexed to the *Political Essay* had been so hurriedly printed that 'those who have seen the originals will bestow but moderate commendation on these *translations*', continued the *Literary Panorama*, making 'translation' synonymous with the derivative and the second-rate. The *Monthly Review* of December 1811 was equally damning. Its critic, Joseph Lowe, acknowledged that the translator was 'a man of education and capacity', but deplored his translation for resolving none of the stylistic and structural problems in the French original.[4] As the *Monthly Review* would reflect some five years later, the great drawback of Humboldt's essay on New Spain was

> the almost total want of method which it betrayed; and which, coupled with the disadvantage of a hasty translation, made it a toil to the English reader to peruse that which, under better management, might have been rendered one of the most entertaining books of the age.[5]

The *Essai politique* had its origins in the Spanish 'Tablas Geográfico-Políticas del Reino de la Nueva-España' [Geographical–Political Tables of the Kingdom of New Spain] (1804), a report written up by Humboldt as a token of his appreciation to the Spanish king for permission to travel through his dominions. Shortly after landing in Acapulco in March 1803, Humboldt had been granted the necessary papers to travel through Mexico for scientific purposes. He was given access to the Mexico City archive, where he consulted its manuscripts, statistics and maps. He also amassed a wealth of other scientific data on his excursions through the country to its mines, factories and towns, before leaving the country in March 1804. As Thea Pitman notes, this material would make the *Essai politique* a fascinating publication precisely because Humboldt was one of very few foreign travellers to Spain's American colonies in the years before they gained independence (Pitman 2007: 211). The 'Tablas', as the title suggests, largely contained tabulated statistics on subjects as diverse as population growth, political and administrative structures or trade. But as Humboldt reworked the brief manuscript, turning it into French, and expanding it considerably by adding maps and extending its historical depth, it morphed into an account sprawling across some 2,000 pages in the French octavo edition of 1811. For all its comprehensiveness, though, the *Essai politique* cannot be divorced from its modest beginnings. Mary Maples Dunn rightly stresses its 'overall limitation of locale' and peripheral feel, compared with the far more comprehensive character of later works resulting from Humboldt's voyage (Humboldt 1988: 8–9).

Even a narrative that focused solely on Mexico was of immediate interest to British readers. Their curiosity had already been piqued in 1805 following the publication of an eighteen-page account by the French naturalist Jean-Claude Delamétherie (1804) of Humboldt and

Bonpland's voyage to the Americas in the 1804 edition of the *Journal de physique, de chimie, d'histoire naturelle et des arts* [*Journal of Physics, Chemistry, Natural History and the Arts*]. The 'Notice' of their travels not only described their itinerary starting from Cumana (in today's Venezuela) in July 1799, down the Río Apure to the Orinoco, back up to the coast and on to Cuba, south again through Colombia and Ecuador to Lima, then northwards via New Spain, up the North American coast to Philadelphia and back across the Atlantic, arriving in Bordeaux in August 1804. It also promised new findings on the geology of the Andean *cordilleras*, the botany of the Río Sinú, temperature and electric charge at the volcanic crater of Pichincha, the passage of Mercury (as seen at Lima) and the rich seams of silver ore at the Valenciana mines. Delaméthrie's account also whet the appetite of the general reading public by describing the physical hardships that the travellers had endured, plagued by biting insects under the burning South American sun, fearful of encountering the belligerent Guaicas Indians, and exposed to the bitter cold in the permanent snows of the Andean peaks. Although the 'Notice' was never translated into English, by 1810 Humboldt was already a familiar name in journals as diverse as the *Anti-Jacobin Review*, the *Literary Panorama*, the *Farmer's Magazine* and the more urbane *La Belle Assemblée: or Court and Fashionable Magazine*.

Black was not merely imagining the 'impatience of the public' (*PE* I, xiii) as he set to work on the *Essai politique*. British reviewers had already paraphrased or translated from it at such length that Black's justification of his own translation on the grounds that a 'considerable part of the Essay on New Spain has not yet arrived in this country' (*PE* I, xiii) was somewhat defensive. Black was careful to offer a full translation of Humboldt's central narrative, complete with tables, as well as its paratextual material (including the dedication to His Catholic Majesty Charles IV of Spain, Humboldt's footnotes and index). But in other respects, his handling of the text was anything but cautious. His own vast footnote apparatus queried, corrected and criticised Humboldt's text. Readers of the English narrative could routinely find six or seven brief additional footnotes appended to just one side of text, while longer annotations easily ranged across a good three pages. These footnotes gave the narrative another, highly audible, voice that was in constant dialogue with Humboldt's own: these paratextual interventions would be abruptly silenced only in the third English volume. Humboldt was already well aware of translation's deforming and transforming powers, having been compelled to translate the first parts of the German edition (1809–14) of the *Essai politique* himself, after paying a translator who produced a first draft so unsatisfactory that Humboldt's ruthless revisions practically obliterated it (Humboldt 2009: 97, 119). Horrified at Black's appropriation of his text in this way, Humboldt fumed to Pictet in April 1811, that he had been mistreated in the translation by a man whom he considered 'd'une bêtise amère' [resentfully stupid] (Humboldt 1869: 72). Black tells the public, seethed Humboldt, that he has added the notes for his own enjoyment – *'pour s'amuser'* – and clearly works himself into a lather over printing errors that any child could have guessed were such (Humboldt 1869: 72). Joseph Banks mollified the peeved Prussian by writing to him 'd'une manière qui prouve que tout le monde là-bas ne pense pas sur moi comme M. Black' [in such a manner as to prove that not everyone over there thinks the same of me as Mr Black] (Humboldt 1869: 72).

This chapter focuses primarily on the footnoting that Black appended to the translation, a form of paratextual intervention exploited so intensively to articulate his own ideas and opinions that the 'voice' of the translator could scarcely be ignored. It starts by exploring how Humboldt's *Political Essay* constituted new knowledge for a British readership and which stylistic forms it adopted to convey this information. It then examines why Black felt motivated to add such weighty paratextual apparatus (not least given the time pressure he faced), and how translating Humboldt's *Essai politique* fitted into Black's translation agenda. Discussions of 'voice' in translation highlight how Black was able to use footnotes to subvert the traditional power differential between Humboldt as author and himself as translator and make of the *Political Essay* a piece critical of the very source text that it apparently reproduced. Finally, this chapter looks beyond the first edition of Black's translation to Longman's later, third edition of the *Political Essay* (1822) and extracts that appeared in John Taylor's edited *Selections from the Works of the Baron de Humboldt* (1824), to explore the further definition that the *Political Essay* gained in successive versions.

The *Essai politique*: Rewriting the New World

Humboldt's account of his travels to Mexico was, as he wrote to his German publisher Johann Friedrich Cotta in 1809, the last of the four volumes into which he had intended to organise the material gathered during his five-year voyage with Bonpland to the Americas (Humboldt 2009: 103). His initial plans to split up his travels into four sections – on the Orinoco, New Granada, Quito and the River Amazon, and on Mexico – would come to fruition only in part, however. The *Relation historique*, intended to be the comprehensive account of his journey, stopped abruptly before the travellers embarked on a trip down the Río Magdalena in Western Colombia in March 1801 and never arrives at a discussion of the Mexican leg of the journey. Humboldt's decision to complete the Mexico work before he finalised the *Relation historique* probably hinges on the fact that his report to the Spanish monarch served as a useful basis for revision and elaboration, and that the *Essai politique* was reasonably contained, both chronologically and geographically. The lengthy 'Geographical Introduction' countered the impression given by the title that this work focused solely on the political and administrative characteristics of Mexico. As Hanno Beck notes, this work was as much a political as a geographical overview of New Spain (Beck 1966: 34), organised around the itineraries of three journeys made: from Guanaxato up the Río Bravo to the ancient capital city of Tula in the central–eastern part of the country in the company of young nobleman Carlos Montúfar; from the mouth of the Río del Norte to the Sea of Cortez; and from Mazatlán to the city of Altamira.

The six books into which the main body of the *Essai politique* was organised indicate the vast thematic range of Humboldt's project: the 'General Considerations on the Extent and Physical Aspect of the Kingdom of New Spain', the 'General Population of New Spain', the 'Particular Statistical Account of the Intendancies of which the Kingdom of New Spain is Composed', the 'State of the Agriculture of New Spain', the 'State of the Manufactures and Commerce of New Spain' and the 'Revenue of the State, Military Defence'. It was essential for rebutting assertions made by previous chroniclers about the differences between the 'Old World' and the 'New World'. As Antonello Gerbi observes, these had cast the Americas as 'an immature or impotent continent' in some way inferior to Africa, Asia and Europe (Gerbi 1973: xv). Buffon had concluded that the animal species in South America were in many cases smaller, weaker and more cowardly than those to be found in Europe or Africa, due to the hostile natural environment in America. Physically, too, America had been portrayed as a new world that, having emerged only relatively recently from the sea, was too underdeveloped to nurture civilised peoples.

David Hume, in his famous essay *Of National Characters* (1748), considered those who dwelt 'beyond the polar Circles or betwixt the Tropics' to be 'inferior to the rest of the Species', thus casting the natives of America living in these regions as culturally insignificant, according to his own mode of geographical determinism (Hume 1748: 279). The Abbé Raynal had similarly described the inhabitants of America as degenerate and decadent, lacking the moral fibre of the Europeans, while Cornelius De Pauw considered the American savages uncultivated, uneducated and indolent. Humboldt's writings on the Americas would thus occupy a central place in the European dispute on the New World, by demonstrating that the assumptions and assertions made by Buffon, Raynal and De Pauw ill represented the reality of the exuberant proliferation of life in America's tropics, the level of civilisation that its ancient peoples had attained or the intrinsic value of its present culture. As Humboldt remarked of Mexican miners:

> The appearance of these robust and laborious men would have operated a change in the opinions of the Raynals and De Pauws, and a number of other authors, however estimable in other respects, who have been pleased to declaim against the degeneracy of our species in the torrid zone. (*PE* I, 125–6)

And he was quick to chide Robertson, Raynal and Pauw, who 'disfigure the names of cities and provinces', for failing to appreciate the subtleties of Aztec orthography and, more generally, the complexities of ancient Mexican culture (*PE* I, cv).

To an Anglophone audience, the *Political Essay* – in Black's rendering – was the first of Humboldt's works to engage critically with such questions.[6] It also took a decisive stance on colonisation and slavery. During his visit to New Spain in 1803 and 1804, he witnessed the advantages that Spanish colonisation had brought to Mexico, not least in the field of education. He sat in on examinations at the School of Mines in Mexico City, where he also lectured on the geological and mining data acquired on his journey and composed his *Essay de Pasigraphie* [*Essay on Pasigraphy*] on how geological formations could be represented in a 'universal', diagrammatic language. His enthusiasm for the benefits brought by contact with Spain was tempered by the oppressive hierarchies that accompanied it. Openly ambivalent about Spanish colonial treatment of the native peoples, he concluded the *Political Essay* by impressing upon his readers

> that the prosperity of the whites is intimately connected with that of the copper coloured race, and that there can be no durable prosperity for the two Americas till this unfortunate race, humiliated but not degraded by long oppression, shall participate in all the advantages resulting from the progress of civilization and the improvement of social order! (*PE* IV, 282)

But as Humboldt complained in aggrieved letters to Pictet in early 1806, such questions were not of the slightest interest to his British publishers. What they really wanted to get their hands on were the statistical tables in the Mexico account. Sir Andrew Snape Hamond, head of the Navy Board, had already advised him that these alone were worth £1,000 (Humboldt 1869: 50). As Humboldt acerbically noted, the subtext to his communications with Longman was that the firm essentially sought works of popular or economic interest, but not science. Tales of the South Seas were more up their street, as was his statistical account of Mexico, but only because it enabled readers to find out 'ce que vaut la cochenille en place' [how much cochineal costs at source] (Humboldt 1869: 50).

By the time Humboldt approached them, Longman and company had already begun to establish a niche in non-fictional travel writing, with works like John Pinkerton's multivolume *General Collection of the Best and Most Interesting Voyages and Travels in All Parts of the World* (1808–14) and Lewis Meriwether's compilation *The Travels of Capts Lewis and Clarke from St. Louis by Way of the Missouri and Columbia Rivers to the Pacific Ocean* (1809). They were clearly still interested in more subjective accounts such as Catalina Schuyler's biographical *Memoirs of an American Lady: With Sketches of Manners and Scenery in America* (1809), but may have hoped to press Humboldt's writing more firmly into the mould of Hugh Gray's *Letters from Canada* (1809), which discussed trade and political relations in Nova Scotia. Humboldt's disgust at what he perceived to be Longman's interest only in exotic light adventure or hard economic fact was not something he would quickly forget, even when the *Essai politique* eventually found favour:

> Mon ouvrage sur le Mexique est vers la fin. Vous vous souvenez de ce libraire léopard qui, ne voulant pas de mon astronomie, parce qu'il assurait que ce ne serait que quelque «mexican Guide» réchauffé, s'extasiait sur la Statistique du Mexique, qu'il regardait comme la pierre philosophale. Hé bien! dans ce dernier cahier, ses yeux seront ravis de l'aspect de tant de chiffres, qui tous expriment de l'argent. (Humboldt 1869: 78)
> [My Mexico work is nearing completion. You will remember the leopardic (=British) publisher who, not interested in my astronomy, since he declared that it would be little more than some reheated 'Guide to Mexico', was in ecstasy at the Statistical Account of Mexico, which he considered on a par with the Philosopher's Stone. Well! In this last section his eyes will delight at the sight of so many figures that all talk money.]

Humboldt therefore realised early on that his concerns to impart scientific knowledge to a general public would need to accommodate the interests of the British book trade, if the translation of the *Essai politique* was to appear in print.

Visible Traces

How did John Black, the son of a Perthshire pedlar, come to translate Humboldt's *Essai politique*? What motivated him to embark on this undertaking and why was his visibility in the translation so important? One answer can be found in Black's humble origins, which instilled in him a driving ambition for self-improvement and intellectual display. Another lies in his radical politics. Black started his working life as a clerk, attending lectures at Edinburgh University in his spare time, and supplementing his income with translation. Besides honing his skills at Latin and Greek, he learned German from an Austrian musician employed in the theatre and Italian from another musical colleague, as well as French (Walford 1856: 270). Black's earliest commissions include translations from the German for the Scottish physicist and mathematician David Brewster's *Edinburgh Encylopædia* (1808–30), which counted among its more illustrious contributors Charles Babbage, Thomas Carlyle, Robert Gordon and Thomas Telford. Black's experience in translating non-fictional, factual, matter was balanced by an interest in the arts: from 1807 to 1809 he wrote articles on Italian drama and German literature for his friend William Mudford, editor of the *Universal Magazine* in London. In *Cobbett's Political Register*, Black and Mudford engaged in an energetic 'battle of the books', Black defending classical study and his opponent advocating the benefits of modern learning (Jones 1989: 79).

Black came down to London in 1810 to translate foreign journal articles for the *Morning Chronicle*, which had already acquired a formidable reputation as a Whig paper. As Black turned his hand to reporting political debates from the House of Commons, he earned a reputation for being radical and relentless, and the nickname 'Flying Scotchman' for the speed at which he penned his columns.[7] The English philosopher and social reformer John Stuart Mill declared him the 'first journalist who carried criticism & the spirit of reform into the details of English institutions [...] Black was the writer who carried the warfare into these subjects' (Mill 1996: XV, 979). To the utilitarian thinker Jeremy Bentham he was the greatest publicist Britain had produced (Escott 1911: 159). Although Black energetically encouraged debate, the publishing historian James Grant remarked on how his insistent argumentation tended to excess: when he 'took up any question of the day, he not only worked it threadbare, but positively hammered away at it long after it had been all but forgotten by the public' (Grant 1871: I, 281–2).

Black's appointment in 1819 as principal editor of the *Morning Chronicle* compelled him to focus his energies solely on journalism. But his brief career as a translator – spanning not even a decade – would nevertheless see him acquire an international reputation as a translator of (scientific) travel writing, literary memoirs and political treatises.[8] After completing Humboldt's *Essai politique* in 1811, he moved on to the German geologist Leopold von Buch's 1810 account of his Scandinavian explorations, which appeared as the *Travels through Norway and Lapland During the Years 1806, 1807, and 1808* (1813) with notes and illustrations by Robert Jameson, Professor of Natural History at Edinburgh. He then turned to the Swedish chemist Jöns Jacob Berzelius's *Attempt to Establish a Pure Scientific System of Mineralogy* (1814), followed by August Wilhelm Schlegel's key text on German Romantic aesthetics, the *Course of Lectures on the Dramatic Arts and Literature* (1815). Black then returned to his own specialism of travel literature, working on Isaac von Gerning's *A Picturesque Tour along the Rhine, from Mentz to Cologne* (1820; original 1814), before rounding off his career with a characteristically radical flourish by translating Johann Joseph von Görres's politically controversial *Teutschland und die Revolution* [*Germany and the Revolution*] (1819; trans. 1820), which had been banned in the German states. Black therefore brought to the Humboldt translation the ability to work at speed and a sound knowledge of foreign languages, coupled with a passionate interest in current affairs. He also brought a propensity for active self-marketing.

The notion of 'visibility' in translation has attracted extensive attention since Venuti's discussion of the 'illusion of transparency', which has long governed how people have thought about the translator's situation. As Venuti stresses, a translated text 'is judged acceptable by most publishers, reviewers and readers when it reads fluently, when the absence of any linguistic or stylistic peculiarities makes it seem transparent', and when it creates the appearance 'that the translation is not in fact a translation, but the "original"' (Venuti 2008: 1). Readers, he asserts, play a significant role in constructing this illusion, partly because they read for content and partly because they use fluency as the benchmark by which to gauge the translator's invisibility: 'the more fluent the translation, the more invisible the translator, and, presumably, the more visible the writer or meaning of the foreign text' (Venuti 2008: 1).

Issues of visibility are generally played out on three levels – intratextually (within the text); paratextually and peritextually (for example, on the cover, in the preface, in footnotes); and extratextually (such as in reviews). Black mobilised the first two of these with particular vigour. Common to all his translations was his visibility on the title page, as the named translator, and in a translator's preface or similar 'advertisement'. His urgent need for 'visibility' as he tackled the two hefty quarto tomes of the *Essai politique* is perhaps best explained by the fact that it was the first major piece the twenty-eight-year-old translator undertook, whereas Williams, Ross and Elizabeth Sabine grappled with Humboldt's works as established writers and translators in their forties and fifties. Longman did not baulk at Black's relative inexperience. But Humboldt might have had greater misgivings about his first British translator, had he known more about him. As Maeve Olohan has emphasised, scientific translators embodied a 'central role as gate-keeper and localiser of scientific material' and were key in shaping 'some of the material and social contingencies' that determined the appearance of a scientific work on the publishing market (Olohan 2013: 433). Humboldt had originally intended, upon his return from the Americas, to entrust supervision of English editions of the *Essai politique* to Pictet. That this work would be translated by a man whom Humboldt did not know and who sought no contact with him other than his authorisation to undertake its translation into English meant that the genesis of the *Political Essay* was largely taken out of his hands.

Opening Moves

The *Essai politique* was a demanding text to put into English. On rare occasions, the translator was called upon to embrace Humboldt's enthusiasm for the exotic landscape of the Americas ('the delicious country which nature has enriched with the most varied productions'), marvel at the 'uncommon beauty and strength' of the vegetation of the *cordilleras* or portray the 'rough and angry' seas off Acapulco (*PE* IV, 76; I, 78; IV, 60). More often, though, he was confronted with pages of measurements and tables that scarcely offered that combination of objective observation and subjective reflection central to later works such as the *Personal Narrative*. Leask is correct in describing the *Essai politique* as 'more of an assemblage of statistical and naturalistic data [. . .] than a travel book proper', in which the personality of the traveller and his narrative 'survive only vestigially beneath the systematic presentation of statistical and geographical information' (Leask 1999: 185).

Humboldt's translator therefore needed to be proficient in the technical vocabulary used across a range of different disciplines, including social and physical geography, agriculture, business and industrial technology (particularly mining) to produce a high-quality translation. It is not difficult to sympathise with Black, struggling to find English equivalents for mineralogical terms such as 'hyalite mammelonnée' ('mammeloneous hyalite' or 'müllerisch-glass'), anatomical words such as 'os occipital' (occipital bone), and economic terminology like 'balance of commerce' – not to mention words from the Aztec language Nahuatl, neatly embodied by the 'problematical reptile called *axolotl*' (*EP* II, 585; I, 90; I, 12, 168; *PE* III, 303; I, 155; I, 18; II, 17). Translating the source text also required fluency in French and Spanish, given Humboldt's comments on the *provincias internas* of New Spain and the plains or *llanos* of Mexico. Before considering how Black handled these various problems, it is worth pausing to analyse how he constructed an identity for himself as Humboldt's translator. An examination of the translation solutions he employed to circumvent these problems then enables us to investigate how these approaches affected the reception of Humboldt's text by an Anglophone reading public.

'It is observed by a popular French writer, Bernardin de St. Pierre, that by far the most valuable and entertaining part of modern literature is the department filled up by travellers,' stated Black in the opening to his eleven-page 'Preface by the translator' in the *Political Essay* (PE I, iii). By associating Humboldt with the author of the best-selling *Paul et Virginie* (1788) – who was also a respected botanist in his own right – Black was immediately affirming the appeal of Humboldt's travel writing to a general audience. Emphasising the modernity of this genre and its ability to encapsulate the 'civilization', 'refinement and enterprize [sic] of the inhabitants of the west of Europe', in contrast to the savagery of more remote corners of the world, Black neatly allied Humboldt's *Essai politique* with that Enlightenment watchword, 'progress' (PE I, iii). But if Black applauded the considerable contribution that authentic, impartial travel accounts had made to the advancement of knowledge, he also registered the complaints voiced, from the mid-eighteenth century onwards, about the indiscriminate and insipid accounts flooding the book market that could make no such claims to novelty and originality. Black therefore used the opening of his Preface to suggest an easy familiarity with the genre of travel writing and to signal his awareness of what constituted a 'good' travel narrative. In acknowledging the principles governing this genre, he would, however, be compelled later in the Preface to adopt a more critical stance towards the text he was otherwise presenting so confidently to a British public.

In his brief survey of European travel writing, Black wrote Humboldt into an impressive pantheon of figures, some of whom would be familiar to a general readership:

> M. de Humboldt belongs to a higher order of travellers [...]. We must place him beside a Niebuhr, a Pallas, a Bruce, a Chardin, a Barrow, and a Volney; and his works will probably be long consulted as authorities respecting the countries which he describes. (PE I, iv–v)

In charting the geographical range of these travellers – the Danish pioneer Carsten Niebuhr's voyages to Egypt, Syria and Arabia or Peter Simon Pallas's travels through Russia and Siberia – Black was subtly demonstrating how Humboldt's narrative on Mexico would make an important new contribution to the existing corpus of travel writing. And in mentioning travellers who would be familiar precisely because their accounts existed in English translation, Black also implicitly emphasised the importance of translation in the transmission of texts on foreign lands, peoples and cultures.

While Black used the opening of his Preface to reflect on what European travel writing had achieved over the previous half-century, by the third page his critical, dissociative stance towards his source text had acquired firmer definition. Black had stressed in the previous paragraphs the important role that authentic travel accounts played in acting as 'proper sources of information' (*PE* I, v–vi). He now pointed up what he perceived as an essential conflict of interests in Humboldt's narrative:

> The work of which a translation is here offered to the public was submitted to a very severe trial: the sketch of it was freely communicated to the natives of New Spain, and underwent the examination of the Spanish government. It may be doubted, however, whether the accuracy and fulness [*sic*] of information which such a measure has a tendency to procure might not be counterbalanced by seemingly unavoidable disadvantages. We never talk of our friends so candidly before their faces as behind their backs. [. . .] Even Dr. Johnson, with all his bluntness, would have hesitated to read his Tour to the Hebrides to his Scotch landlords. (*PE* I, v–vi)

Clearly, Black could not wholly endorse the publication of an account that had initially been destined for the Spanish government, with all the political complications this entailed. Above all, he was suspicious that the narrative underpinning the *Essai politique* had been written to discharge a number of personal debts Humboldt had incurred in the Americas. 'We accordingly find him exceedingly prone to give favourable accounts of all the individuals of that country whom he has occasion to mention,' Black noted, adding that Humboldt was 'profuse in his compliments to their learning, science, and their other good qualities, and nothing ever appears to shade the picture' (*PE* I, vii). Black did not so much take issue with the effusive compliments Humboldt paid to those whose help he had appreciated, as wonder at the lack of balance in an account in which few individuals or institutions were mentioned that incurred Humboldt's disapproval. Conscious of the high expectations that critics would have of a travelogue that claimed the authority of a Bruce or a Barrow, Black sought to parry any attacks critics would make on the obvious impartiality of Humboldt's narrative by admitting this weakness from the outset.

He was not only critical of the limitations to the information that Humboldt had gathered, but also found fault with the style of its presentation. While Black conceded that the great strength of the *Essai politique* was its 'precise data on a very great variety of important subjects', this information was poorly packaged, with repetition and prolixity among the greatest shortcomings (*PE* I, viii). In response to Humboldt's inordinately long discussion of the importance and value of the banana, Black observed that it was to be regretted 'that the author could not throw occasionally more rapidity into his descriptions, and give somewhat more condensation to his materials' (*PE* I, viii). Black did not, however, put these problems down to personal stylistic idiosyncrasy. Rather, he associated them with those notions of 'national' style that commentators such as De Quincey would point up a couple of decades later:

> This failing is not peculiar to M. de Humboldt, but is common to him with too many authors, and particularly those of his own country, Germany. Indeed the faculty of selecting the more important and leading features of an object is, perhaps, the rarest and most valuable which any writer can possess. It is this which communicates such a charm to the history of Hume, and arrests so strongly our attention in the travels of Volney. (*PE* I, viii)

Black was therefore keenly aware of differences in narrative style and structure that could mark this text as potentially 'un-English', although he did not address the complications of Humboldt's own linguistic situation: namely, that he was a native speaker of German composing a text in French. Rather, he seemed to suggest that writing style was an innate, nationally discrete affair, the characteristics of which even pervaded authorship in a foreign language.

As Humboldt's translator, Black set a critical distance between himself and his author that was to become far more marked in the extensive footnotes crowding below the main narrative. He signalled from the outset that he was not prepared to manipulate the text by tacitly omitting here and there a few needless repetitions or irrelevant passages to resolve the problems. In any case, abridgement was scarcely an option, given that an English audience already knew in some detail what the work contained. Moreover, by editing the translation to produce an abridged version, Black would have opened the door to competition, since a rival translation could swiftly purport to be better because it was more comprehensive.

There were two exceptions to Black's fixation with fidelity that he was at pains to justify in the Preface: his changes to spelling and his addition of footnotes. 'Who', he queried, 'could find out *Washington* in *Ouachinnetone*?', as he brought Humboldt's orthography into line with standard English usage (*PE* I, xi). With regard to his footnotes, Black initially made light of them by remarking that they had been 'occasionally thrown in' and were of little importance, intended only to make the work more easily comprehensible to the Anglophone reader (*PE* I, ix). In the 'Advertisement' to the third volume, in which Black largely silenced his own critical voice and continued to use footnotes only to give the conversions of weights, measures and currency, he was, however, more forthright about justifying the role that annotation had played in the first two volumes:

> It is hardly possible for a Translator of the most obtuse intellect not occasionally to perceive a vulnerable point in his original; and what the present Translator perceives or imagines he perceives, he is at no time very willing to keep locked up from others. . . . (*PE* III, n.p.)

Here Black neatly voices the predicament of reflective translators: that they are not in a position to speak, yet unwilling to keep their insights 'locked up from others'. Black was also querying the tacit acceptance that translators should have no voice in their translations and take a solely deferential stance towards the source text and its author – a role that Black would adopt, under duress, only in the second half of the *Political Essay*.

Translation and Annotation

In his seminal work on the footnote, Anthony Grafton has observed that in the Enlightenment it achieved 'a high form of literary art' that served both scholarly and polemical ends (Grafton 1997: 1–2). Historical works – notably Edward Gibbon's *History of the Decline and Fall of the Roman Empire* (1776–89) – used footnotes to anchor the narrative within extensive scholarly bodies of research but also to engage in fierce intellectual disputes that included vitriolic, satirical characterisations of fellow historians (Palmeri 1990: 249). Other subversive ways of making the footnote a place of controversy rather than seemingly impregnable superiority had been employed much earlier in British fictional writing. Jonathan Swift's *A Tale of a Tub* (1704) used extravagant textual apparatus, including ample footnotes, to deflate the intellectual mystique associated with the book, and to provoke readers into approaching the written word with a more questioning and discerning eye. Swift's goal was not to reject textual authority outright but to insist that it be 'vested in the proper places' and foster greater interpretative acumen (Mueller 2003: 208). A couple of decades later, Alexander Pope in his *Dunciad Variorum* (1729), the second edition to his mock epic *The Dunciad*, likewise appended an enormous apparatus of footnotes to the text, many attributed to the scribal persona of Martinus Scriblerus, which were deployed as a way of satirising Grub Street authors and scholarly pedants. A vilification of dusty antiquarianism, false scholarship and a misplaced faith in fact, Pope's footnotes were themselves a thinly veiled attack upon the idea that the footnote was valid testimony to the inherent truthfulness of the text it accompanied. They also parodied contemporary misuses of the footnote such as excessive length, thematic irrelevance or inclusion of patently nonsensical content (Strang 1992: 214–15). Pope and Swift therefore poked fun at the seductive hold that footnotes could have over author and reader alike as narrative devices that were seemingly 'objective' and critically aloof, yet imparted information of questionable value.

Footnoting in scientific writing had been the subject of fierce debates in the eighteenth century that revolved primarily around factuality – what facts signified and how they could be deployed in non-fictional writing. Lorraine Daston has observed that long before the word 'objective' had come into common use, the Royal Society had appealed to its members to adopt a language of impartiality, closely allied with notions of 'civility' in communication, to avoid personal squabbles openly impinging upon intellectual debate (Daston 1991: 337–63). As an indispensable tool in bolstering a scientific text's apparent objectivity, the footnote offered the reassurance that the arguments in the main body of text depended upon verifiable objective fact. The author of the text therefore appeared in two different personae: in the central part of the narrative as the 'interested pleader of the text who does not conceal his subjective interests as the proponent of a specific idea' and in the marginal part of the text, the footnote, as the 'objective gatherer and presenter of evidence' (Cosgrove 1991: 132). Grafton goes further in arguing that it now signals that a work is bound up 'with the ideology and the technical practices of a profession', indicating the specialised training scholars have undergone to achieve the approval of their peers (Grafton 1997: 5).

Humboldt had himself used extensive footnoting to disarm scientific critics, to alert readers to related publications in the field, to embed his writing within wider historical, philological and literary discourses, and to position himself within the rapidly expanding web of scientific communication. But they also became his Achilles heel. Footnotes hampered revisions as he struggled to include the most up-to-date references across a bewildering range of disciplines, publications and languages. They also anchored the production of the text to an even more specific moment in the development of scientific knowledge. By the time the manuscript, wrenched from Humboldt's hands by an exasperated publisher, had appeared in print, the footnotes would have immediately betrayed that it was out of date.

Forster was one of the first writers of scientific travel literature to harness the footnote's full rhetorical potential. He deftly deployed annotation in his two-volume *Voyage Round the World* (1777) to consolidate his position professionally, fire a few political broadsides at the British establishment and give his own account a central position within modern European travel literature. He also frequently used footnotes to cross-reference his own narrative with writing by more established travellers, notably John Hawkesworth's authoritative narrative of Cook's first circumnavigatory voyage *An Account of the Voyages Undertaken ... for Making Discoveries in the Southern Hemisphere* (1773), the Welsh naturalist Thomas Pennant's *British Zoology* (1766) and the Scottish explorer Alexander Dalrymple's *Historical Collection of the Several Voyages and Discoveries in the South Pacific Ocean* (1770). Leask has observed that Forster's work would become a 'milestone for romantic period travel writing, establishing the principles which would increasingly be demanded from scientific travel writers over the next half-century' (Leask 2002: 41).

It is important to recall that Forster's *Voyage* also served as a prime example of the many ways in which annotation could be deployed in the genre, particularly in the translation of travel accounts. Forster started out by using footnoting to advertise his own publications (notably the *Voyage Round the World*) – but swiftly developed his skills as a translator and annotator, to use footnoting in more complex ways (Martin 2006; Martin 2007). In the *Nachrichten von den Pelew-Inseln in der Westgegend des stillen Oceans* (1789), a translation of George Keate's exotic tale of maritime disaster, the *Account of the Pelew Islands Situated in the Western Part of the Pacific Ocean* (1788), he established a provocative presence in the text that affirmed his status as both an Anglo-German cultural ambassador and a European authority in the area of scientific travel. While Keate had himself given the English version thirty-one footnotes, Forster added a further forty-one to his German translation, some covering three or four pages in length, querying, correcting and supplementing the information provided by Keate. Rather than carefully manipulating the text as he translated it, and diplomatically reshaping or even omitting the passages with which he disagreed, Forster deliberately used translation to engage in provocative debate with the author of his source text, aggressively deploying annotation to undermine the status of the very work he was putting into German.

Translators therefore potentially share common ground with

historians and scientists who are also keen to convey, clarify and criticise the ideas of others. As Esther Allen argues, throughout history translators have also been 'called upon to gloss, annotate, comment upon and provide source references for the texts they translate' (Allen 2013: 211), and translation therefore implicitly carries an editorial component. More fundamentally, though, translation and annotation share a 'mutual orientation towards a clearly delimited subset of readers', since annotation 'anticipates the lacunae, requirements and areas of expertise of a specific group of readers', as does translation when it considers how best to make a text intelligible to a foreign culture (Allen 2013: 213). Black's addition of a critical and extensive footnote apparatus to Humboldt's *Essai politique* was not particularly unconventional.[9] But the stridency of his editorial voice certainly was. By refusing to keep his opinions 'locked up from others', Black was articulating what Jacques Derrida has proposed is an inherent double bind in footnoting: 'I am forbidden to speak' and yet 'I cannot remain silent' (Derrida 1991: 192). Footnoting, Derrida suggests, reinforces a legitimised distribution of space on the page that assigns hierarchical relationships to the so-called 'principal' text that is higher (symbolically and spatially) on the page than the footnote that is lower and more marginal (Derrida 1991: 193). The hierarchy between footnote and principal text is, above all, swiftly upturned in those cases where the annotator is operating in a polemical context. A riposte is actually most often read, Derrida argues, when it is in a footnote, since the footnote gives comments a paradoxical freedom and autonomy by dint of being perched at the very edge of the text, neither quite within nor outside it (Derrida 1991: 194).

Hermans's work on 'voice' in translation offers a slightly different perspective on how translators use translated texts to articulate their own opinions about the subject matter. Translation is a complex form of quotation, the reuttering of someone else's ideas in another language (Hermans 2007: 69–74). As translators respond to what they are 'quoting', they can choose to position themselves towards it in terms of detachment and accountability. 'The suspension of personal views is not always easy for translators,' Hermans suggests, 'even when they are in sympathy with what they translate' (Hermans 2007: 57). As we shall see, Black's longer annotations were rarely appended to passages where he was 'in sympathy' with Humboldt. Rather, they were used by Black to demur, show detachment from the original, and signal a refusal to be held responsible for its contents. Hermans terms this resistance by translators to the very text they translate a form of 'ironic' translation, an act of distancing that 'springs from a critical value judgement' (Hermans 2007: 79). In tracing the contours of Black's 'ironic' approach to translating the *Essai politique* in what follows, we can explore not only how it threw his own identity as translator into sharper relief, but also what image of Humboldt it presented to an Anglophone audience.

Between Domestication and Defamiliarisation: Translating the *Essai politique*

What is most striking about Black's prose is his grammatically literal translation of the French original, and his routine use of English cognates for French terms. For example, the map of the Gulf of Mexico, published by order of the King of Spain in 1799, had been 'retouched' ('retouché') in 1803, and Humboldt's own reworking of that map 'exhibits in a coup d'œil all the territories which depend on the viceroyalty of Mexico' (*PE* I, xviii). The expressions 'retouched' and 'coup d'œil' could be found in contemporary English writing, but other phrasings, such as 'revised' and 'at a glance', were also common and would have created a more natural-sounding English text. Using definite or indefinite articles in the English text in the same way as in the French also generated awkward expressions such as the 'cultivation of the sugar cane' (*PE* III, 2). Black therefore did not always do a good job of 'domesticating' Humboldt's prose in his translation, and did not convincingly sustain any illusion that this was a home-grown text.

Literal renderings also abound in the translations of the many, sometimes quite specialist, scientific terms in the *Political Essay*. From the outset, Black tried to resolve this problem by offering a tentative solution in English, immediately followed by the original foreign term in parentheses. In a section on the Anahuac region of central Mexico, Black observed that the 'undulations of the surface (*mouvemens du terrain*), the form of the mountains, their relative height, and the rapidity of the declivities, can only be completely represented in vertical sections', while in a later discussion of the relative aridity of the table-land of New Spain, Black's translation records that 'the springs are rare in mountains composed principally of porous amygdaloid, and fendilated (*fendillé*) porphyry' (*PE* I, cvii, 76). In a chapter on the mines of Old California in northwestern Mexico, Black was confronted with German mineralogical terms as Humboldt reflected on how 'these *masses* (*gangausfüllungen*) are partly the same with those which are exhibited in the veins of Saxony and Hungary', and remarked that the veins of native gold were most frequently found in the province of Oaxaca, 'either in gneiss or micaceous slate (*glimmerschiefer*)' (*PE* III, 129, 134, 148). Black's 'double entries' helped to ensure there was no loss of meaning and gave an intriguing insight into the different terminological systems that were emerging, but they also repeatedly pointed up the slightly uncertain mediated quality of the *Political Essay*.

At several junctures, even Black's belt-and-braces approach to tackling terminology failed to hold. In a section on modernisation and technology in the mines of Mexico, Humboldt had discussed in some detail the forms of labour used, the health of the workers and the structuring of mines to ensure that air circulated properly through the pits and galleries. He concluded, in Black's translation, 'In proportion as the mines of New Spain resemble more and more those of Freiberg, Clausthal, and Schemnitz [sic], the miner's health will be less injured by the influence of the *Mofettes**, and the excessively prolonged efforts of muscular motion', the asterisk referring to a footnote in which Black admitted that the translator 'professes his ignorance of the meaning of this word' (*PE* I, 127). This annotation was singled out for particularly vicious treatment by Humboldt, who exclaimed in his letter to Pictet about the British translator's annotations: 'il assure qu'en Angleterre personne ne peut lui dire ce que c'est que *moffette*' [he asserts that no one in England is able to tell him what a *moffette* is] (Humboldt 1869: 72). But in his irritation at Black's apparent laziness, Humboldt failed to understand that this term (denoting the exhalation of carbon dioxide from a fissure in the earth) was still relatively uncommon in early nineteenth-century English texts. It had been employed by William Hamilton in a letter published in his *Observations on Mount Vesuvius, Mount Etna, and Other Volcanos* (1772) – albeit spelt 'mofetes' – and it could also be found in the third edition (1797) of the *Encyclopaedia Britannica*, but here it was buried deep in an article on 'Damps' or 'noxious exhalations issuing from some parts of the earth', rather than commanding an entry in its own right (V, 656).

Black was doubtless wrestling with his own lack of scientific expertise, a reluctance to consult specialists in the field, and pressure of time, as he skated over complex terminology. Worse still, some terms quite simply did not exist in English at that time. Linguistic inventiveness was the only solution. If the historical etymologies in the *Oxford English Dictionary* (*OED*) are anything to go by, then Black's translation was the first popularly available English-language text to use words such as 'aerolith', 'cargador', 'greenstone', 'melastome', 'phonolite' and 'porphyritical'. A further twenty or so entries in the *OED* testify to the importance of Black's *Political Essay* in offering new or striking examples of how words such as 'American', 'black', 'civilization', 'criollo' or 'oriental' could be understood, thus emphasising the linguistic contribution this work made to early nineteenth-century discussions of society, race and ethnicity. On this basis, then, the *Political Essay* had a greater influence on the English language in terms of linguistic innovation than the far longer *Personal Narrative* or the magisterial *Cosmos*. The difficulties encountered by Black essentially served only to emphasise the impressive range of Humboldt's scientific and linguistic expertise, as much as the relative inaccessibility to the lay reader of some parts of the *Essai politique*, particularly in its excursions into scientific detail. They also demonstrated how quickly language needed to evolve to keep pace with scientific developments. But those readers who took at face value Black's assertions that the main text of the *Political Essay* was an unadulterated version of Humboldt's original would have been disconcerted by its foreign-sounding turns of phrase, its dual presentation of scientific terminology in different languages, and its failure to buy into the 'illusion of transparency' that conventionally characterised translation.

Annotation, Authorship and Authority

If Black sensed, and occasionally signalled, his lack of scientific expertise, he was merciless in using translation to (re)gain control over the translation and present himself paratextually as the confident mediator of Humboldt's writing to a British reading public. His initial footnotes served simply to domesticate the text for an English audience by giving the imperial equivalents of heights, weights and volumes. By page eighteen of the English translation, however, Black had begun to append notes that would query Humboldt's method, correct his calculations or enter into open (and sometimes fierce) debate, in moments of unabashed intellectual self-promotion. Black's dissociative deployment of annotation began in a section where Humboldt was discussing trade and communication between the American colonies and their European mother countries. As Humboldt had remarked, in Black's rendering:

> At a time when the New Continent, profiting by the misfortunes and perpetual dissentions of Europe, advances rapidly towards civilization; and when the commerce of China, and the north-west coast of America, becomes yearly of greater importance, the subject which we here summarily discuss is of the greatest interest for the *balance of commerce**, and the political preponderancy of nations.
> *It may be necessary to inform the reader, that he is indebted for this *term*, at present in some sort of disrepute from the proscription of political economists, however much the *idea* may still haunt the wise heads of our commercial men, to the author and not to me. *Trans.* (PE I, 18)

Black's asterisked intervention indicated that he was not prepared to be responsible for what he considered Humboldt's tenuous discussion of international commerce based on balance of trade figures. As the economic commentator Thomas Mortimer had outlined in his *Lectures on the Elements of Commerce, Politics and Finances* (1801), establishing what a country's balance of trade really amounted to was an almost impossible task, given that custom-house entries on foreign imports gave only a partial account of incoming goods, since contraband, smuggling and clandestine import by post automatically skewed the figures (Mortimer 1801: 151). Add to this bad markets, foreign bankruptcies, confiscations of goods abroad and losses at sea, and it was similarly impossible to come up with a true figure for the income a country made from its exports (Mortimer 1801: 151). Black therefore swiftly set the tone for both this and the next translated volume, as he unambiguously distanced himself from such terms by attributing them 'to the author and not to me', and making repeated use of his signature mark '*Trans.*' to indicate the presence of two different authors in this text.

On occasion, though, this earnestness would lapse into pedantry of the sort that had so incensed Humboldt. Talking of the difficulties of travelling through Mexico, uphill then downhill, through biting cold and shimmering heat, Humboldt remarked how problematic it was to transport goods across such terrain. While carriages could use the road to Acapulco, this was impossible for other routes across the plains of Mexico. 'On the contrary, of the 84,5 [*sic*] leagues from the capital to the port of Vera Cruz, 140 belong to the great plain of Anahuac,' translated Black and swiftly added, 'Here is evidently a mistake, 140 cannot be a part of 84,5. *Trans.*' (*PE* I 59–60). This was just one of seventy printing errors recorded in the errata at the back of the first of the French quarto volumes, and thirty-six at the end of the second – the version of the text from which Black must have worked to encounter these mistakes in the first place, given that they had been corrected in the French octavo edition. Whether Black translated from the printed edition of the 1811 quarto volume (complete with errata, which he ignored) or from its proofs (not fully copy-edited) is unclear: either way, he wilfully used such errors to downplay the significance of Humboldt's ground-breaking work on Mexico.

Black not only criticised factual errors in the *Essai politique*. He also attacked what he perceived as Humboldt's highfalutin style. In a discussion of the apparent immunity of Europeans or those of white descent to yellow fever, Humboldt had observed, 'Les individus de la race du Caucase ne paroissent pas exposés à ce typhus mortel, tandis que, d'un autre côté, la fièvre jaune ou le vomissement noir n'attaque que très-rarement les Indiens mexicains' (*EP* I, 352). Black translated this as:

> The individuals of the race of Caucasus* do not appear subject to this mortal typhus, while, on the other hand, the yellow fever or black vomiting very seldom attacks the Mexican Indians.
> *Who are the individuals of the race of Caucasus? The Europeans. So at least we learn from the context where they are opposed to the Mexican Indians. This involves the theory of the mountains of Asia being the nursery of the old continent. Every one however will not so easily be able to understand Europeans by this denomination. Such attempts to elevate the style, at the expense of perspicuity, can never enough be reprobated. *Trans*. (*PE* I, 117)

Black had got the wrong end of the stick, though. As Humboldt groaned to Pictet, 'le mot *race du Caucase* (nom d'une variété de Blumenbach) lui paraît une élévation du style' [the expression *race of Caucasus* (name of a race used by Blumenbach) appears to him to be a stylistic exaltation] (Humboldt 1869: 72). The pitfalls in translating Humboldt's writing therefore lay not only in finding the English equivalents for a raft of different scientific terms, but also in actually recognising that they were scientifically significant and knowing in which discipline they originated. The expression 'race du Caucase' (Blumenbach 1803: I, 74), used in the 1803 French translation of the German physiologist Johann Friedrich Blumenbach's *Handbuch der Naturgeschichte* [*Manual of Natural History*] (1779–80), had been employed by Humboldt to make an implicit link between his account of travel to Mexico and early work on racial difference, thus adding anthropology to the range of discourses into which the *Essai politique* could be embedded.

This Blumenbach gaffe had essentially been a criticism levelled at Humboldt's style – a subject on which Black would launch further assaults in the course of his translation. Humboldt had observed (in Black's rendering) that, according to the sixteenth-century historian Diego Muñoz Camargo, the Mexicans 'were so astonished at the address of Alvarado, that on seeing him make his escape, they bit the earth (a figurative expression which the Tlascaltec author borrowed from his language, and which signifies being stupefied with admiration)' (*PE* II, 75). Black noted in a footnote:

> There is such a thing, perhaps, as explaining too much. Few of M. Humboldt's readers, I dare say, will be led to conceive that the Mexicans fell literally to the eating of earth. There are bounds to commenting, which a salutary dread of prolixity should impress on every writer, but which, unfortunately, the countrymen of M. de Humboldt (Germans) seem seldom to have a clear conception of. I shall make myself sufficiently understood when I allude to

the prolixity of their most celebrated writers, their Herders, Gentzes, and Wielands. *Trans*. (PE II, 75–6)

Rehearsing the assertions made in the Preface that Humboldt's style was a particularly 'German' cultural product, rather than characteristic of his personal idiom and idiosyncrasies, Black lumped Humboldt together here with a collection of German Enlightenment figures, the philosopher Johann Gottfried Herder, the political journalist Friedrich von Gentz and the poet Christoph Martin Wieland. Black had few qualms about pigeonholing Humboldt's style of explication as apparently 'German' and, by implication, verbose. Yet Black seemed blissfully unaware that his own translator's footnotes frequently 'explained too much' in shows of petty pedantry. Daston remarks that while science might seem to exemplify rationality and facticity, it is guided by its own 'moral economies' – values shared by individuals working in science that influence their 'subject matter and procedures, [. . .] sifting of evidence, and [. . .] standards of explanation' (Daston 1995: 6). Black's misreading of the moral economy of scientific knowledge meant that instead of casting himself as the trustworthy, impartial translator of Humboldt's work, he presented himself as an ungentlemanly, immodest and sometimes uninformed purveyor of information.

By far the longest notes that Black appended to the *Political Essay* were those that contradicted the sources that Humboldt had used, compared them with other narratives or supplemented them with further commentary. In passages like these, Black seems to be less a translator than a historian, whose task it is to 'examine all the sources relevant to the solution of a problem and construct a new narrative or argument from them' (Grafton 1997: 4). Humboldt had already located his account of Mexico within different networks of knowledge, from contemporary scientific travel writing to historical accounts of the Americas, histories of other ancient civilisations and modern works on the international political economy. Black's own additions would not only offer glimpses into other related works but also display the extensive background knowledge he brought to his annotation and translation of the *Essai politique*. Two sources on which he drew most heavily were the Jesuit missionary José Gumilla's *El Orinoco illustrado y defendido* [*The Orinoco Illustrated and Defended*] (1741) and the astronomer Antonio de Ulloa's *Noticias Americanas* [*American Observations*] (1761). De Pauw had scorned the Jesuit writers for painting a livelier, more sympathetic portrait of the American Indians whom they had converted to Christianity, compatible with the image of the native peoples that Humboldt (and Black) aimed to convey (Gerbi 1973:

64). On rare occasions, Black used their accounts to confirm the assertions that Humboldt had made as he borrowed from those 'established' travellers. Where Humboldt quoted what Volney had reported in his *Tableau du climat et du sol des Etats-Unis* [*View of the Climate and Soil of the United States*] (1803) – namely, that the children of the Canadian Indians were born as white as Europeans, Black used Gumilla's account to develop this point further by adding that Indians reportedly remained white 'for several days after they are born, with the exception of a small spot [...], of an obscure colour' (*PE* I, 146). More often, though, the narratives of Gumilla and Ulloa were employed to contradict or cast doubt upon statements that Humboldt made. While Humboldt had asserted that the native American Indians rarely turned grey with old age, Black immediately remarked that this differed from Ulloa's view, 'who says expressly that the symptoms of old age among the Indians are *grey hairs* and a *beard*', added a few lines of Spanish quotation for good measure, complete with source reference, and closed the footnote with the remark that 'Father Gumilla gives an account somewhat similar to Ulloa's' (*PE* I, 150–1). As Humboldt continued his description of the Mexican Indian, describing him stereotypically as 'grave, melancholic and silent, so long as he is not under the influence of intoxicating liquors', Black again drew on Ulloa's *Noticias Americanas*, this time to refute this assertion of Indian taciturnity.

The religious oppression of native peoples that accompanied colonisation was a theme to which Black repeatedly returned as he openly criticised Catholic missionaries for the dubious means by which they obtained the conversion of the Mexican Indians. Humboldt's description of how in Pasto, on the ridge of the Andean *cordilleras*, he had seen Indians masked 'and adorned with tinkling bells, perform savage dances around the altar, while a monk of St. Francis elevated the host' demonstrated how native peoples had adapted Christian belief to their existing traditions (*PE* I, 168). The striking incompatibility of these images was, to Black, a perfect example of how senseless such attempts at conversion were. Contrasting the methods adopted by Anabaptists, Methodists and Presbyterians with those used by 'our more volatile catholic brethren', he argued that the success of Catholic missionaries lay in giving weapons as incentives to convertees, a method that Black neatly termed 'hatchet bribery' (*PE* I, 168–9). And as Black cynically remarked a little further on, the Spanish clergy had few qualms about extracting large sums of 'earthly treasure' from those Indians who had converted to Christianity so as to ensure them 'heavenly felicity' (*PE* I, 187). Not all Black's criticisms were levelled solely at the Catholic Church, however. Placing in a wider context Humboldt's remark that oppression corrupted morals,

Black offered a range of more specific examples to illustrate this, noting that West Indian slaves, when freed, became as cruel as their masters, Greeks collecting taxes from the Turks were more exacting than their Turkish counterparts, and the Hindu could expect to suffer most at the hands of 'his own brethren armed with foreign authority' (PE I, 170). Black's plea for the end to such tyranny and oppression amplified the political message of Humboldt's account, giving it a sharper critical edge than Humboldt had permitted himself in the *Essai politique*.

The anti-imperial, abolitionist stance articulated in these footnotes was just one agenda that Black pursued in his annotations as he responded to the diverse range of topics addressed in Humboldt's *Essai politique*. A subtle patriotism underpinned some of Black's footnotes as he either reminded Humboldt of British (or specifically Scottish) issues or achievements he felt had been overlooked, or came to the defence of British figures whom he considered had been hard done by. In his discussion of how difficult it was to assess the population of a newly discovered country, Humboldt had taken issue with a series of accounts by British travellers, not least Cook:

> The celebrated Cook estimated the number of inhabitants of Oteheite at 100,000; [. . .] I cannot allow myself to believe that these differences are the effect of a progressive depopulation. The maladies with which the civilized nations of Europe infected these once happy countries must, no doubt, have caused a depopulation; but it could never have been so rapid as to carry off in forty years nineteen-twentieth parts of the inhabitants*. (PE I, 93–4)

In his long footnoted riposte to such assertions, levelled even at the 'celebrated Cook', Black was quick to remind the reader that 'Captain Cook may have somewhat exaggerated the number of inhabitants of Otaheite; but [. . .] he did not form his estimate so much from conjectural circumstances as from having seen the whole population of the island', adding that he was 'in general extremely sober and moderate in his judgments' (PE I, 94). While these footnotes scarcely suggest an impartial and detached 'ironic' stance towards the source text, they do highlight Black's commitment to accountability, which he understood as the translator's duty to present the reader with accurate, not arguable, fact. As Jeanne Fahnestock has cogently argued, scientific authors have a particular responsibility in this regard: 'the distinction between the sciences and other fields depends less on the forms of argument used initially than on the degree of accountability demanded eventually, and in the sciences the degree of accountability is high' (Fahnestock 1999: 43). Black's corrective footnote strongly indicates that he presumed his readership to comprise men of science, rather than a general audience. These readers would be particularly critical of factual detail and the arguments that turned upon them. By stressing that Cook had not made his statement 'from conjectural circumstances', he was arguing that it could withstand rigorous scrutiny.

Scottish Enlightenment

If Black would not brook criticism from Humboldt on the authority of Cook's account, he responded still more fiercely to comments made by Humboldt on Scotland. As Humboldt tried to assess the level of cultivation that the ancient peoples of Mexico had attained, he reflected that it was possible to observe even in Europe that the 'peasant of Brittany or Normandy, and the inhabitant of the north of Scotland, differ very little at this day from what they were in the time of Henry the Fourth and James the First' (*PE* I, 157). Black countered that this assertion about the Highlands of Scotland might have held true half a century earlier but was certainly not the case at the time of writing, since the obstacles to improvement and civilisation presented by the barren and mountainous terrain of this part of the world had 'seldom been more successfully overcome than in the highlands': proof, were it needed, could be found as much in statistical accounts of Scotland as in the high moral character of the Highland regiments (*PE* I, 157). These annotations pointed up Humboldt's dated knowledge of Black's homeland, and cast Scotland in a more favourable light as a country beginning to embrace modernity and free itself from the backwardness implicitly attributed to it in Humboldt's remarks.

Although Humboldt's own references to Scotland were at best sporadic, Black was always eager to draw on Scottish examples to signal discord with Humboldt or simply vaunt his own learning. Humboldt had argued that the increase in the population of New Spain might be due to improved cultivation and a shift towards agricultural colonisation. Black noted:

> †The author may be very probably in the right; yet it is but an indifferent proof that the population of the whole kingdom has increased, because, in those places where shepherds have given place to agriculturists, the population has been rapidly increasing. By a similar mode of reasoning, it may be concluded that the population of Britain is on the decline, because the population of the highlands of Scotland, converted from agriculture to sheep farming, is on the decline. *Trans.* (*PE* I, 95)

Black was therefore critical of generalisations Humboldt had made as he tried to explain the reasons for population expansion across the territory as a whole, based on changes in land use in just one area. Whether Black was justified in drawing a parallel between the situation in Mexico and in Scotland is debatable, but in transposing Humboldt's argument to a Scottish context, he was trying to make the essence of Humboldt's argument easier to grasp for the British reader, even if this doubtless misrepresented and even invalidated the point that Humboldt had wanted to make.

Black used other references to Scotland to gesture towards the intellectual achievements of the Scottish Enlightenment. Humboldt had observed that the 'civilised' Mexican Indians reasoned 'coolly and orderly' but never manifested 'that versatility of imagination, that glow of sentiment, and that creative and animating art which characterize the nations of the south of Europe, and several tribes of African negros'. Black exclaimed to the British reader:

> What must our brethren of the northern part of this island, who have attained no small reputation for a pragmatical and metaphysical disposition, and who are so much disposed to give metaphysical superiority a precedence over all the other human faculties, feel, when they find that, most probably, their future rivals are not to spring up in any of the rival colleges of the south, or even in any of the great German universities, but among the beardless tribes of the Mexican mountains, and the banks of the Orinoco! (*PE* I, 170–1)

Homing in on the Indian's clear-headed and logical way of thinking, Black shifted the focus away from Humboldt's admiration of fiery-tempered Mediterranean vivacity towards a recognition of the achievements of Scottish pragmatists – figures such as Francis Hutcheson, Adam Ferguson and David Hume. Footnotes such as these might have been deliberately targeted by Black at London intellectuals to goad them into recognising achievements north of the border. They certainly presumed a group of readers with wide-ranging interests that spanned the philosophical writing of the Scottish Enlightenment, as much as the political, economic and geographical themes addressed by Humboldt's *Political Essay*.

To what extent did the targeted audience of Black's annotations belong to a delineated subset of readers? Although his footnotes signal an eager desire for self-promotion through debate and discord, some voiced his concern to make his translation appeal to a general audience. He certainly posited a wide reading audience in notes declaring, for example, '[e]very English reader will recollect the fine passage in Gulliver's Travels on this subject' (*PE* II, 453). By contrast, other annotations quite patently excluded a 'popular' audience, not least in the language that they used. The quotations from the Greek, which Black inserted into a footnote relating to a discussion of the types of grain used in bread-making in Mexico, are a striking example of this. Humboldt had made a hasty comment on Homer's use of the term for spelt, about which Black took him thoroughly to task. A prime example of Black's pedantry, the correction scarcely affected Humboldt's enumeration of the products available on the Mexican markets. Yet for all that Black himself acknowledged this to be 'an affair of small consequence', he still felt the need to give three quotations, complete with line numbers, from books five and eight of Homer's *Iliad* and four of the *Odyssey* (see Figure 2.1).

These quotations obviously illustrated Black's competence in Greek

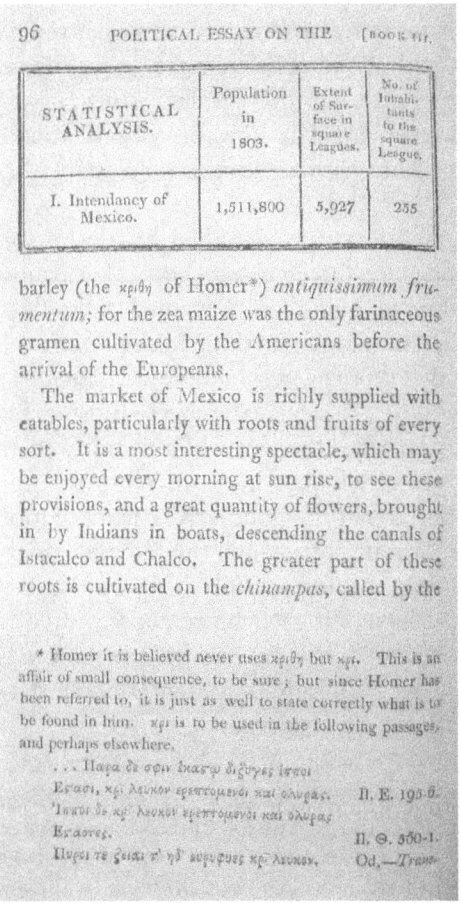

Figure 2.1 Alexander von Humboldt, *Political Essay on the Kingdom of New Spain*, trans. John Black, 4 vols (London: Longman, Hurst, Rees, Orme and Brown, 1811), II, p. 96.

and his close familiarity with Homer's writing, as much as his desire to impress his readers through intellectual display. In demonstrating that he was fluent in Greek, in a footnote that covered almost as much of the page as Humboldt's principal narrative, Black signalled his need to belong to that class of 'gentleman' readers who had enjoyed a classical education. Snide sideswipes elsewhere in his annotations at 'the refuse of talent carefully preserved in the cells of some fat old university' or the permanent absence of the professors at Mexico's Tridentine college, who indulged in 'the praiseworthy custom of considering their chairs as sinecures', suggest that Black had an axe to grind with those who considered universities the true seats of learning (*PE* I, 218, 220). Self-improvement, self-education and self-promotion were all embodied in Black's interweaving of quotations from Homer into his annotations. He was also obviously keen to locate learning and a knowledge of the natural world in the classical past – a subtle reminder of where he stood in the 'battle of the books'. And, in measuring himself against the achievements of the privileged aristocrat Humboldt, Black sought to demonstrate that intellectually he could hold his own.

If Black's Greek quotations signalled his desire for inclusion in the British intellectual elite through his show of learning, they were also highly exclusionary in that they prevented a whole swathe of British readers from understanding everything in the *Political Essay*. Most obviously, they excluded women, who might have acquired a knowledge of modern, but rarely of classical, languages. Indeed, Black generally did not appear to bear in mind the needs of a female audience in his annotations. Either he presumed that his readership would largely comprise men, or he considered that the 'scholarly' nature of footnotes made them a decidedly male domain into which the female reader would not stray. One annotation that would certainly have appealed to liberal-minded women, but might have left their more conservative counterparts cold, was in Black's discussion of why gender imbalances exist in any given society. Prompted by Humboldt's presentation of data for the cities of Puebla and Valladolid, he observed:

> It would be interesting were we enabled to discuss thoroughly the influence of the diversity of casts in the proportion of the sexes to one another. I saw, from the enumeration in 1793, that in the city of Puebla and at Valladolid there were among the Indians more men than women, while among the Spaniards or the white race there were more women than men. The intendancies of Guanaxuato and Oaxaca exhibit in all the casts the same excess of men*.
> *This hardly makes in favour of John Rheinhold [*sic*] Forster's theory, embraced with so much ardour by the far-famed Mary Wollstonecraft in her

Rights of Women [*sic*], that the sex of the offspring is determined by the side on which the preponderance of ardour lies in the sexual intercourse. Hence, says she, 'there are more females than males in the east; for the females being deprived of their just share in that intercourse, have consequently a more than ordinary share of ardour. Yet here we see these beardless Indians, who are cool enough in all conscience, and to whom their women prefer any thing that comes in their way, black or white, beget more males than females. *Trans*. (*PE* I, 247–8)

It is a popular misconception that Mary Wollstonecraft's *Vindication of the Rights of Woman* (1792) was greeted with horror and derision by a late eighteenth-century audience (Janes 1978: 293). It was more William Godwin's publication of the *Memoirs of the Author of a Vindication of the Rights of Woman* on her death in 1797, which revealed that she had borne a child out of wedlock and twice attempted suicide, that would bring about Wollstonecraft's subsequent demonisation. But as Anne K. Mellor argues, even if by 1800 it had become dangerous for a woman to identify openly with Wollstonecraft as a person, many would still continue to espouse her ideas (Mellor 2002: 145).

Black may have rightly assumed that unlike the *Personal Narrative*, which had been so avidly read by women like Lucy Aikin, the *Political Essay* would appeal less to a female audience. Although it could broadly be categorised as scientific travel writing – a non-fictional genre that did appeal to women, and from which they read out loud to their children (Flint 1993: 198) – it could not be marketed in quite the same way as the *Personal Narrative*, which carried a far stronger appeal to individual experience and promised a certain linearity of description. The *Political Essay*, with its important insights into trade, commodities and prices, did not offer the same vicarious reading experience to the armchair traveller that the *Personal Narrative* would do just a couple of years later, and this may have encouraged Black to draw the conclusions he did about its potential market and readership.

Echoes and Afterlives

The significance of the *Political Essay* to a nineteenth-century Anglophone audience is probably best gauged by the ways in which Black's text was recycled in the fifteen years or so following its publication. A second edition of Black's translation appeared with Longman in London in 1814 and a third in 1822, both of which were practically exact reproductions of the original, complete with Black's footnote apparatus. Longman wrote to Humboldt in August 1822 to inform him that the 'new edition of the Essai Politique will be ready in a few days and we shall have the pleasure of sending you a copy before publication so that we can make any alterations you may suggest either in the form of an Addenda or an Errata as you may consider best'.[10] Humboldt had noted to Cotta a decade or so earlier how translating the *Essai politique* into German had enabled him to correct several errors in the original, thus acknowledging translation's inherent potential to improve a text, rather than just distort and disfigure it (Humboldt 2009: 97). Humboldt does not, however, appear to have responded with any major changes. While it seems surprising, given his explosive reaction a decade earlier, that he did not have Black's annotations removed, either he had reconciled himself to his British translator's visibility in the *Political Essay* or the publisher was willing to make changes only in the form of an additional page of errata. Humboldt himself would produce a freestanding second edition of the *Essai politique* (1825–7) in French that was a significantly revised edition of the first.

Where Black's translation of the *Essai politique* was to find other 'afterlives', beyond these later editions, was in works that selected two aspects of its content for further reproduction: the statistical information that it gave on Mexico, and the detailed account of how silver and other ores were mined. One Baltimore edition, entitled *Humboldt's Statistical Essay on New Spain Abridged* (1813) and shortened by 'A Citizen of Maryland', reclaimed and repackaged parts of Black's text (minus its footnotes) for an American readership, even if Black was never mentioned at any point as Humboldt's British translator. The focus of this abridged account was simply on the lists of the fifteen intendancies of Mexico and their population, position, main towns and geography. Even though the information it gives is largely statistical, it shares sufficient turns of phrase with Black's rendering of the *Essai politique* for it to be clearly modelled on his translation. That being said, the anonymous author–editor's Preface explicitly recast the work for an American audience, as the opening lines demonstrate:

> The interest which the revolutions now taking place in the vice-royalty of Mexico, and the provinces [. . .] comprising the kingdom (so called) of New Spain, excites in the breasts of the American people, from its contiguity to the U. States, cannot but render a geographical knowledge of that vice-royalty and captaincy-general highly useful to the American reader; (Anon 1813: v)

Humboldt's Mexico account was all the more relevant to North American readers because New Spain had hitherto been only 'very imperfectly and inaccurately described by our English geographers and gazetteers' (Anon. 1813: v). In a revolutionary spirit, this writer called for New Spain to 'throw off the trammels of European thraldom', although the reasons for doing so were essentially commercial, given that this writer seemed more interested in the trading possibilities offered by Mexico than with anything else (Anon. 1813: v). Humboldt's cynicism at what he considered a British obsession with commerce – memorably embodied in the tussle that he had with Longman over publishing the *Essai politique* – was slightly misplaced, given that interest on an international level seemed to revolve more around the economic information it had to offer than the new light it cast on the culture, history and languages of the peoples of Mexico.

Where this American edition essentially homed in on the commercial detail of Humboldt's account, other British publications placed greater focus on the geological and mineralogical information it contained. The *Selections from the Works of the Baron de Humboldt, Relating to the Climate, Inhabitants, Productions, and Mines of Mexico* (1824), published with Longman by John Taylor, Treasurer of the Geological Society, melded text from Black's translation of the *Essai politique* with sections from Humboldt's *Geognostical Essay*, which had also appeared with Longman the previous year. Taylor worked highly selectively, drawing only on Humboldt's reports of six intendancies – those relevant for mining – and jettisoning Black's footnotes completely. He had, Taylor acknowledged, been permitted 'free use' of Black's translation and had cut all marginal references, suggesting that any reader who wished to 'verify a fact which appears to him suspicious, will not think it too much trouble to consult the original, or Mr. Black's translation' (Taylor 1824: n.p.). Taylor's edition of extracts from the *Essai politique* made of the translation a seemingly monologic piece in which the translator's overt presence was eradicated and the author's voice restored to its full authority. In taking this step, Taylor set a precedent for more modern re-editions of the *Political Essay* – notably the 1988 republication for the University of Oklahoma Press by Maples Dunn. All measurements had been silently 'corrected' from imperial to metric and foreign currencies converted into pounds sterling, so that the foreign origins of the *Political Essay*, so frequently marked in Black's translation, had now been erased. Taylor, a mining specialist, was also better placed than Black to hunt down the correct terminology with a technical audience in mind: 'I have adapted the technical language more to the comprehension of my English mining readers, than that of the translation would be found to be' (Taylor 1824: iv). The notes that Taylor did append to this edition were anything other than confrontational. Indeed,

Taylor noted early in his Introduction that he had been in contact with Humboldt to ask for further clarification regarding some of the mines of Mexico, to which he had received highly detailed and knowledgeable replies. The revised version of the *Political Essay* that Taylor published in the *Selections* may well have righted a number of wrongs committed by Black a decade or so earlier, in that Taylor had ironed out terminological problems, removed his footnotes and also informed Humboldt of the publication of this new edition, restoring to him the illusion of a sense of control over where his works were being translated and by whom.

Conclusions

Black's restyling of Humboldt's *Essai politique* for a British audience was, then, essentially a two-part process, characterised in the first stage by the highly literal rendering of the original French text into English, and in the second by the annotation of this translation with footnotes that complemented, criticised and contradicted it. Translation, for Black, was not about subservience or invisibility. It was about accountability. The *Political Essay* became an intellectual sparring partner for its British translator as he used it to point up Humboldt's factual inaccuracies and to distance himself openly from such errors. While Black may have intended that such interventions in the text would affirm his own intellectual achievements, they called his own integrity into question as he lost sight of Humboldt's greater achievements behind a barrage of petty quibbles. Black's translation was a striking example of the way in which the footnote could be wrenched away from objective aloofness and made to participate in the rhetorical processes of debate and disagreement that underpin the making of knowledge. The footnotes to the *Political Essay* could be seen as a productive addition to Humboldt's narrative to the extent that, in combination with Humboldt's own footnotes, they more explicitly situated his account within the contemporary 'Dispute on the New World' and the discussions it fostered about the history and cultural identity of the Mexican Indians and the uses and abuses of colonial power.

While Black's footnotes embedded Humboldt's work more firmly in a range of relevant political, cultural and scientific discourses, they also defamiliarised the text by constantly reminding the reader of its mediated, hybrid nature. It would be equally easy, then, to cast Black's annotations as displays of intrusive pedantry that undermined Humboldt's authority, blunted the terminological sharpness of his scientific descriptions and played out in the margins irrelevant controversies that distracted the reader from the central arguments. It is also not difficult to understand Humboldt's irritation at the unwarranted injection of such overt multivocality in his own text. Yet in one crucial aspect Black's footnotes did Humboldt no disservice. By emphasising the mediated nature of the work, they forcibly illustrated that the *Political Essay* was not an innocent rendering of Humboldt's French original into English, directly reflecting the author's personality or intention. Despite Humboldt considering the *Political Essay* a disastrous début on the British book market, it must have been productive in compelling him to reflect on the kind of relationship he would wish for with future translators. Mutual respect, close collaboration and productive exchange were more what he sought than Black's solitary, confrontational and self-promotional approach. As the following chapter shows, cooperation was what Humboldt both sought and found in the working relationship that underpinned the English translation of his next work, the *Relation historique*.

Notes

1. *New Universal Magazine: Or Miscellany, of Historical, Philosophical, Political and Polite Literature*, 12 (1809), p. 177; *Critical Review; Or Annals of Literature*, 22 (1811), p. 207.
2. *Monthly Review*, 66 (1811), p. 353.
3. *Literary Panorama and National Register*, 9 (1811), p. 235.
4. *Monthly Review*, 66 (1811), p. 353. For the identification of Lowe as the author of this review, see Nangle 1995: 38–9.
5. *Monthly Review*, 79 (1816), p. 1.
6. Ette (2004) rightly argues that the *Vues des Cordillères* was the first piece by Humboldt to engage in this debate, but in the context of Humboldt's British reception, the *Political Essay* appeared before Williams's translation of the *Vues*, which would not be published until 1814.
7. *The Gentleman's Magazine*, n.s., 44 (1855), p. 210.
8. Black's death did not go unrecorded in German journals. The *Leipziger Repertorium der deutschen und ausländischen Literatur* describes him as someone who worked his way up to being one of the most respected and influential journal writers in England (see volume 13 (1855), p. 294), while the *Deutsches Museum* described him as one of the most active and most worthy journalists in Britain, under whom the Whig *Morning Chronicle* flourished (see volume 5 (1855), p. 119).
9. This spatially and intellectually deferential relationship between paratext and main text has been routinely difficult to maintain in translations of the *Essai politique*. Even when the translator's additional comments have been tucked away in endnotes, their sheer volume has still been hard to overlook. Hensley C. Woodbridge's 1957 translation of just the first three chapters of

Book 1 includes a further 106 notes that cover thirty pages, almost as long as the translation itself (Humboldt 1957).
10. University of Reading, Special Collections, hereafter 'URSC', MS 1393 Longman Letters I, 101 no. 296c.

Chapter 3

'A Colossal Literary and Scientific Task': Helen Maria Williams and the *Personal Narrative of Travels to the Equinoctial Regions of the New Continent* (1814–1829)

In 1814 translations of two further works by Humboldt went on sale in Britain. One was the English edition of the *Vues des Cordillères et monumens des peuples indigènes de l'Amérique*, entitled the *Researches Concerning the Institutions and Monuments of the Ancient Inhabitants of America*. The other comprised volumes one and two of the *Personal Narrative of Travels to the Equinoctial Regions of the New Continent during the Years 1799–1804*, which was gradually appearing on French booksellers' shelves as the *Relation historique du voyage aux régions équinoxiales du Nouveau Continent*. Translated by Helen Maria Williams (1761–1827), both pieces corresponded to the seventh and eighth works that Humboldt had planned to publish in collaboration with Bonpland, deriving from their journey to the Americas between 1799 and 1804. The *Researches* comprised chapters giving visual and textual descriptions of Mexican and Peruvian antiquities – architecture, sculpture, historical paintings and hieroglyphics – interspersed with picturesque scenes of the Andean *cordilleras*. The *Personal Narrative*, which would eventually span seven volumes, described Humboldt and Bonpland's journey across the mountains, plains and tropical rainforests of Venezuela, down the river systems of South America (notably the Orinoco), to the Catholic missions of Spanish America and the sugar *haciendas* of Cuba. While the two-volume *Researches* was relatively compact, the illustrations having been pared down from sixty-nine to nineteen plates, the *Personal Narrative* ranged across almost 2,000 quarto pages in the French original and over 3,900 in the seven-volume octavo English translation. A pivotal work in Humboldt's œuvre, the *Personal Narrative* was as much the embodiment of Humboldt's struggle to combine aesthetic and scientific representation in an all-encompassing travelogue as it was a testament to the staying power of his translator.

By 1814 British critics were already showing signs of fatigue. 'It may be doubted whether the method of publication adopted by M. de Humboldt is that in which either his interest or his reputation has best been consulted,' declared John Barrow, Fellow of the Royal Society and author of travelogues on China and South Africa, in the *Quarterly Review*.[1] While many reviewers (including Barrow) readily paid homage, in reviews routinely thirty pages long, to Humboldt's international reputation as a scientific traveller, they were equally quick to criticise what they perceived as his poorly managed material, the dryness of the scientific sections and the crippling slowness with which the seven-volume *Personal Narrative* started to appear. The *Monthly Review* protested that the complete imbalance between scientific modes of representation and literary narration left readers 'wading through an endless variety of wearisome details' and fighting the 'philosophical phantasmagoria' that floated through Humboldt's work.[2] While the *Edinburgh Review* conceded that 'so extensive a collection of facts, reasonings and opinions' could scarcely be embraced in a single-volume work, the *British Critic* was less conciliatory.[3] In 1819 it complained of Humboldt's accounts that 'we found ourselves at the end of them exactly in the same place in which we were at the beginning', and by 1821 it had ominously concluded that he would die before the task of writing his travel account had been completed.[4]

If Humboldt's management of his material was an easy target, his translator was potentially another. Williams, an outspoken supporter of the aims of the French Revolution, was one of the most prominent of Humboldt's British translators and also one of the more 'visible' women of her time (Figure 3.1). Her *Letters from France* (1790–3) had made her a household name, with some contemporaries praising her as 'a warm friend of civil liberty' (Anon. 1823: III, 615). Others, like Hester Lynch Piozzi, used memorably fiercer epithets, calling her a 'wicked little Democrate [sic]' (Knapp 2005: 158). The *Letters*, which Vivien Jones has aptly described as a 'triple violation' for anti-Jacobin propagandists of sexual morality, generic decorum and national political loyalty, appeared in Britain at a time when feelings about the Revolution were running extremely high (Jones 1992: 179). Richard Polwhele's poem *The Unsex'd Females* (1798), a vitriolic attack on women who departed from the course of 'proper' female writing, was quick to highlight what he saw as Williams's apparent volte-face from penning innocuous poems that 'charmed the moonlight vallies' to becoming an 'intemperate advocate for Gallic licentiousness' (Polwhele 1798: 19). The *Letters* also famously caused James Boswell to expunge the word 'amiable' from his reference to Williams in the second edition of his *Life of Johnson*,

Figure 3.1 Stipple engraving of Helen Maria Williams (London: Dean and Munday, 1816).

incensed at a work 'written in favour of the savage Anarchy with which France has been visited' (Boswell 1828: IV, 251).

Reviewing the fifth volume of the *Personal Narrative*, Barrow swiftly censured other 'liberal' aspects of Williams's life, besides her politics. Here, he noted, was displayed a 'specimen of the descriptive powers of our author, which however suffers not a little in the verbose and languid translation of Helen Maria Williams, alias Mrs. Stone'.[5] This poisonous jibe at her extramarital relationship with the Paris printer John Hurford Stone demonstrates just how closely the British reception of Humboldt's *Personal Narrative* and *Researches* was bound up with Williams's own biography, reputation and public image. Few critics overlooked the significance of Williams as their translator. The *Augustan Review* admired the 'spirit and freedom which might be augured from the earlier productions of its translator', clearly locating Williams's rendering of the *Relation historique* and *Vues des Cordillères* within her own translation œuvre.[6] The *Edinburgh Review* focused more on her authority as its translator by dint of her long-term residence in France, noting that this 'is the work of a lady well acquainted both with the language *from* which, and the language *into* which, her translation is made'.[7] But linguistic expertise was not all that reviewers sought in a translator. The *Monthly Review* implied that she should have edited the material as part of the translation process:

> We cannot but regret that Miss Williams, who has executed the office of translator in a style much superior to that which is generally seen in this department of literary labour, did not deem herself authorised to take the liberty of new-modelling the arrangement of the materials

and suggested that 'the publication would have gained largely on being re-cast by her hands'.[8]

This chapter focuses on Williams's contribution to the translation and presentation of Humboldt's writing within the context of British scientific and literary culture. Leask's observation that Williams's translation of the *Personal Narrative* should be considered 'a *collaboration* with Humboldt' is of signal importance in understanding the genesis of this work (Leask 2001: 235). No other English rendering of Humboldt's work would be collaborative to quite the same degree. As Ulrike Leitner (1997) notes, Williams worked directly from the French print proofs, rather than from the published volumes, to produce her first draft, which Humboldt meticulously corrected before returning it to her for further reworking. The nature of their shared undertaking repays further examination. It was shaped as much by how Humboldt – proficient in English – considered that British editions of his work should read, as by what kind of a stylistic imprint Williams aimed to leave on his work.

Elizabeth Bohls describes Williams's writing as 'closer to *précocité* than to the universalist rhetoric of Revolutionary reason', emphasising her overt, effusive sentimentality, and echoing what even contemporary admirers such as Mary Wollstonecraft described as her 'affected' manner (Bohls 1995: 125). In his 1995 retranslation of the *Relation historique* for Penguin Classics, Jason Wilson has rightly highlighted the stylistic disparities between Humboldt's 'curiously flat, scientific and modern' French, which contrasts with Williams's translation: her rendering plays on an appeal to the heart, and 'when Humboldt enthused', Wilson complains, 'then his translator interpreted and exaggerated' (Humboldt 1995: lix). Others have viewed her stylistic repositioning of Humboldt's text in a more positive light. Leask considers the *Personal Narrative* to be an 'outstanding', 'excellent translation', but gives little indication of the rationale behind such judgements (Leask 2002: 281; Leask 2001: 225). Williams herself modestly cast it as 'my imperfect copy of a sublime model', suggesting that her translation was in some way inferior to the original (*PNW* I, xii).[9]

Leask has stressed that the *Personal Narrative* is integrally rather than peripherally related to other works in which Williams had a creative input, but has focused primarily on its relationship to her non-fictional travelogues (Leask 2001: 219). This chapter contends that parallels drawn with Williams's other translations and her earlier fictional works are equally important in understanding how Humboldt's writing acquired its distinctive literary flair in English. Moreover, it suggests that Humboldt was so keen for Williams to undertake the translation of the *Relation historique* and the *Vues des Cordillères* precisely because he was aware that her style would enhance the literariness of his own writing in English translation and thus increase its appeal to a more 'popular' audience. By viewing her translation work as an essential element in the continuum of Williams's writing life, this chapter also asks where we can locate the *Personal Narrative* within her own creative œuvre. In what follows, the focus will be less on her translation of the *Vues des Cordillères* than of the *Relation historique*, given the latter's considerable length and Williams's greater translational visibility in it. Hitherto overlooked archival material, notably Humboldt's corrections of her translation, enables us to reconstruct the opening chapters of the *Personal Narrative* in their draft form as a pre-published version, or '*avant-texte*', and analyse the genesis of the translation and the processes underpinning the 'Englishing' of Humboldt's account. The remainder of this chapter concentrates on comparing the published versions of the original French text and its English rendering by Williams, to see how she left her own stylistic mark on the *Personal Narrative*, in the exhausting undertaking she termed 'a colossal literary and scientific task' (Kennedy 2002: 186).

Translation, Identity and the Salon

Williams's salon had its heyday in the first decade of the nineteenth century, when it was one of the most important places of enlightened exchange and intellectual sociability in Paris. Throughout the period in which she was translating Humboldt's works, she would continue to host these regular Sunday evening meetings of minds, which contributed productively to her translation activities. Indeed, Leask has argued that the key to her 'multifaceted involvement' with Humboldt lay in the contribution her salon made to 'the construction of knowledge in early nineteenth-century Paris' (Leask 2001: 220). Like other intellectual women of the period, such as Madame de Staël, her daughter the Duchess de Broglie, or the Duchess de Duras (all of whose salons Humboldt frequented), Williams used her role as a *salonnière* to encourage intellectual, political and cultural debate, and to foster a culture in which merit overrode differences of class or gender. Joan B. Landes notes that in such salons 'men of the aristocracy mingled with writers, artists, scholars, merchants, lawyers, and officeholders' (and we could add to that list 'scientific explorers'), to create a place of exchange 'between educated men and literate, informed women who functioned not just as consumers but as purveyors of culture' (Landes 1988: 22). As a translator, Williams's salon enabled her to foster contacts with individuals from a wide range of disciplines, which would prove essential as she came to tackle the scientific aspects of Humboldt's writing and draw on outside expertise. It was also a space in which she could exchange books and periodicals, as much as ideas, and it helped her to remain in indirect contact with Britain some thirty years after she had left it for good.

In 1790, enthralled by the 'sublimer delights of the French Revolution' and the 'blessed expectation of this glorious spectacle in Paris', Williams had set out for France (Woodward 1930: 32). A 'loss to the country', as Anna Seward generously described her departure from Britain, Williams was quick to find her feet in the French capital (Knapp 2005: 103). Her first salons in Paris were lively, busy affairs, drawing a wide variety of visitors. The American diplomat, poet and politician Joel Barlow asserted in 1802 that she received as many as fifty guests every night and that 'Helen really runs us down with her great parties' (Woodress 1958: 222). Schiller's sister-in-law, Caroline von Wolzogen, reports, more realistically, that Williams received every other evening (Woodward 1930: 149). Visitors to Williams's salon in its early years included Humboldt's fellow traveller Bonpland, the natural philosopher Georges Cuvier, and influential political writers and activists such as Charles James Fox, Thomas Paine and Benjamin Constant. Mary A. Favret has aptly termed Williams 'an exile playing hostess to a world in revolution' (Favret 1993: 276), emphasising her performative role as a woman very much at home in the theatre of post-revolutionary politics. Members of her salon came from many walks of life, but contemporary reports of Williams's gatherings repeatedly stress the literary persuasions of her many guests (Woodward 1930: 147). Coleridge's friend, Thomas Poole, remarked, 'I have been three times to Helen Marie [sic] Williams' *conversations*. You meet here a very interesting society. Many of the *literati*' (Sandford 1888: II, 290). Williams's erstwhile friend, Hester Lynch Piozzi, with whom she was now decidedly out of favour, noted contemptuously that her apartment in the early 1800s was 'the resort of a *Literary Coterie, all malecontents*' (Knapp 2005: 248).

The American writer and Hispanist George Ticknor gives us a snapshot of Williams's salon a good decade later in 1817 – when translation of the *Relation historique* was well under way – which confirms that it continued to have a literary and political character. He gives us a rare insight into how Humboldt used these meetings to air his first-hand knowledge of South America (and his republican views):

> In the evening I was – as I usually am on Sunday eve – at Miss Williams's, and was amused to hear Humboldt, with his decisive talent and minute knowledge of the subject, show how utterly idle are all the expectations now entertained of the immediate and violent emancipation of South America. Without knowing it, he answered every argument Mad. de Staël had used, this morning, to persuade me that the fate of the South was as much decided as the fate of our Independence was at the capture of Yorktown. (Ticknor 1876: I, 138)

Ticknor was also overjoyed to meet the poet Robert Southey quite by chance at Williams's salon, but was less impressed by their hostess: 'Miss Williams is evidently waning. Her conversation is not equal to her reputation, and I suspect never was brilliant' (Ticknor 1876: I, 130). For some, the fading *salonnière* was a figure of fun. The Irish writer Thomas Moore noted in 1821, the year Williams turned sixty, that 'Rogers joined us, after a visit to Miss H. M. Williams, and gave us an amusing account of it; the set of French Blues assembled to hear a reading of the "Mémoires de Nelson" which R. was obliged to endure also' (Clayden 1889: I, 310–11).

Although Williams was naturalised as a French citizen in 1817, her salon remained an enclave of Englishness, which she sustained until her move to Amsterdam in 1824 to be with her nephew, the preacher Athanase Coquerel, returning finally to Paris in 1827, the year of her death. This linguistic and cultural *dépaysement* was an experience she shared with Humboldt. Probably the last letter Humboldt ever sent Williams is defined by a deep sense of loss and nostalgia. 'Vous êtes séparée de tant d'objets qui Vous étaient chères aussi' – 'you are separated from so many objects that were dear to you too' – Humboldt wrote, drawing parallels between her deterritorialised position as an English translator distanced from Britain, and his own constantly peripatetic lifestyle as a Prussian who would spend almost a third of his professional life in Paris.[10]

During the time that they were working on the translation of the *Relation historique*, Williams and Humboldt profited from her salon as a space of intellectual, cultural and political exchange. In it, they seem to have shared what Venuti terms 'simpatico', an affinity felt by the translator with the author's ideas, tastes and persuasions, which can make of translation a 'recapitulation of the creative process by which the original came into existence' (Venuti 2008: 237). On a geographical level, too, it was a location well situated for such exchanges. Its relative proximity to his apartment in Paris – he lived at 67 rue d'Enfer from 1811 to 1813 then 3 quai Malaquais from 1813 to 1816 – facilitated swift communication between them as they worked on the translations of the *Vues des Cordillères* and the *Relation historique*. Living on the Left Bank more or less opposite the Louvre, Humboldt was not more than a good half an hour's walk from Williams's home in the rue d'Hauteville in the tenth *arondissement*. While their letters remain formal in register, never venturing into the familiar second-person pronoun 'tu', Humboldt's highly informal dating of the correspondence (only giving the weekday but no month or year) indicates just how direct and uncomplicated these communications really were. Humboldt's comment in a note to Williams that 'l'éloignement rend difficile les moindres choses; lorsqu'on a des intérêts communs' [distance makes even the smallest things difficult, when one has common interests] emphasises the extent to which he prized proximity to his translator spatially, but also intellectually.[11]

Williams as Writer and Translator

Williams's effusive sentimentality had already won her literary acclaim by the late 1780s, when she was barely twenty. Burns enthused over 'the unfettered wild flight of native genius and the querulous, *sombre* tenderness of time-settled sorrow' in her work, while Wordsworth published that same year his 'Sonnet on Seeing Miss Helen Williams Weep at a Tale of Distress' (Woodward 1930: 21). Williams swiftly came to embody this image of feminine sensibility, her weeping state ('deliciously distressed', as Janet Todd has described it) presenting women as sentimental, susceptible and spontaneous (Todd 1986: 63).

This was sustained in Williams's first main prose work, *Julia, A Novel; Interspersed with some Poetical Pieces* (1790). It appealed to what Gary Kelly has termed a 'sentimental aesthetics of expressivity', in which the manner of narration was more important than its matter, and its plot-line, centred around its eponymous heroine, automatically echoed Jean-Jacques Rousseau's epistolary work of 1761, *Julie, ou la Nouvelle Héloïse* (Kelly 1993: 33). *Julia* is relevant to my analysis of the *Personal Narrative* not just because it embodies Williams's early (and lifelong) preoccupation with sentimentality. Her novel abounded with direct quotations from British canonical works. Near the beginning, Julia's grandfather tells her 'tales of his battles in Germany, he "shewed how fields were won"', she later watches the setting sun 'till the bright vision gradually dissolved, and "Twilight grey, had in her sober liv'ry all things clad"', and when nature offers restorative calm, 'every gale is peace, and every grove is melody' (Williams 1995: I, 61, 81, 223). These quotations from Goldsmith's *Deserted Village* (1770), Milton's *Paradise Lost* (1667) and Thomson's *The Seasons* (1726–30) are just a few examples of how Williams wove snatches from widely respected authors into her text. Others came from the Bible and Shakespeare, Goldsmith's *Vicar of Wakefield* (1766), Charlotte Smith's *Elegiac Sonnets* (1789) and Gray's *Elegy* (1751). Non-fictional texts included Cook's *Voyage Toward the South Pole, and Round the World* (1777), John Moore's *Various Views of Human Nature* (1789) and Boswell's *Life of Johnson* (1791). As Stephanie Hilger rightly comments, these references not only enhance certain messages in Williams's work, but they also permit her to display her familiarity with contemporary writers, casting herself as a well-read author (Hilger 2009: 47–8). While Deborah Kennedy has suggested that Williams 'presented her credentials as those of the heart, not the head, consciously excluding herself from the more intellectual arena reserved for talented and educated men' (Kennedy 1991: 79), in *Julia* the opposite is the case. Williams adopted a highly self-confident role by writing herself into the British literary tradition, juxtaposing her own lines with that of Shakespeare, Milton and Goldsmith, and actively mobilising their writing for her own ends.

Williams's enthusiastic appropriation of texts by other authors was not dampened by her move from writing to translation. The English rendering of Jacques-Henri Bernardin de Saint-Pierre's enormously popular *Paul et Virginie*, which appeared with the London publishers Robinson in 1795, derives its multivocality from the eight sonnets of her own making with which she interspersed his text, presenting them as if written by one of the main characters, Madame de la Tour. A tragic romance that drew on the rhetoric of Rousseauvian sentimentality, *Paul et Virginie* took an island paradise (modelled on Mauritius) as the setting for Bernardin de Saint-Pierre's mordant criticism of slavery and the corrupting influence of aristocratic culture upon natural harmony. A scientist and, later, secretary to the philosopher and botanist Jean-Jacques Rousseau, he had spent two years on the tropical volcanic island of Mauritius in the late 1760s. His vivid descriptions of the heady scent of orange-flowers and mouth-watering succulence of pawpaws captivated contemporary readers. A work that Bonpland and Humboldt read to each other as they floated down the Orinoco, *Paul et Virginie* 'throbs with natural life', as Dassow Walls aptly puts it, in its rich scenes of tropical nature (Dassow Walls 2009: 244). Michael Dettelbach has noted how Humboldt's construction of the tropics derived from the observation and recording of his own inner states and owed much to the sentimental narratives of Rousseau and Bernardin de Saint-Pierre (Dettelbach 2005: 46). Indeed, as Humboldt was contemplating early in 1807 how best to present his many findings to a German public, he remarked that he need look no further than the 'unnachahmliches Muster' – the unsurpassable model – provided by *Paul et Virginie* (Humboldt 2009: 79). Since Humboldt's world view was informed as much by the collation of scientific fact as by his own physiological responses towards nature, he remained enthralled by this work long after his return from the tropics.

Scholars of translation studies have focused on the Preface and additional sonnets in Williams's *Paul and Virginia* to understand this work as what Anna Barker has termed 'a fusion of Williams' talents as creator and re-creator' (Barker 2011: 68). Williams's ten-page Preface, from June 1795, describes how she worked on it 'amidst the horrors of Robespierre's tyranny' during her own imprisonment in the Luxembourg Palace, and indicates that putting *Paul et Virginie* into English offered 'the most soothing relief [...] from my own gloomy reflections' (Bernardin de Saint-Pierre 1795: iii, vi). Williams used translation as a means to sustain her literary activities – despite the oppression of the Jacobin Terror – through the seemingly derivative and uncreative practice of translation. In so doing, she was employing it as what Kelly terms 'an act of political defiance' (Kelly 1993: 56). Sherry Simon's assertions (1996), that translation allows women to activate political concerns in cultural exchange, are particularly relevant here. Barker, in a similar vein, has suggested that this exemplifies a 'mode of *engagement* with literature, a kind of literary activism' (Barker 2011: 68). The Preface also gives us a strong sense of Williams's awareness as a translator of how readable the end-product should be, 'omitting several pages of general observations, which, however excellent in themselves, would be passed over with impatience by the English reader, when they interrupt the pathetic narrative' (Bernardin de Saint-Pierre 1795: iii, vi). English readers, she argued, expected events to be portrayed in rapid succession, unlike their French counterparts, who could happily listen to 'long philosophical reflections' while dramatic action was held in suspense (Bernardin de Saint-Pierre 1795: ix–x). Williams's attempts to be stylistically domesticating in her translation therefore echoed Schleiermacher's discussions about whether the foreignness of the text should be preserved in translation, or whether a text should be made more accessible to the translation's audience. Emphasising her central position between the two cultures – an 'otherness' already well flagged by references to her imprisonment by post-Revolutionary forces – she further highlights her significant mediatory role in bringing Bernardin de Saint-Pierre's work before a British public.

By the time Williams came to work on Humboldt's narrative in the early 1810s, she had also acquired a highly politicised identity through other texts she had translated. In a contract drawn up with Williams for her translation of a speech by Robespierre, the Irish translator Nicholas Madgett, employed at the French foreign ministry, described her as 'one of the leading pens in England', whose English writings on the French Revolution 'were incontestable evidence of her principles'.[12] The three-volume *Political and Confidential Correspondence of Lewis the Sixteenth; with Observations on Each Letter by Helen Maria Williams*, published in London in 1803, was, like her *Letters from France*, political dynamite. A collection of seemingly authentic letters attributed to the late Louis XVI, who had been executed at the start of the Reign of Terror in 1793, it was published as a bilingual volume. Williams supplied the English translations, exploiting the opportunity to express her republican, anti-Catholic sentiments in observations appended to each letter. Williams produced a highly literal rendering, perhaps due to pressure of time, which contained such awkward formulations as: 'You long balanced whether you should come' for 'vous balançâtes long-tems à venir' and 'questions the most absurd are agitated in the Assembly' for 'on agite maintenant, dans l'assemblée, les questions les plus absurdes' (Williams 1803: I, 43, 49; II, 30, 32). Some of her translations and commentaries also drew heavily on the language of sensibility. She gave the far more lyrical 'I stem alone the stormy torrent' for 'je fais seul tête à l'orage' [I alone face the storm], while the original 'je porte dans mon cœur' [I carry in my heart] became 'I feel engraven on my heart' (Williams 1803: I, 200, 203; II, 92, 94). Williams's lexical choices therefore allied Louis XVI's correspondence in English translation fairly and squarely with the language of sensibility that was her own stylistic hallmark.

But the letters were a scam, and Williams was not alone in being fooled by them. Even Francis Horner, the Scottish Whig and critic for the *Edinburgh Review*, thought them genuine, finding only later that two 'impudent' men came forward to avow they had composed them as a 'fair literary enterprise one morning after breakfast' (Horner 1843: I, 228). Even prior to this revelation, Horner's review of the *Correspondence* had been withering. Writing to the Scottish publisher John Murray II, he had noted, 'I have taken [. . .] for the Review, Miss Williams's correspondence of Louis XVI,' adding, 'I shall probably adore the unfortunate prince, and flagellate the conceited heartless woman' (Horner 1843: I, 228). He did just that. Williams, whom he disparaged as a 'foreign refugee authoress, whose reputation in her own country has scarcely reached beyond the customers of the circulating libraries', had adulterated the original with her additional commentaries by padding them out to 'a large bulk by a mixture of rubbish'.[13] In the translation, the King's style and natural expression had been lost, he opined, to the 'tawdry bombast, and the chilling affectation, of Miss Helen Maria Williams' (Horner 1843: I, 214).

Her final translation project, done at Humboldt's suggestion, was an English version of Xavier de Maistre's *Le Lépreux de la cité d'Aoste* [*The Leper of the City of Aoste*] (1811, trans. 1817). In her Preface she registered an acute nostalgia for the language of her youth, which she had not spoken in her home country for over twenty years: 'Although long habit may render a foreign tongue as familiar as our own, we have best to weep over sorrows recorded in that language in which our earliest emotions were felt, and our first accents were uttered' (Maistre 1817: v–vi). Willams's high degree of cultural assimilation through years spent in France meant she presumably understood the French of Humboldt's *Relation historique* very well. But this prolonged exposure to a foreign language also appears to have been detrimental to her English. Woodward comments that her short *Letters on the Events Which Have Passed in France Since the Restoration in 1815*, published four years later in 1819, was a work Crabb Robinson was obliged to free of Gallicisms before he could approach a London publisher on Williams's behalf (Woodward 1930: 187). And Williams's chatty letters to Murray betray a tendency to mix English and French – 'Being of this moment extremely busy and hurried, I am forced to reply quite à la hate,' she wrote to him in December 1815 – suggesting that she generally cultivated a hybrid, frenchified form of English, which also pervaded her translations.[14]

Humboldt was evidently familiar with the range of Williams's creative writing and translation work, and his correspondence reveals admiration for her political travel account *A Tour in Switzerland* (1798), her *Narrative of the Events Which Have Taken Place in France* (1815) and her pamphlet *On the Late Persecution of the Protestants* (1816).[15] In correspondence with his publisher Cotta in the 1840s, he repeatedly described Williams as the famous literary writer – 'die berühmte Dichterin' (Humboldt 2009: 224, 407) – looking back on her career as an author of fiction and neatly sidestepping her controversial reputation as a political commentator. But perhaps by then he remembered most clearly her final works, notably the collection *Poems on Various Subjects* (1823), which included 'To the Baron de Humboldt, On his Bringing Me Some Flowers in March'.

Money, Misery and the *Medusa*

Correspondence between Humboldt and Hurford Stone, and, more importantly, between Humboldt and Williams, allows us to pick up some of the threads by which to reconstruct how the *Personal Narrative* was managed as a business undertaking. Humboldt had already lost money (to the tune of 35 Louis d'or, he complained) on Black's translation of the *Essai politique*. Since Hurford Stone had bought the rights to the *Voyage de Humboldt et Bonpland* for the tidy sum of 100,000 francs, Humboldt was prepared to place the English translation in his seemingly capable hands (Stern 1980: 349). As Humboldt wrote in a most obliging tone:

> Depuis que Vous, mon excellent ami, avez acquis la propriété de mon ouvrage, je Vous ai abandonné toute l'affaire de la traduction. C'est Votre affaire seule, Vous l'avez mis entres les meilleurs mains possibles. Je ne veux y être que pour autant que Mademoiselle Williams veut bien m'employer et c'est à Vous et à Elle que je m'abandonne volontiers. . . .[16]
> [Since your acquisition, my excellent friend, of the rights to my work, I have completely ceded the business of translation to you. It is your business alone, you have placed it into the best possible hands. I only wish to be there should Miss Williams need my services and it is to you and to her that I willingly surrender myself. . . .]

Humboldt's trust in him would not have seemed misplaced. An established member of the British expatriate community in Paris, Hurford Stone had started out in 1794 by asking his brother to look into 'the means of getting books of merit which may come out, to translate from the English into French', particularly those that 'would have a speedy sale: of this sort are travels' (Stern 1980: 320). He then founded the 'Imprimerie Anglaise', or 'English Press', which issued Paine's *Dissertation on Government* (1786) and Barlow's epic, *The Vision of Columbus* (1787).

It also produced Jean-Baptiste Say's translation of Williams's Swiss travelogue as the *Nouveau Voyage en Suisse* (1798) and her *Political and Confidential Correspondence of Louis XVI* (Stern 1980: 339). Hurford Stone was therefore an established and capable printer by the time he took on Humboldt's *Vues des Cordillères*, the *Essai politique*, the *Recueil d'observations de zoologie et d'anatomie comparée* and the *Atlas géographique et physique du royaume de la Nouvelle-Espagne*.[17] These lavishly illustrated volumes – the *Atlas géographique* alone had twenty engraved plates, of which five were double-spread or large folded sheets – were to be his downfall because they required far greater capital investment than he had originally imagined. In 1812 he resigned from his directorship of the firm (Stern 1980: 350).

Williams and Humboldt were also crippled financially by the publication of these works. Lady Morgan describes Williams's demise in the 1820s as a decline 'into absolute indigence some time before her death, a circumstance which, in her independent spirit she endeavoured to conceal till all further concealment was impossible' (Morgan 1830: I, 362–3). This is difficult to rhyme with the figure who had arrived in Paris some thirty years earlier a relatively wealthy woman, following the success of *Edwin and Eltruda* (1782) and *Poems* (1786), which had boasted over 1,500 subscribers. Humboldt would likewise see his fortune slip away as he prepared the first works for publication that derived from his expedition to the Americas. In January 1829, the year in which the last volume of the English translation of the *Relation historique* was published, Humboldt complained in a letter to the German general and diplomat Count Georg von Cancrin that he had spent everything he had inherited, estimated at 100,000 Prussian thaler (Humboldt 1869: 43).

The financial straits into which both Humboldt and Williams sank seem, however, to have strengthened rather than sundered the relationship between author and translator. Together, they fended off queries and complaints from Longman, and worked as an effective team to ensure the successful completion of the *Relation historique* and the *Vues des Cordillères* in English translation. As Williams reflected in a letter of 25 March 1819 to Henry Crabb Robinson:

> the only thing in heaven or earth that M. Humboldt does *not* understand, is business, but he had given me a great proof of his devotedness in 'binding up his loins to such a feat', he had done his best, and therefore I acquiesced in conditions with which I was little satisfied. (Woodward 1930: 184)

Humboldt was never very mercenary in his dealings with publishers, and Williams herself conceded to Murray, 'you will have perceived that I am little skilled in bargaining, or in commercial speculation'.[18] But the main problem was less that Longman and Murray were mean-spirited than that Humboldt worked with meticulous care and, as they saw things, infuriatingly slowly. Another dangerous side-effect of his determination to use the material he had gathered exhaustively was that the number of projects on which he was working continually multiplied, becoming many-headed and Gorgon-like – a reminder of the immense, fragmentary nature of what Ette has described as Humboldt's 'work in progress' (Ette 2001a: 35). In January 1819, in an uncharacteristically downbeat letter to Williams, Humboldt even forecast 'le triste spectacle du naufrage de la Méduse' [the sorry spectacle of the shipwreck of the Medusa].[19]

Since the *Personal Narrative* was financially such a bottomless pit and printing and publishing it something of a nightmare, why did Williams never contemplate pulling out? One reason surely lies in Humboldt's constant encouragement of her activities and admiration for her achievements. His letters abound with reassurances that he had their best interests at heart, expressions of appreciation for her work and promises that he would continue to present an optimistic front to their frustrated publisher, Longman. Humboldt also made a point of reminding Williams that others greatly esteemed her work, relaying to her from his travels abroad that the publisher Mr Rees was 'full of admiration' for her and the Scottish novelist, Jane Porter, 'speaks of you and your poetry as I would wish'.[20] Writing from London in June 1814, he described how much her translation was valued by the English physician and chemist Sir Charles Blagden, and noted that while staying at the Duke of York's residence, Oatlands Palace in Surrey, he had found almost all Williams's works in the library there and everyone was impressed by her talent.[21] In September he again reported back to her on how her translation was faring in England, writing in a quaint mixture of French and English, '[v]otre belle traduction est en vente à Londres. Un Mr. Cory qui vient d'arriver l'a vue dans tous les Shops' [your fine translation is on sale in London. A Mr. Cory who has just arrived has seen it in all the shops].[22] Ironically, then, it was not Williams who kept track of the translation's fate in Britain, but Humboldt. And it was in England – from which Williams had dramatically 'eloped' thirty years previously – that their financial salvation seemed to lie.

While Humboldt fended off much of the criticism from Longman about the slowness with which the *Personal Narrative* was appearing, Williams too must have played her part in placating the publishers. 'We are happy to find that the Personal Narrative is proceeding, for the sooner this is completed the more productive it will be to all interested,' pressed Longman in a letter to Williams of 5 June 1817.[23] Longman's accounts ledgers show that large payments were going out quite regularly to Williams in the period between 1818 and 1820 – as much as £78 on 28 December 1818 – as she completed volumes three, four and five of the *Personal Narrative*.[24] But by January 1824, relations between Williams and Longman had become less cordial, and the publishers contacted Humboldt to ask if he would 'oblige us by a few lines informing us of the progress of the Relation Historique as we have not heard from Miss Williams for a considerable time'.[25] In May a keener wind was blowing. In response to a letter from Williams requesting more money, Longman regretted that 'the sale of the work does not warrant any increase of remuneration beyond the Agreement', and added reproachfully that '[t]he delay of the work has been a serious injury to the sale of the translation, the public having lost all confidence as to a completion'.[26] He drew to her attention to the fact that 'as we have now no demand for the French edition in our hands, that we can no longer Insure that stock, or incur any further expences [*sic*] on it', adding that 'the stock may be removed from our Warehouses'.[27] By the end of summer 1824, communication between Williams and Longman had completely broken down. She had not responded to their notice regarding insurance on the property in which unsold stock was being stored, and they were now demanding acknowledgement of receipt 'by the next post'.[28]

Although it is relatively easy to chart the beginnings of the *Personal Narrative*, it is harder to assess how the translation was concluded. Williams had been dead for two years by the time the last volume appeared in London in 1829, so Longman made the final payment of £84 1s 6d to her inheritors.[29] It seems most likely that Charles Coquerel, Williams's nephew, guided her mammoth translation project to a conclusion. If so, he was careful to sustain Williams's style throughout this volume, since it does not read as if it were the work of a different hand. Despite obvious misgivings, Longman was right to see the *Personal Narrative* through to the end. Without the seventh volume, the English version of the narrative would have seemed still more fragmentary and incomplete than the *Relation historique* already was.

First Steps: Correcting the *Personal Narrative* and the *Researches*

Humboldt's letters to Williams, covering a period from 1810 to 1824, give us a clear insight into which aspects of her translation skills he valued most, how their working relationship evolved and how they developed an efficient *modus operandi*. Humboldt's introductory letter to Williams of 10 May 1810 indicates that it was she who had taken the initiative to start work on a translation:

> Un de nos communs amis, littérateur distingué, [...] m'a appris que Vous avez commencé à Vous occuper de la traduction de mes ouvrages. Je ne puis Vous exprimer assez vivement combien cette nouvelle a flatté mon amour propre. Je connais assez Votre langue et le charme du Style de Miss Helena Williams pour sentir combien sa plume éloquente donnera d'attrait à tout ce qui est susceptible d'émouvoir l'imagination.[30]
> [A mutual friend of ours, a distinguished man of letters, [...] has informed me that you have begun to busy yourself with the translation of my works. I cannot express emphatically enough how much this news has flattered my self-esteem. I know well enough your language and the charm of Miss Helena Williams's style to sense how much her eloquent pen will render appealing everything that is capable of inspiring the imagination.]

Two points stand out here: Humboldt's familiarity with Williams's writing style and his interest in the imaginative potential of the language of her translation. Still something of an *ingénu* when it came to managing his translators, Humboldt's attitude towards Williams in this letter was one of immediate cooperation. He would place at her disposal all the manuscripts she required, he promised, adding that there were certainly plenty of works to go at. He reeled off the *Géographie des plantes*, the *Tableaux de la nature* and the *Vues des Cordillères* as good starting points – doubtless to the slight bewilderment of his new translator.[31]

Charm, flattery and praise were all part of Humboldt's rhetorical arsenal. While his repeated adulation of Williams's 'admirable' style could be seen merely as an extension of this linguistic diplomacy, the frequency with which he applauded it suggests a real respect for her achievements. One letter written to Williams at two in the morning (he usually corrected her translations at night) records his impressions of an early passage of the *Personal Narrative*: 'I cannot thank her enough for the care she has been so good as to invest in this work. Everything is elegant, spirited, exact, precise.'[32] He praised in similarly glowing terms the breadth of her linguistic talents, and the precision and good taste exhibited in her translation.[33]

Humboldt was acutely aware of the contribution that Williams made to his work, not just in the basic 'Englishing' of his text, but also in her complex recasting of his writing in her own idiom. Thanking Williams effusively for the most recent pages she had sent, he commented, 'Je sais combien mon ouvrage a souvent gagné sous ses pinceaux!' [I know how much my work has often been improved by the strokes of her brushes!].[34] By presenting Williams as an artist in her own right, he emphasised her investment in the visually powerful, imaginatively engaging elements of his work. In allowing her 'brushstrokes' to be an identifiable component of the English translation of his *Relation historique*, Humboldt signalled his acceptance that her own creativity should be apparent in his work. In some senses, then, a faithful translation of the poetic elements of the French original was never Humboldt's goal. As he exclaimed, 'Que Vous savez mettre de la grâce dans tout ce qui échappe de Votre plume. [. . .] Tout est senti, embelli par Vous' [How well you know to render graceful everything that issues from your pen. [. . .] Everything is felt and embellished by you].[35] This was surely the ultimate proof that he endorsed the stylistic changes she made to her translation – which critics would later term 'exaggerations' – as she deliberately heightened the imaginative appeal of the *Relation historique* to an Anglophone readership.

Hidden among the Alexander von Humboldt papers at the Royal Archive in The Hague are twenty-nine pages of scribbled comments, corrections and queries – some with quirky sketches to aid visual clarification – made by Humboldt in response to the opening sections of the translation in its first draft. They are deeply revealing of the various lives Humboldt's *Personal Narrative* enjoyed prior to publication, as a draft passed to and fro between translator and author. The very existence of any pages of corrections is surprising: they should have landed in Williams's wastepaper basket some two centuries ago, discarded in the translation process as draft supersedes draft, and the final, polished version emerges. Once deciphered, reordered and positioned alongside source and target text, they bear first-hand witness to how the collaborative enterprise of translating the *Relation historique* was managed in practice.

Most importantly, they enable us to adopt a process-oriented approach to translation, investigating the various stages a translation undergoes, as well as the input and influence of different agents at various junctures. Rather than focusing on the 'reception end', Hélène Buzelin argues that by following translation projects 'in the making', we can more easily examine 'the link between linguistic/stylistic decisions and those pertaining to the product's features and its mode of dissemination' (Buzelin 2007: 165). These attempts to reconstruct the translator at work share similarities with the call by Jerome McGann (1992) to acknowledge pre-publication documents as important traces of a dialogue between author, publisher and editor (and, we could add, translator). Representatives of genetic criticism, or *critique génétique*, notably Almuth Grésillon, are likewise interested in a text's process of evolution, its development from 'private', rough draft to 'public', final text (Grésillon 1994: 109). Such archival fragments are the embodiment of a dynamic process of rethinking and rewriting. They enable a study of the traces left behind by the writer, which emphasises the power and weakness of language, the moments of inspiration and failure, the instability of the text and the precariousness of language. Above all, though, they gives us a glimpse of the 'aesthetics of production', the processes of creation and deletion, omission and addition, rejection and acceptance, which enable a text to be honed to a point at which it is deemed ready to enter the public sphere (Grésillon 1994: 209).

Two of the twenty-nine pages of comments in the archival material relate to the descriptions accompanying plates I to VI of the *Researches*. The remainder refer, with one exception, to the Introduction and first three chapters of volume one of the *Personal Narrative*.[36] It may simply be coincidence that only these sheets pertaining to the opening volumes survived. It is equally possible that Humboldt rapidly rethought the way in which he went about reworking Williams's draft. If, as he claimed several decades later to Cotta, he actually revised Williams's translation orally by having her read it out to him, then these pages indicate not only how the text itself changed over time, but also how the author and translator adjusted their own working practices to accommodate the pressures of publication (Humboldt 2009: 224).

While he had little to cavil at in the Introduction, Humboldt very swiftly recognised one major area in which Williams's translation skills were inadequate. From the outset she had immense difficulty grappling with the scientific terminology and expressions. It was precisely the factual detail, with its concomitant terminology, that wrong-footed Williams time and again. Bret's overview of scientific translators in nineteenth-century France suggests that the majority were specialists in the area in which they were translating, some at the beginning of their career, some already well established in their field (Bret 2012: 948). Williams did not fall into these categories. Humboldt had to tell her that the correct term for 'calorique rayonnant' was actually 'radiant heat' and that sulphur crystals were 'of conchoid fracture', that 'plantes grasses' should be 'succulent plants' and 'reconnoissance' was a maritime expression signifying that a particular landmass had hove into view.[37] Place names and geographical designations were also not without their pitfalls. He frequently corrected Williams's rendering of 'la mer du Sud' from 'southern sea' to the 'South Sea' (the expression he had found in Cook), and the Straits of Magellan should, he reminded her, be in the plural.[38]

These examples merely gesture towards the immense range of geographical, mineralogical, botanical and maritime terms in which the *Personal Narrative* abounds. While Humboldt did not expect Williams's knowledge of scientific terminology to be encyclopaedic, he had clearly presumed she would research the correct translations of these terms. He queried early on the term 'formulary', suggesting that 'method' would be more appropriate, and begged Williams to seek help from 'Mr. Warden' in such matters.[39] David Baillie Warden, for whom Hurford Stone had printed work in 1813, nurtured interests that ranged across medicine, chemistry and mathematics. He had studied natural history, anatomy and theology at the University of Glasgow and was therefore well placed intellectually to advise Williams on a range of issues. He was also someone whom Humboldt trusted. His letters to and from Humboldt in the period from around 1809 onwards indicate the valuable contributions that Baillie Warden made to Humboldt's understanding of the Gulf Stream (Humboldt 2004b: 128–30). Humboldt made sure he had to hand the tables of temperature sent him by Baillie Warden in 1816 when he was drawing up his landmark essay on isotherms and heat distribution across the globe in 1817 – a sign of the esteem in which he held the scholar's research (Humboldt 2004b: 24).

Warden's expertise was called upon to confirm that a 'squall' was the correct marine term for the 'grain de vent' that battered Humboldt and Bonpland as they sailed from Santa Cruz en route to Cumana, rather than the 'weak storms' in Williams's first draft.[40] Warden also stepped in to advise that 'angles or bearings' was the right translation for 'relèvemens', where Humboldt was describing how difficult it was to calculate a ship's position if the coast was so far off that its outline was difficult to make out.[41] Williams should not be satisfied with mere guesswork, Humboldt chided:

> J'attache beaucoup de prix à l'extrême justesse des termes techniques. Rien [n']est plus fait pour inspirer de la méfiance que le vague des expressions de Physique et/d'Astronomie nautique![42]
> [I set great store by the total accuracy of the technical terms used. Nothing is more likely to arouse mistrust than the vagueness of expressions pertaining to physics and nautical astronomy!]

The defensive tone of his comment implies that he was afraid that the language, rather than the structure or ideas in the *Personal Narrative*, could become the Achilles heel of his work and be detrimental to his scientific reputation. But it was not just Humboldt's reputation that was at stake. As he commented apologetically, 'Vous direz que je suis bien insolent. Il y va de Votre bien et Vous savez pardonner' [You will say that I am rather insolent. It is for your own good and you know how to forgive].[43] This may have been a conscious strategy on Humboldt's part to goad Williams into more meticulous translation. By appealing to her *amour propre* and implying that her own public image could be tarnished by language sloppily used, he also emphasised just how closely her own fortunes were bound up with this undertaking.

On one occasion, when Williams took the initiative of drawing on external advice, this backfired quite disastrously. Describing the crossing from Tenerife to South America, Humboldt related how he and his companions entered the torrid zone and saw for the very first time stellar constellations unknown to them in the northern hemisphere. These included the Southern Cross, so poetically described in *Paul et Virginie*. Williams erroneously added that this belonged to the group of stars known as 'Sobieski's Shield'. Humboldt threw up his hands in horror:

> hélas! qui Vous a inspiré la parenthèse the Sobiesky Shield!! C'est comme si nous disions Andromède ou la petite ourse. Je fais une guerre à morte à Votre conseiller. C'est notre ennemi! L'écusson de Sobiesky se voit en Suède,

et la Croix du Sud ne se voit qu'aux Indes. Il en étoit fait de ma réputation d'automne à Londres et l'on devoit croire que la parenthèse étoit de moi.[44]
[alas! who has inspired you to add in parentheses the Sobiesky Shield!! It is as if we were to speak of Andromeda or the Little Bear. I will challenge your advisor to a fight to the death. He is our enemy! The Sobiesky Shield can be seen in Sweden, and the Southern Cross can be seen only in the Indies. It would have ruined my reputation in London this autumn and they would have believed that the information in parentheses came from me.]

With self-styled 'Prussian frankness' and good humour, he alerted Williams to the dangers of being careless and inaccurate in her translation.[45]

Humboldt not only relied on other Anglophone scientists working in Paris for help in correcting Williams's mistranslations of technical terms. In some cases, he combed works in his own scientific library – particularly the Royal Society's *Philosophical Transactions* – for the right solutions. Two whole pages of his handwritten notes are filled with lists of equipment, geological formations or plant structures that Humboldt had sifted from scientific texts to facilitate Williams in her translation work. It was 'feldspar', he wrote, and not 'feldspath', 'calcareous carbonate' for 'carbonate de chaux', and 'geodetical' rather than 'geodesical' instruments.[46] Some terms – 'Variation Compass' for 'boussole de déclinaison' or 'Dipping Needle' for 'boussole d'inclinaison' – are transcribed letter for letter, rather than in Humboldt's characteristically crabby hand, to avoid orthographical errors creeping in.[47] The *Philosophical Transactions* were also a useful source of geological terminology. In an 1804 communication from George Watt (son of the steam engine developer, James Watt) about the structure of basalt, read before the Royal Society in May, Humboldt had found the term 'distinct concretions' for 'pièces separées', which Williams needed for descriptions of basalt formations.[48] The expressions 'isolated mountains' and 'relative age', which Humboldt drew from the Scottish naturalist Robert Jameson's *System of Mineralogy* (1808), also to hand in his library, were relevant for the discussion in the *Personal Narrative* of different types and formations of porphyries. Work by the historian and geographer John Pinkerton (most likely his *Petralogy* of 1811) aided Williams and Humboldt in solving the problem of how 'lamelleux' [lamellar] and 'schisteux' [slaty] should be translated in a passage on the geological structure of the hills encircling the peak of Tenerife.[49]

Williams was stumped not only by astronomical and mineralogical terms. Humboldt also drew on medical terminology to describe how, as they approached the Antilles, an illness of near epidemic proportions rendered even the most robust delirious. Without any supplies of cinchona bark on board, 'being more occupied with our instruments than our health', he noted ruefully, they had no means of treatment but fresh air (*PNW* II, 23). Yet this seemed to do the trick, more than 'all the asthenic remedies' of bleeding or evacuating the body (*PNW* II, 24). Williams had described the remedies as 'palliative', whereas Humboldt indicated that he was looking for 'debilitating' or 'asthenic' – the latter term, he added, being more commonly used in English, since he had read it in publications by the physician John Brown of Edinburgh.[50]

One other difficulty would have flummoxed even the most competent scientific translator: the fact that it was sometimes impossible to visualise what Humboldt had seen or know what he was describing. The popularity of exotic travel writing turned on precisely this: confronting the armchair traveller with the strange, the intriguing and the extraordinary. As Dettelbach comments, 'every traveller worth his or her salt knew that few things were as effective in stimulating interest in a voyage as a few carefully chosen engravings of new and spectacular plants', the 'palm tree, that emblem of the tropics' chief among them (Dettelbach 2005: 46). Although the English translation of the *Relation historique* was accompanied by seven plates, these were geographical or geological. Taken from illustrations already prepared for the *Atlas géographique* (1808–34), they included maps of the course of the Orinoco and the Río Meta, of Colombia and the eastern part of the province of Verina, and of Humboldt and Bonpland's journey towards the summit of Chimborazo – all of which would have been of little help to Williams as she struggled to translate some of Humboldt's more unusual botanical finds.

Although gardens such as the Jardin des Plantes in Paris or London's Kew competed fiercely to hold the first of any exotic species in their collections, few would have ostentatiously displayed what interested Humboldt most on the crossing to Venezuela – seaweed. As they neared the island of Margarita, off the coast of Venezuela, on 15 July 1799, Humboldt noticed a species of the genus *Fucus* floating in the water, its stems carrying 'appendices extraordinaires en forme de godets et de panaches' [extraordinary appendages in the form of little cups and plumes] (*RH* I, 218; *PNW* II, 37). Williams left the word as 'godet' in English – perhaps because she was unable to translate it, perhaps because she knew of its rather archaic use to mean a drinking tankard. But Humboldt must have sensed that Williams was on unsure ground and had taken a word that the general British reader might not recognise. It invited his response, 'de grace [sic] faites disparaitre [sic] godets – n'est-ce pas little cups [?]. La forme d'une petite tasse, un gobelet' [please remove godets – is it not little cups [?] – The form of a small cup,

'A Colossal Literary and Scientific Task'   97

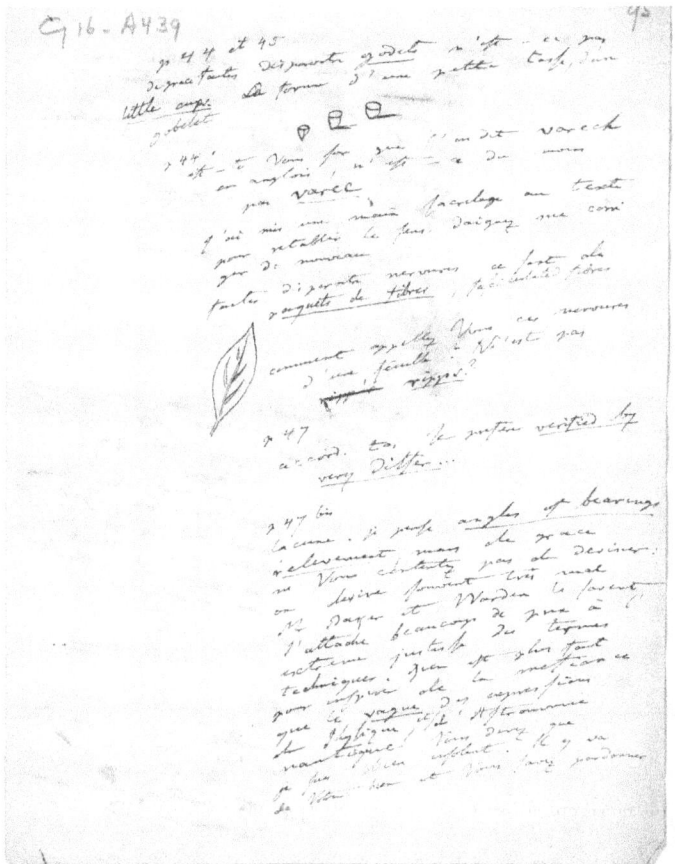

Figure 3.2 Humboldt's handwritten feedback on Williams's draft translation of the *Personal Narrative*: Top: 'de grace faites disparaitre godets – n'est-ce pas little cups'.

a beaker], accompanied by a sketch of three small curved vessels (Figure 3.2).[51]

A second drawing on the same page of Humboldt's feedback was intended to help Williams visualise a different aspect of the plant he was describing – its leaves. Here too he demanded that she replace her solution 'nervures' with a set of terms to describe the 'paquets de fibres' [bundles of fibres] surrounding the cellular tissue of the seaweed.[52] The printed text confirms that she adopted his suggestion of the technical 'fasciculated fibres' (*RH* I, 218; *PNW* II, 37).

Another sketch was made by Humboldt to illustrate a different species of the same genus, this time the *Fucus vitifolius*, illustrated on the left in plate 69 of the second volume of the *Plantes équinoxiales* (Figure

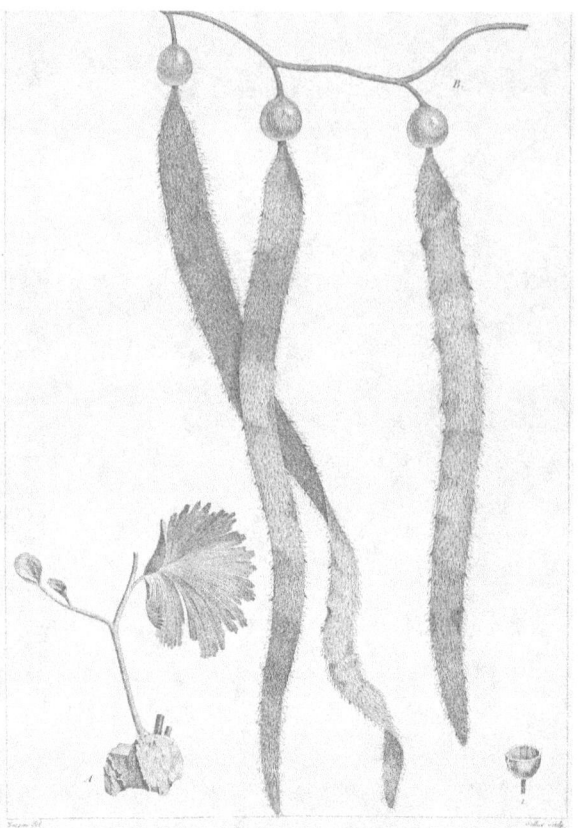

Figure 3.3 'Plate 69: *Fucus vitifolius*', Alexander von Humboldt and Aimé Bonpland, *Plantes équinoxiales*, 2 vols (Paris: Schoell, 1808–9) II, unpaginated.

3.3). The leaves of this were of a 'vert tendre: elles sont membraneuses et striées commes les feuilles des Adiantes et du Gingko biloba' [light green: and they are membranous and streaked like those leaves of the adiantums and the Ginkgo biloba] (*RH* I, 85; *PNW* I, 85). While Williams might have been familiar with the *Ginkgo biloba* tree, she clearly had difficulties visualising the 'streaked' patterning of the leaves, or picturing how the leaves could be 'concave and envelloped [sic] one in the other' before they opened (*PNW* I, 85).[53] Since the *Plantes équinoxiales* had appeared well before she started her translation of the *Relation historique*, it seems odd that Humboldt did not draw her attention to this plate, but perhaps getting this large, heavy volume to his translator was impractical, and Humboldt could swiftly supply the term he felt was appropriate.

A final sketch in Humboldt's feedback, designed to aid his bemused translator in selecting the correct term, was that of part of a banana plant. In a luxuriantly descriptive passage, Humboldt conjured up the exotic fare espied in Indian canoes as their ship the *Pizarro* approached Cumana, having survived the hardships of the Atlantic crossing:

> They gave us some fresh cocoa nuts, and very beautifully colored fish of the chætodon genus. What riches to our eyes were contained in the canoes of these poor Indians! Broad spreading leaves of vijao covered bunches of plantains. (*PNW* II, 44)

Williams had encountered difficulties in conjuring up in her mind's eye the 'régimes de bananes', which would eventually be described in the English translation as 'bunches of plantains'. Humboldt sought to resolve this problem by drawing a generously sized hand of bananas (albeit with the fruit curving outwards) and added by way of explanation 'c'est presque grappes' [it is almost bunches], thus creating a parallel with the French 'grappe de raisins' or 'bunch of grapes' (Figure 3.4) (*RH* I, 222).[54]

Some of Williams's translations were not always that far off the mark, while others displayed a significant tendency to exaggerate. She initially described a 'cross wind', which had caused barometer readings to become irregular, as a 'whirlwind', and she cast de Saussure's travels around Geneva in her first draft as an 'expedition', causing Humboldt to counter that it was more a 'journey' (*PNW* II, 6; *PNW* I, xlii).[55] Other corrections that Humboldt made were more far-reaching in the effect they had on the tone of his writing. In a paragraph at the start of the *Personal Narrative* – before, in fact, the voyage had even got under way – Humboldt had begun to describe relations between Spain and 'the new continent', the speed with which news reached the 'mother-country' from the outposts of its colonies and the means of communication between the two. This appeared in the published version as:

> The circulation of ideas is become more expeditious; the complaints of the natives reach Europe with more facility, and the supreme authority has sometimes succeeded in repressing vexations which, from the distance of the place, would have remained for ever unknown. (*PNW* I, 24)

Humboldt's corrections indicate that Williams had originally put 'acts of tyranny' for the French words 'vexations', which ended up as the English cognate in the published translation: Williams's solution had been too strong and Humboldt demanded that she tone it down (*RH* I, 52).[56] Her initial suggestion would have been stridently critical of the colonial administration in Spanish America. Although he shared Williams's political leanings, Humboldt must also have been conscious

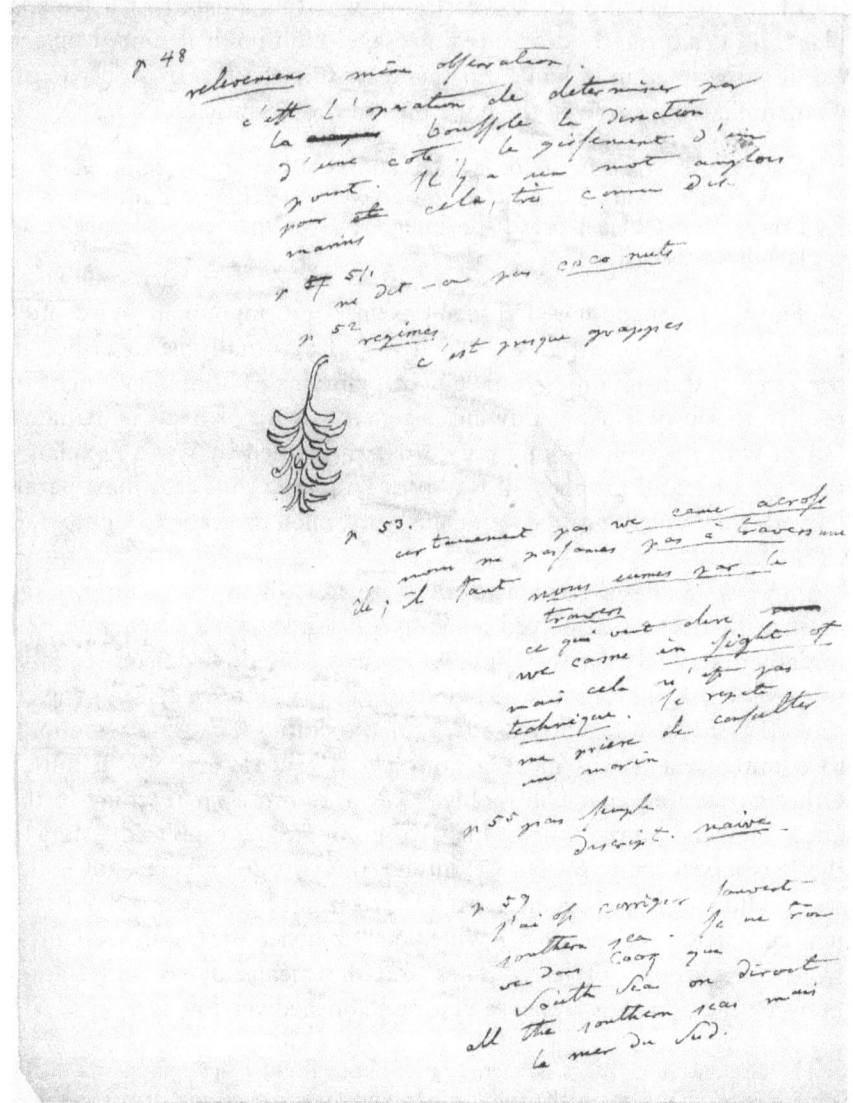

Figure 3.4 Humboldt's feedback on Williams's draft translation of the *Personal Narrative*. Middle: 'c'est presque grappes'.

that her radical enthusiasm could be potentially damaging to his work and reputation.

Many of Humboldt's recommended alterations were definite improvements upon Williams's original. He had suggested in the *Relation historique* that captains of merchant ships could best judge how much longer it would take them to get to New York, Philadelphia or Charleston, once they had reached the 'edge' – rather than Williams's suggestion of 'brink' – of the Gulf Stream.[57] He also proposed the word 'drapery' in a lyrical description of the vegetation on the sides of Niagara Falls in the *Researches*: 'The fall of Niagara, placed beneath a northern sky, in the region of pines and oaks, would be still more beautiful, were its drapery composed of heliconias, palms and arborescent ferns.'[58] But not all his suggestions hit the mark. She decided against 'naive description' to refer to the way in which a local pilot had represented various species to Humboldt and Bonpland in the creole language, preferring to stick with the 'simple description' of her draft – perhaps because of the patronising associations made with 'naive'.[59] Even in scientific passages, she did not automatically take up Humboldt's suggestions. In his discussion of the merits and demerits of the whalebone hygrometer versus Deluc's hygrometer, Humboldt had noted that the latter would respond only with extreme slowness to changes in humidity, hindering a swift reading of the instrument at the summit of a mountain: 'we are often uncertain whether we have not ceased our observations before the instrument has ceased its movement' (*PNW* II, 85–6). Humboldt's initial suggestions of 'progress' and 'march' (for what eventually became 'movement') – echoing the French 'cessé de marcher' – would have made it sound as if the hygrometer were itself tramping up the mountain with them.[60]

Humboldt read the English translation from various perspectives and with different audiences in mind – with the hypercritical eye of the scientific reader looking to cavil at inappropriately used scientific terminology, and through the lens of the general reader seeking literary amusement and imaginative 'transport'. This discussion of Humboldt's queries, suggestions and corrections has necessarily focused on his proposed alterations to Williams's first draft. It is worth noting, though, that long stretches of her translation went untouched. The very language of his queries to Williams, with its fair sprinkling of question marks, tentative 'is-it-nots' and pleas to be forgiven for his impertinence, point to a conscious desire not to deal heavy-handedly with her translation. The genesis of the English version of the *Relation historique* was therefore founded on the robust exchange of opinions – a discursive process between author and translator rooted in openness, honesty and mutual respect.

Final Versions: The *Personal Narrative* and the *Researches*

The title pages of both the *Personal Narrative* and the *Researches* indicate how much thought their author and translator – and possibly also their publisher – gave to the associations that these would evoke on the British book market. Neither the English title of the *Relation historique* nor that of the *Vues des Cordillères* is a simple, literal rendering of the original. The *Vues des Cordillères*, fully titled the *Researches Concerning the Institutions and Monuments of the Ancient Inhabitants of America, with Descriptions and Views of Some of the Most Striking Scenes of the Cordilleras!*, may have been thus named to call to mind the celebrated *Asiatic Researches* of the British Asiatic Society of Calcutta. The English rendering of the *Relation historique* as the 'Personal Narrative' does not appear to be modelled on any previous titles of a similar nature, but it certainly set its own trend. This word-pair rapidly became a constituent element in titles of non-fictional travelogues and historical biographies in the wake of Williams's translation, growing in popularity through the 1820s to peak in the 1830s.[61]

Although subsequent travel writers would therefore exploit the success of this formulation, reviewers had actually been quite disapproving of it. In the *Edinburgh Review* of 1815, one critic had noted that the title of a 'Personal Narrative of Travels, may not, perhaps, convey to every one [sic] a very precise notion of the work to which it is prefixed', insisting that Humboldt's was nothing more nor less than an 'ordinary book of travels'.[62] Over the next decade, critics' patience would wear thin as they ploughed through the endless pages of statistical tables, paragraphs peppered with measurements and readings, and footnotes bulging with the encyclopaedic references that characterised the later volumes of the *Personal Narrative*. On the appearance of the penultimate volume in 1826, the *Eclectic Review* snapped that 'the title of the work is, in fact, a complete misnomer: it is any thing [sic] rather than a "personal narrative"' and complained that the 'slender thread of narration' by which Humboldt attempted to keep the rather heterogeneous parts together (presumably constituting what the reviewer understood by the account's 'personal' qualities) was constantly being broken off.[63] The *Literary Gazette* of the same year concurred: 'we have as little of personal narrative as can well be fancied', the chief merit of the volume being its statistical tables.[64] While the *Personal Narrative* did indeed become increasingly short on individual reflection, Humboldt (or indeed Williams) could not have forecast how the

Relation historique would evolve, as they deliberated on its English title some twelve years earlier.

What is also striking about the title page of the *Personal Narrative* is Williams's prominence on it as the translator, her name given in a font size and type identical to that of Humboldt. The presentation of the author to the British public also repays scrutiny. In the mid-century English translations of the *Relation historique*, *Kosmos* and the *Ansichten der Natur*, he becomes 'Alexander von Humboldt'. However, in Williams's rendering, he is still 'Alexander de Humboldt'. This shift over the thirty years in which Humboldt's writing was first published in English neatly charts his own biographical journey from Parisian *savant* to celebrated Prussian scientist. Perhaps Williams opted for the French 'de' because 'Alexandre de Humboldt' graced the title page of the French edition and had also been used by Black in the *Political Essay*. But despite the anglicisation of 'Alexandre' to 'Alexander', this still made him seem frenchified and foreign to English ears, as if deliberately rejecting his Germanic roots.

Williams's eight-page translator's preface to the *Personal Narrative*, described by the *Monthly Review* as 'marked equally by taste and sensibility', made her presence in the text immediately felt.[65] The Preface, which opens with the spotlight firmly on Williams (Humboldt's name appears only in the second half), shows her subtly admitting a scanty knowledge of science and conceding that her English might have a foreign ring. She also seeks to deflect English critical opinion by asserting that, '[l]ong a stranger to my country', she hopes for impartial treatment from its critics (*PNW* I, xi). But if she appears to distance herself from her own volatile biography, her preface is far from apolitical. Travel writing, she asserts, gives respite from the 'terrible page of our history' just turned, acquaints us with 'new systems of social organisation' and brings 'civilisation' to wilder parts of the globe (*PNW* I, vi–viii). As Kelly has perceptively remarked, the *Personal Narrative* did not constitute a change in direction for Williams away from political writing. Rather, its Preface is shot through with allusions to Robespierrean reason and Napoleonic imperialism, which make of Humboldt a 'scientist for a post-Revolutionary age, feminizing his subject with feeling and imagination to give it a fully human character and appeal' (Kelly 1993: 208–9). Despite Williams's self-deprecating stance, she writes herself into the *Personal Narrative* not as Humboldt's 'translator' but as his 'interpreter' – for, as Kelly also notes, she avoided calling the text a 'translation' (Kelly 1993: 209). By casting her English rendering of Humboldt's work as 'my pages', she also aligns translation with authorship, echoing the opportunities for self-empowerment and self-promotion (as well as manipulation of the original) in her version of *Paul et Virginie*.

Williams's relegation of Humboldt to the second half of the Preface was deliberate. But it was less a calculated move by her than by Humboldt. Kurt-Reinhardt Biermann has remarked that Williams's Preface was essentially not her own work at all (Biermann 1982: 60). She sent Humboldt a first draft for inspection, which provoked the following response from him:

> Je sens combien Vous avez d'amitié pour moi: l'élévation de Votre style en est la preuve la plus flatteuse. Elle sera durable, car ces lignes dictées par les sentimens les plus purs et les plus nobles, ne peuvent rester que parmi les papiers qui me sont les plus chers. Il est impossible, Mademoiselle, que cette lettre soit publié. Qu'y a-t-il de plus dangereux que ces éloges au moment ou le public doit juger par lui-même de l'ouvrage auquel j'ai mis le plus de soin.[66]
> [I sense how much friendship you feel towards me: your elevated style is the most flattering proof of it. It will last, because those lines dictated by the most pure and most noble sentiments can only remain among those papers that I value the most. It is impossible, Mademoiselle, for this letter to be published. What could be more dangerous than these paeans of praise at a moment when the public has to judge of its own accord the work to which I have applied the utmost care.]

Her eulogies of praise, though well meant, were ill placed. In a subsequent letter, Humboldt suggested she extend her comments on Cook's voyage (doubtless with its appeal to English patriotic sentiment in mind) and he wrote an additional passage, now covering the best part of pages eight to ten of the English Preface, which she could translate for inclusion in her piece.[67] This she dutifully did, reducing her adulation of Humboldt to a mere two or three sentences.

On the printed page, then, the textual collaboration between Williams and Humboldt starts even before we embark on the translated narrative of his voyage to the Americas. Williams's Preface is in fact a subtly orchestrated piece of self-advertisement by Williams and Humboldt that obliquely registers her compliance with his wishes to steer a more modest course. We can only now surmise what Williams had originally penned, but Humboldt was right to be cautious. Barrow, in the *Quarterly Review*'s first piece on the *Personal Narrative*, commented that it was 'not the fault of M. de Humboldt, though it may be to his misfortune that he has fallen into the hands of injudicious friends, who speak of his pretensions in a strain of exaggerated panegyric that must pain a modest man, and shame a wise one'.[68] If Barrow curled his lip at the excessive idolisation of Humboldt in the published version of Williams's Preface, he would surely have responded with greater vitriol to the wording she originally had in mind.

Although Williams's effusions were problematic in the Preface, Humboldt was far more receptive to them in the main text of the *Personal Narrative*. Some of the changes she introduced made her translation strikingly different from the French original. Luise von Flotow proposes, in *Translating Women* (2011), that it is time to move away from considering female translators as trapped within systems that censor their work, reduce them to quasi-invisibility or simply disregard their contribution to knowledge transmission. Rather, Flotow suggests that translation scholars should revisit ideas about the discursive and performative nature of gender, which would allow us to look more productively at how women actively mobilised translation for their own ends (Flotow 2011: 7).

Williams boldly reconfigured Humboldt's descriptions of the natural world in ways that echo the intertextual approach to writing and translation that characterised her understanding of authorship. As his party left the Canary Isles, Williams records in her translation that we 'quitted this solitary place, this domain where Nature towers in all her majesty' (*PNW* I, 190). The French had been far less effusive, referring simply to 'ce site dans lequel la nature se montre dans toute sa majesté' [this area in which nature presents itself in all its majesty] (*RH* I, 144). Making nature an imposing, threatening or overwhelming force was a leitmotif in Williams's translation. Throughout the narrative, mountain peaks and the crowns of trees were never simply 'high' but always 'lofty'. Shadows are never 'dark' but routinely 'gloomy', and barren landscapes almost without exception 'austere and savage'. She not only echoed the language of Romantic landscape aesthetics. As Humboldt had noted at the end of an early chapter on the Canary Isles:

> l'on peut espérer qu'un jour les îles Fortunées où l'homme éprouve, comme partout, les bienfaits et les rigueurs de la nature, seront dignement célébrées par un poète indigène. (*RH* I, 197)
> [we may hope that one day the Fortunate islands, on which man experiences, as elsewhere, the benefits and hardships of nature, will be celebrated in a fitting manner by a native poet.]

Williams rendered it like this:

> we may be led to hope, that, at some future period, the Fortunate islands, like every other climate of the Globe, either where man reposes on the lavish bounties of Nature, or shrinks from the severity of her frown, will inspire the muse of some native poet. (*PNW* I, 289)

In her translation 'les bienfaits [. . .] de la Nature' became 'the lavish bounties of Nature' and implicitly imbued the natural world with a greater richness, generosity and fecundity than in the original. The 'rigueurs' were reformulated to become the 'severity of her frown', sustaining Williams's personification of Nature as a force that could even inspire fear in man as he 'shrinks' away from her displeasure. Reflecting on the wholly absorbing richness of the natural world, Humboldt commented:

> Au milieu d'une nature imposante, vivement occupé des phénomènes qu'elle offre à chaque pas, le voyageur est peu tenté de consigner dans ses journaux ce qui a rapport à lui-même et aux détails minutieux de la vie. (*RH* I, 28)
> [Amidst the impressive power of nature, deeply absorbed in the phenomena that it offers at every step, the traveller is little tempted to record in his diaries what relates to himself and to the minute details of his own life.]

Williams's exuberance in the English translation seemed to know no bounds:

> Amidst the overwhelming majesty of Nature, and the stupendous objects she presents at every step, the traveller is little disposed to record in his journal what relates only to himself, and the ordinary details of life. (*PNW* I, xxxviii)

She added the notion of 'stupendous', and heightens 'imposing' to 'overwhelming majesty', intensifying wonder at Nature's power. She also edged the focus away from the interested traveller, 'vivement occupé' in the source text, to place greater emphasis on the grandeur of the natural world.

A common feature of her translation was the heightening of sense impressions – particularly the visual – in relation to the natural world. In a lyrical passage recounting a nocturnal scene off Lanzarote in the Canary Isles, Humboldt had described how the moon cast its light over this volcanic landscape and the phosphorescence of the ocean enhanced the glow, before clouds passed across the sky and threw the scene into darkness. These elements – scientific and imaginative appeal – were key to defining the mood of the scene that night. This blend of the aesthetic and the scientific recalls the drawings of artist William Hodges, made on board the *Resolution* during Cook's second voyage, which distinguished themselves from traditional modes of landscape depiction by their treatment of the effects of light, cloud and other meteorological phenomena (Greppi 2005: 27). Williams, alert to the poetic qualities of this piece, amplified such contrasts. Humboldt had noted that Antares (the large star in the constellation of Scorpio) 'brilloit près du disque lunaire' [shone near the disk of the moon], but in the English translation it 'threw out its resplendent rays near the lunar disk' (*RH* I, 82; *PNW* I, 80). By using the turn of phrase 'threw out its resplendent rays' Williams made the star much more of a spectacle, distorting the aesthetic balance of light sources that Humboldt seemed to consider so important, and even personifying it to cast its own rays of light out into the night sky.

If the different quality of light was one aspect that Humboldt had emphasised as he set out on his voyage to the Americas, the richness of the vegetation was another. He had divided the flora of Tenerife into five zones, distinguished from each other by the height at which certain plants normally grew (and the temperature and aridity they could correspondingly withstand). Of the second zone, where woody plants such as laurels could be found, he noted, 'c'est aussi la région des sources qui jaillissent au milieu d'un gazon toujours frais et humide' [this is also the region in which springs gush forth amidst grass that is always fresh and damp] (*RH* I, 185). Williams rendered this as 'the region of the springs that rise up amidst a turf always verdant, and never parched with drought', casting the landscape as rich, vibrant and life-giving, enhancing the visuality of the scene through the word 'verdant' and using the construction 'always . . . never' to reinforce the contrast (*PNW* I, 267). Given that Williams had not travelled further than Switzerland, the flora of Tenerife that Humboldt described here must have conformed to her romanticised image of exotic nature. Her revisioning of Humboldt's scene cast it in the language of natural luxuriance informed by pre-established modes of viewing associated with grandeur, variety and self-sufficiency.

It was particularly in the opening chapters of the *Personal Narrative* that Williams reinforced this exoticised image of nature. In a description of the volcanic peak of Tenerife, Humboldt gave the reader an impressionistic sketch of the lively scene they witnessed from the sea:

> Malgré le grand éloignement, nous ne distinguions pas seulement les maisons, la voilure des vaisseaux et le tronc des arbres, nous voyions briller aussi, des plus vives couleurs, la riche végétation des plaines. (*RH* I, 139)
> [Despite the great distance, we were able to make out not only the houses, the sails of the vessels and the trunks of the trees, but also, resplendent in the brightest colours, the rich vegetation of the plains.]

Here again, Williams took advantage of the range of meanings associated with the verb 'briller' to heighten the power of Humboldt's description: 'Notwithstanding the great distance, we distinguished not only the houses, the sails of the vessels, and the trunks of the trees, our eyes dwelt on the rich vegetation of the plains, enamelled with the most vivid colouring' (*PNW* I, 181). The term 'enamelled' lent the colours a jewel-like quality, enhancing the vibrancy of the scene in the mind's eye of its English reader. Luciana Martins and Felix Driver remark that when travellers such as the British botanist William Burchell first set eyes on tropical nature, they searched for metaphors in literature, poetry and the visual arts with which to describe it (Martins and Driver 2005: 74). While, in this case, Humboldt did not do so overtly himself, his translator had clearly understood the associative power of such language, which she deliberately mobilised in her English version. By taking the word 'enamelled' to describe natural scenery, she was echoing its use by writers such as Milton, in 'O'er the smooth enameled green' from line 55 of his *Arcades* (1645), or, closer to her own time, the Quaker poet John Scott, who enquired in his *Elegy* (1786) 'Blows not a flow'ret in th' enamel'd vale [?]' (l. 29).

Not all of Williams's more creative solutions relied simply on emotionally engaging or visually appealing language. Her reformulations of the source text were sometimes so unusual that a critical reader comparing original and translation might initially wonder how she came by them. In the Introduction to the first volume of the *Relation historique*, Humboldt had compared the savages of America with the South Sea islanders on whom Volney had recently reported. The state of half-civilisation that characterises these islanders, Humboldt noted, lent their customs a particular charm: 'tantôt c'est un roi qui, accompagné d'une suite nombreuse, vient offrir lui-même les fruits de son verger, tantôt c'est une fête funèbre qui se prépare au milieu d'une forêt', [at times a king, accompanied by a large entourage, comes himself to offer the fruits of his orchard; at others a funeral feast is prepared in the middle of a forest] (*RH* I, 33). Williams rendered this as: 'Here a king, followed by a numerous suite, comes and presents the fruits of his orchard; there, the funereal festival imbrowns the shade of the lofty forest' (*PNW* I, xlv). She added her characteristic 'lofty', read in the notion of 'shade' and, most strikingly, introduced the the term 'imbrown'. Her use of the word 'imbrown' in relation to shadow and foliage recalls 'The unpierc't shade Imbround the noontide Bowrs' from Milton's *Paradise Lost* (iv, 246), *Autumn* from Thomson's *Seasons*, in which the autumn woods 'the country round/Imbrown; a crowded umbrage, dusk and dun' (ll. 951–2), and Scott's *Rokeby* (1813), in which 'No deeper clouds the grove embrown'd' (III, ix, 115).

Another example of Williams's urge to tap into the language of sensibility can be found in her translation of the following reflections by Humboldt on the delights of Tenerife:

Ces avantages ne sont pas uniquement l'effet de la beauté du site et de la pureté de l'air; ils sont dus surtout à l'absence de l'esclavage, dont l'aspect est si révoltant aux Indes et partout où les Colons européens ont porté ce qu'ils appellent leurs lumières et leur industrie. (*RH* I, 109)
[These advantages are not solely due to the effect of the beauty of the site and the purity of the air; they are above all due to the absence of slavery, the sight of which is so disgusting in the Indies and everywhere that European settlers have brought what they term their enlightenment and their industry.]

Williams had translated this passage thus:

These advantages are the effect not of the beauty of the site and the purity of the air alone; the moral feeling is no longer harrowed up by the view of slavery, the appearance of which is so revolting in the West Indies, and in every other place, whither European planters have conveyed what they call their civilization, and their industry. (*PNW* I, 127)

Once again, she deliberately selected a more literary solution – 'harrowed up' – for the really rather simple 'dus . . . à', which she could just as well have translated as 'due to' or 'can be put down to'. The phrase 'no longer harrowed up by the view of slavery' echoes Milton's 'Amaz'd I stood, harrow'd with grief and fear' from *Comus* (l. 565), 'It harrows me with fear and wonder' from Shakespeare's *Hamlet* (I, v, l. 16) and Blake's 'But now my Soul is harrow'd with grief and fear' (l. 27) from *The Warrior and the Daughter of Albion* (1793).

Many of Williams's literary translations echo canonical pieces of British fiction, but some simply reflect the idiom of the period. Contemplating the political upheaval in South America since his travels, Humboldt had noted that internal dissent was likely to have its effect upon the natural world as well as on the scientist's ability to access the regions he himself had visited.

Cette circonstance ajoute peut-être à l'intérêt d'un ouvrage qui présente l'état de la majeure partie des colonies espagnoles au commencement du dix-neuvième siècle. Je me flatte même, en me livrant à des idées plus douces, qu'il sera encore digne d'attention, lorsque les passions seront calmées. (*RH* I, 37)
[This circumstance perhaps adds interest to a work that describes the state of the greater part of the Spanish colonies at the start of the nineteenth century. I even flatter myself that in contemplating more pleasant ideas, it will still be considered worthy of attention when peace has been restored.]

Williams's translation read thus:

This circumstance may perhaps add to the interest of a work, that portrays the state of the greater part of the Spanish colonies at the beginning of the 19th century. I may even indulge the hope, under the influence of more soothing ideas, that this work will be thought worthy of attention, when the passions shall be hushed into peace.... (*PNW* I, l)

Her choice of the phrase 'hushed into peace' for Humboldt's plain 'calmées' [calmed, soothed] is rather free, but it was a formulation not uncommon in Williams's day. A standard expression found in hymns, British sermons and evangelical writing, it occurs in George Walker's *Collection of Psalms and Hymns for Public Worship* (1788), Mark West's *Sermons on Different Subjects* (1789) and Vicesimus Knox's *Sermons* (2nd edn 1793). Williams was therefore deliberately deploying a religiously inflected language to convey what had been formulated more neutrally in Humboldt's original.

Williams's recourse to an overtly literary vocabulary sometimes enabled her to convey information more concisely. Solutions such as 'compact short-swarded turf', 'keen-edged weapons' or 'athwart the clouds' drew on word formations common in poetry of the period (*PNW* I, 190; *PNW* I, 217; *PNW* II, 33). Others, such as 'in the lapse of ages', gave her writing a slightly archaic, even biblical, air that would reinforce the sense that she was writing in an established literary tradition (*PNW* III, 342). But not all her attempts to meld the scientific and the literary were entirely successful: 'bedewed with sulphurous acid' is one of her more startling turns of phrase (*PNW* I, 176).

Williams also cast Humboldt's cooperation with his fellow traveller Bonpland in a slightly different light. It is noticeable, for example, that Bonpland is written out of her translation of the Introduction, so that while the French read 'Il me sera par conséquent permis de rappeler ici les travaux que nous avons antérieurement publiés, M. Bonpland et moi', the English simply had, 'I may therefore be permitted to enumerate in this place all that we have hitherto published' (*RH* I, 15; *PNW* I, xx). Although the 'we' did, of course, implicitly refer to Bonpland, the omission of his name at this point more generally reflects his disappearance from Humboldt's larger publishing project. Dassow Walls is right to note that Humboldt seldom mentioned Bonpland except where the narrative required it (Dassow Walls 2009: 41). It is therefore all the more telling that on a rare occasion where he is named in the original (and here it is a key moment of introduction scarcely fifteen pages into the *Relation historique*), he should be missing from the translation. Humboldt's critics were equally quick to pick up on this. Barrow, in his first review of the *Personal Narrative*, used Bonpland's absence from the collaborative undertaking as a perfect opportunity for a barbed comment on Humboldt's alleged homosexuality, by describing Bonpland, Humboldt's inactive co-author, as 'a kind of *sleeping* partner'.[69]

Williams took great pains – certainly in the first two volumes of her translation – to give the *Personal Narrative* an overtly literary feel that 'domesticated' the narrative and facilitated its integration into the corpus of scientific literature in Britain. Yet as the translation progressed, it became increasingly literal and un-English in its formulations. These 'foreign turns of phrase', which Ross would use as justification for a new, modernising, translation some three decades later, may have derived from Williams's need to work at greater speed, or quite simply her increasing inarticulacy in her native language (*PNR* I, v). The most commonly un-English expressions were those where English verbs were used reflexively, echoing the French formulation. Williams happily asserted in the second volume that 'The island of Tobago presents itself under a very picturesque aspect,' in the third volume that '[w]e reposed ourselves at the foot of the cavern', and that 'the yellow fever declared itself' (*PNW* II, 27; *PNW* III, 81; *PNW* III 391). Other awkwardly literal renderings of the original show Williams thrown off course by the indefinite pronoun 'on' in 'on est d'autant plus difficile sur le choix que le luxe de la végétation est plus grand', which became the garbled 'we become difficult in our choice in proportion to the luxury of vegetation' (*RH* I, 616; *PNW* III, 522). These solutions reflected as much the 'colossal' nature of the undertaking as the pressures that it placed on both Humboldt and Williams, as they desperately sought to bring it to a satisfactory conclusion.

Onward Journeys: MacGillivray's *Travels and Researches*

The volumes that made up the first edition by Longman and Murray came out at such a snail's pace that it was not for those impatient to acquaint themselves with the *Relation historique* in English. Although later editions did appear, the second one stopped after the second part of volume five, and in the third edition of 1822 after only the first two volumes. Nor was this translation for those on a moderate income: it set eager purchasers back by at least 123 shillings. Humboldt's *Personal Narrative* therefore remained a work inaccessible to a large proportion of the British public, beyond the volumes they could read in libraries and reading rooms or garner from reviews.

The Scottish ornithologist and natural historian William MacGillivray remedied this. His 428-page *Travels and Researches of Alexander von Humboldt*, which appeared with Oliver and Boyd in Edinburgh in 1832, was aimed at a more 'popular' market and set out to offer a condensed narrative of Humboldt's journeys through the Americas and Asiatic

Russia. Compressed into one smart volume, it appeared as volume ten in the 'Edinburgh Cabinet Library' series, priced at five shillings. It conformed to the 'instructive and amusing' genre that was typical of children's literature in the early part of the nineteenth century and which carried it with an air of moral didacticism (Fyfe 2000: 276). The *Travels and Researches* did not, MacGillivray warned, recount '[r]omantic incidents' or 'perilous adventures': it contained matter that would be of interest to those drawn by 'the great facts of creative power and wisdom' (MacGillivray 1832: 5–6).

The *Travels and Researches* would have appealed to adolescent boys wishing to model themselves on the frontispiece engraving of a dapper young Humboldt (by then, in fact, over sixty).[70] Illustrations of the basaltic rocks at Regla, a rather melancholic 'Jaguar, or American Tiger' and the majestically broad-trunked Dragon-Tree enhanced its visual interest. The one-page map of Humboldt's route up the Orinoco gave a ready overview of the geographical coordinates of his journey and an index turned it into a useful small-scale reference work for the serious reader. Footnote references to other works such as Joseph Andrews's *Journey from Buenos Aires Through the Provinces of Cordova, Tucuman, and Salta* (1827), George Francis Lyon's *Journal of a Residence and Tour in the Republic of Mexico* (1828) and Lyell's *Principles of Geology* located MacGillivray's adaptation within the wider corpus of popular scientific and geographic writing of the time, thus tempting its readers to embark on more ventures into natural history and more armchair voyages of discovery.

MacGillivray's *Travels and Researches* made no claim to be a new translation of the *Relation historique* – and indeed it was not. Rather, it was a revised and much abridged version of Williams's translation that, although it still abounded with measurements of latitude, depth and temperature, omitted longer enumerations and many of the detailed tables. Although MacGillivray sought plainer solutions for Williams's 'imbrowned' and 'harrowed up', borrowings from the *Personal Narrative*, such as the 'lofty chain of mountains' in the Empire of Morocco and the 'savage' nature of the country around the Colombian city of Cartagena, reveal the continuing presence of Williams's translation choices (*PNW* I, 8; VII 463; MacGillivray 1832: 21, 313). But the *Travels and Researches* differed from her rendering in two significant ways. Firstly, Humboldt was ousted from his position as central narrator. In MacGillivray's text, the narrative perspective shifted from the first-person plural to the third, so that a phrase such as 'we directed our course to the north west' (*PNW* I, 41) became the more distanced 'they directed their course to the north-west' (MacGillivray

1832: 25). This was compensated, however, by a second change. In the *Travels and Researches*, parts of Humboldt's narration of events were given in direct speech, making these passages seem more immediate and personal. For example, at the point where Humboldt was admiring the beauty of the southern sky, he noted in MacGillivray's edition:

> 'One experiences an indescribable sensation,' says Humboldt, 'when, as he approaches the equator, and especially in passing from one hemisphere to the other, he sees the stars with which he has been familiar from infancy gradually approach the horizon and finally disappear. Nothing impresses more vividly on the mind of the traveler the vast distance to which he has been removed from his native country than the sight of a new firmament.' (MacGillivray 1832: 57–8)

Claire Brock notes how learning lessons 'through effectively placing oneself in the body of the scientist' became popular in children's scientific works in the nineteenth century (Brock 2012: xvii). MacGillivray probably selected this passage for inclusion in the *Travels and Researches* because it invited the young reader to gaze up with Humboldt into the night sky and be enthralled by the constellations slowly passing across it. Aided by scientific knowledge, Humboldt is able to 'read' the heavenly bodies as an indication of his position on the globe. Yet Humboldt's curiosity at the unfamiliarity of this 'new firmament' in the southern hemisphere indicates that there is always more to explore. With youthful determination, this extract suggests, Humboldt's readers might start to acquire the same familiarity with constellations he knew 'from infancy', and chart an intellectual course as spectacular as his. The *Travels and Researches* therefore encouraged a fascination with nature, with scientific achievements, and with the lives of the men who had accomplished them that would instil in Humboldt's young audience a commitment to uncover more of the wonders and mysteries of science.

Conclusions

In Humboldt's *Personal Narrative*, translation was, above all, about 'added value'. Williams's investment in the translation – in terms of money, time and energy – was immense. Her stylistic contribution to the work that the *Relation historique* became in its English edition was likewise substantial. Her voice is less clearly heard in those sections that turn upon the enumeration of scientific facts and the discussion of theories. But she is undeniably present, and stylistically 'visible', in the more esoteric parts, where Humboldt actively encouraged her to use her expertise as a literary writer to create vibrant, engaging and above readable prose. The *Personal Narrative* is a testament to Humboldt's willingness to understand translation not as a result but as a process, and to acknowledge his translator's co-authorship of the English text.

Humboldt – very much at home in the tropics – had cast exotic landscapes as a place of profusion, brilliance and textural richness. His representation of nature, figured in arrestingly visual terms, made a direct appeal to the reader's powers of imagination. Williams, eager to embrace such startling differences between the temperate and the tropical zones, between cosmopolitan and arguably less 'civilised' worlds, between the 'tamed' plants of Europe and the wild profusion of the tropics, was undeniably creative in some of her lexical choices. But where critics have considered such passages exaggerated, this chapter has shown that Williams was writing within an established discourse that viewed exotic landscape as a shifting space of encounter and exchange, as much between Humboldt and the natural world of the Americas as between the British reader and European projections of the tropics.

By employing the lyrical vocabulary of the literature of sensibility, Williams enhanced the sensual appeal of Humboldt's *Relation historique* to an Anglophone audience. Like Lyell, she also instinctively alluded to canonical works to locate the *Personal Narrative* within a British literary tradition. In so doing, she produced an 'imperfect' and 'unfaithful' version of the original – divergences that were, however, essentially authorised by Humboldt. By drawing so insistently and effusively on the language of sensibility, Williams's idiom gave her rendering of the *Personal Narrative* a specific set of literary characteristics and cultural associations, which would ultimately make her translation feel dated sooner than the more restrained language of Humboldt's source text. It was this that would galvanise Thomasina Ross into action some thirty years later, as she reconfigured Humboldt's *Relation historique* for a rather different audience.

Notes

1. *Quarterly Review*, 15 (1816), p. 440.
2. *Monthly Review*, 79 (1816), p. 14; *Monthly Review*, 88 (1819), p. 235.
3. *Edinburgh Review, or Critical Journal*, 25 (1815), p. 87.
4. *British Critic*, 12 (1819), p. 337; 16 (1821), p. 1.
5. *Quarterly Review*, 25 (1821), p. 367.
6. *Augustan Review*, 1 (1815), p. 467.
7. *Edinburgh Review*, 25 (1815), p. 111.
8. *Monthly Review*, 79 (1816), p. 15.
9. When quoting from Williams's *Personal Narrative* I have silently corrected

'A Colossal Literary and Scientific Task'   115

her spelling of 'it's' to the modern 'its', where appropriate, and have also silently added any missing accents from Humboldt's French correspondence with Williams and his corrections to the translation by Williams.

10. Koninklijk Huisarchief, The Hague: Alexander von Humboldt papers, hereafter 'KHA', G016-A439, 137(2), letter of 22 April 1824.
11. KHA, G016-A439, 128(4).
12. Archives of the Ministère des Affaires Etrangères, La Courneuve, Madgett file, Microfilm P6303: 'Au reste Miss Williams est reconnue pour une des premières plumes de l'Angleterre; son ouvrage anglais sur la révolution française et les certificats infiniment honorables de civisme, qu'elle a eu de Rouen, sont une preuve incontestable de ses principes.' Thanks to Susan Pickford for drawing my attention to this.
13. *Edinburgh Review, or Critical Journal*, October 1803, 6th edn (1810), pp. 212–13.
14. John Murray Archive, National Library of Scotland, hereafter 'JMA', Ms.41282, letter of 31 December 1815.
15. KHA, G016-A439, 181 (1); KHA, G016-A439, 225; KHA, G016-A439, 240.
16. KHA, G016-A439, 40.
17. Hurford Stone's press was not the only firm to take on Humboldt. Karl Bruhns underlines the complexity of managing the printing operation by noting that Humboldt's publishers were 'at one time, the firm of Schoell, Gide, Dufour and Maze, at another time Gide alone, then Fuchs, subsequently Gide fils, Gide and Baudry, and Levrault'. See Bruhns 1873: II, 19.
18. JMA, Ms.41282, letter of 28 August 1815.
19. KHA, G016-A439, 247.
20. KHA, G016-A439, 124(2/3): 'qui parle de Vous et de Vos poésies comme je le désire'.
21. KHA, G016-A439, 243(3).
22. KHA, G016-A439, 237(2).
23. URSC, MS 1393 Longman letters I, 100, no. 107.
24. URSC, MS 1393 Longman ledgers, 2A, p. 46a.
25. URSC, MS 1393 Longman letters I, 101 no. 416A.
26. URSC, MS 1393 Longman letters I, 101, no. 448.
27. URSC, MS 1393 Longman letters I, 101, no. 448.
28. URSC, MS 1393 Longman letters I, 101, no. 460.
29. URSC, MS 1393 Longman ledgers, H9 f40r.
30. KHA, G016-A439, 226(1).
31. KHA, G016-A439, 226(1).
32. KHA, G016-A439, 129(1): 'J'ai passé une partie de la nuit avec les MSS de Mademoiselle W[illiams] et je ne puis assez La remercier du soin qu'Elle a daigné mettre à cet ouvrage. Tout est élégant, spirituel, net, précis.'
33. KHA, G016-A439, 155; KHA, G016-A439, 139(1).
34. KHA, G016-A439, 141.
35. KHA, G016-A439, 180(1).
36. The pages referred to are to be found at KHA, G016-A439, 92–100. The first page (KHA, G016-A439, 92) pertains to information missing from what became *PNW* IV.1, 278–88.

37. KHA, G016-A439, 100; KHA, G016-A 439, 98; KHA, G016-A439, 93(1).
38. KHA, G016-A439, 94; KHA, G016-A439, 95.
39. KHA, G016-A439, 94; passage now at *PNW* I, 12.
40. KHA, G016-A439, 93(1), passage now at *PNW* II, 6; *RH* I, 200.
41. Passage now at *RH* I, 220; *PNW* II, 40.
42. KHA, G016-A439, 95.
43. KHA, G016-A439, 95.
44. KHA, G016-A439, 93(3).
45. KHA, G016-A439, 173.
46. KHA, G016-A439, 99.
47. KHA, G016-A439, 99.
48. KHA, G016-A439, 99.
49. KHA, G016-A439, 100; KHA, G016-A439, 98. Passage now at *RH* I, 106; *PNW* I, 122.
50. KHA, G016-A439, 93(3).
51. KHA, G016-A439, 95.
52. KHA, G016-A439, 95.
53. KHA, G016-A439, 98.
54. KHA, G016-A439, 95.
55. KHA, G016-A439, 93(1); KHA, G016-A439, 96.
56. KHA, G016-A439, 94.
57. KHA, G016-A439, 98, now at *PNW* I, 49.
58. KHA, G016-A439, 97, now at *R* I, 77.
59. KHA, G016-A439, 95, now at *PNW* II, 47.
60. KHA, G016-A439, 95, now at *RH* I, 243.
61. My discussion is based on the holdings in the British Library Integrated Catalogue.
62. *Edinburgh Review*, 25 (1815), p. 86.
63. *Eclectic Review*, 26 (1826), p. 290.
64. *Literary Gazette*, 479 (1826), p. 180.
65. *Monthly Review*, 79 (1816), p. 15.
66. KHA, G016-A439, 174.
67. KHA, G016-A439, 219.
68. *Quarterly Review*, 14 (1816), p. 368.
69. *Quarterly Review*, 14 (1816), p. 369.
70. The *Christian Examiner* of 1832 recommended the work to 'enlarge the mental scope, and enliven pursuit after the physical sciences in a young man' (p. 861).

Chapter 4

'A Plain and Unassuming Style': Thomasina Ross and Humboldt's *Travels* (1852–1853)

In 1852 and 1853 a new title appeared in the London publisher Henry Bohn's 'Scientific Library' series. Bound in smart red cloth with 'Humboldt's Travels' in gilt letters on the spine, this second English translation of Humboldt's *Relation historique* to come out on the British book market was the work of Thomasina Muir Ross (1796?–1875). By condensing Humboldt's travel narrative into three 500-page volumes, where Williams's translation had straddled seven tomes and nearly 4,000 pages, Ross provided mid-century readers with a version of Humboldt's voyage to the Americas that was less daunting in size and more agreeable on the pocket. Affordably priced at five shillings a volume – since Bohn had dispensed with maps and illustrations – and with an index for easy reference purposes, this more modern translation swiftly won critical approval. *The Athenæum* considered the translation well executed and Ross's editorial additions pertaining to 'the modern political and moral conditions of equinoctial America' essential in increasing the value of the work.[1]

It was not just the supplementary material that drew the critics' attention. The extensive omissions required to slim the *Relation historique* down to three volumes – the same format as the immensely popular nineteenth-century 'three-decker novel' – were deemed a great improvement. As the *Examiner* noted, 'we observe some judicious curtailment of statistical and political details which have quite lost any supposed importance they may have had when the narrative was published originally'.[2] Putting it more bluntly, the critic in the *Daily News* applauded Ross's omission of those passages in Humboldt's account that recent political changes in South America had made 'obsolete, and – to the general reader, – useless'.[3]

Restoring relevance to the *Relation historique* almost half a century after Humboldt's return from the Americas, and some thirty years after the publication of the French original, was fraught with difficulty. Vivid descriptions of tropical exuberance would always make for inspiring reading, but the collapse in 1808 of monarchical power in Spain had initiated movements towards independence across South America that made Humboldt's account of colonial occupation decidedly passé. Science too had moved on. As the reviewer in the *Literary Gazette* perceptively observed:

> Since the travels were written, however, the nomenclature of many sections of natural history and geological science has undergone considerable changes, whilst many of the facts and phenomena incidentally noticed by the traveller have been developed and elucidated by those who went after him, or by men of science in Europe, working on fresh materials brought from these interesting regions. We should have liked additional notes calling attention to these changes and accessions of knowledge, and such as would link more closely this perennial work with the knowledge of the present day. The book would gain greatly by such a treatment.[4]

If we understand 'modernity', in Michael Cronin's terms, as the 'accelerated flow of goods, signs and people around the planet' (Cronin 2000: 109), then the changes that Ross needed to bring to the *Relation historique* reflected precisely this shift into a 'modern', more globalised world. The language of science was changing rapidly as new phenomena needed to be described and explained, the international scientific community was expanding fast, and knowledge was circulating ever more widely. Ross was therefore forced to engage with change on a number of different levels – linguistic, cultural and intellectual – to make Humboldt's account saleable to a British public in the early 1850s.

The Bohn edition of the *Relation historique*, which bore the full title *Personal Narrative of Travels to the Equinoctial Regions of America, During the Years 1799–1804*, has largely been overlooked by Humboldt scholarship. The Williams translation tends to be considered the authoritative version because it was a collaborative undertaking between author and translator, and a comprehensive rendering of the original into English. Yet it was Ross's version of Humboldt's *Relation historique* that enjoyed far more popularity, remaining in print throughout the nineteenth century and running to at least eleven different editions in Britain and America (Leitner 1997: 109). Secord is understandably critical of scholarly discussions that focus solely on the first version in which key nineteenth-century scientific titles such as Lyell's *Principles of Geology* or Darwin's *Origins of Species* appeared, as if these works were static entities, never changing in later editions: rather, he suggests, they should be understood as 'serial publications, part of a process of constant rereading and revision' (Secord 2000: 152). It is in this light that we will be analysing the second translation of the *Relation historique*, seeing it as a 'rereading and revision' both of Humboldt's source text and of the English version it would supersede.

The *Examiner* praised Ross for her 'plain and unassuming style', which stood in welcome contrast to the 'less agreeable and easy translation' by Williams.[5] Ross did indeed remove some of the more effusive renderings found in the earlier translation of the *Relation historique*. However, she also retains some of Williams's sentimental idiom and Wilson is justified in remarking that the edition by Ross 'still indulges in period flavour' (Humboldt 1995: lix). It does so, though, in intriguing ways that make the later edition of Humboldt's *Relation historique* a complex and multivocal, palimpsestic text. Although Williams's translation can still be glimpsed beneath the surface of Ross's reworked text, the Bohn edition is rather different from what Longman and Murray had issued several decades earlier. Indeed, of all Humboldt's works translated twice for the nineteenth-century British book market, these two versions of the *Relation historique* by Williams and Ross show most stylistic divergence. Ross not only cleansed her translation of Williams's foreign turns of phrase, but also her restyling of the *Personal Narrative* owed much more to its strategy of extensive omission, which was in turn motivated by different ways of presenting scientific knowledge some thirty years on.

This chapter opens by analysing how, through its publication in Bohn's 'Scientific Library', the *Relation historique* was recontextualised for a mid-century reading audience. It subsequently examines why Ross's own habitus as a journalist and translator of non-fictional travel writing meant she had the skills, confidence and experience to handle Humboldt's text in a rather different way from her predecessor. The main body of this chapter then offers a close discussion of the alterations and omissions Ross made to Humboldt's travel narrative as she put it into English, and reflects on the ways in which this altered the image readers gained of him as a member of the international scientific community. Finally, we explore how Ross's abridged version aligned itself within the growing corpus of condensed new editions of Humboldt's *Relation historique* at mid-century.

'From thy translated tap': Henry Bohn's Library Series

'It is difficult to keep pace with the rapidity of Mr. Bohn,' sighed the critic of the *Daily News* in June 1853, whose job it was to review each of the latest publications in Bohn's 'Standard Library', 'Classical Library', 'Antiquarian Library' and 'Scientific Library' series.[6] Henry George

Bohn was one of the foremost publishers of the Victorian period to target a mass market by providing high-quality literature at affordable prices to a readership keen on self-improvement. Francesco Cordasco describes how his Libraries were 'for perhaps a century, the most widely known series of volumes ever to be assembled and distributed in common format' (Cordasco 1951: ix). Bohn's sensitivity to changes in the market, notably the rise in railway reading, coupled with his winning idea of issuing in attractive formats 'cheap issues of works of a solid and instructive kind', made his publishing house one of the most famous cheap reprint enterprises of the nineteenth century (Cordasco 1951: 13). He was also a ruthless publisher. He had no qualms about infringing other publishers' copyrights, particularly when it came to books initially published abroad – still a legal grey area at mid-century.

Bohn, the son of a German bookbinder and second-hand bookseller who had settled in London, established himself in the book trade and in 1846 launched his first series, known as the 'Standard Library'. The following year the 'Scientific Library' was born, together with the 'Antiquarian', 'Classical' and 'Illustrated' libraries, and over the next twenty-nine years another twelve series would come into existence. Each was characterised by the different colour of its binding, reflecting the importance of recognisability and collectability in nineteenth-century publishing. Such series openly drew on what Richard Altick has termed 'package psychology' and 'snob appeal': these functioned on the premise that once readers had purchased a few volumes in a given series, they would be likely to want the rest (the 'whole package'), and that possession of a shelf of books prominently labelled as a 'library' would give their owner a feeling of enhanced status (Altick 1958: 11–12). The works in Bohn's series in particular were 'valued as highly [...] as the Everyman and World's Classics editions would be in the twentieth century' (Altick 1957: 285–6). Bohn's own series were successful partly because of the energy he invested in helping to compile and translate the texts that comprised them, and partly because the many copyrights he had bought up through the remainder trade enabled him to publish a constant stream of works (Cordasco 1951: 14–15).

The Bohn libraries relied heavily on the translation of new titles or the retranslation and revision of older pieces to keep the series evolving. Just under half the works in his library series derived from originals written in other languages. As Kenneth Haynes has suggested, 'it was Henry Bohn, more than any other publisher, whose series actively influenced the formation of a canon of world literature in translation' (Haynes 2006: 8). *Punch* put it more poetically, in a versified obituary on Bohn's death in 1884, written in memory of the:

> Dear renderer of many a learned page
> Into the – rather dryasdust – vernacular;
> True source of many an utterance oracular
> From many a pseudo-pundit, who scarce owns
> To wandering in that valley of dry Bohns.
> [. . .] From CATULLUS down to CRUSOE,
> From Plato, Xenophon, and Aristotle deep,
> To GOETHE, SCHLEGEL, SCHILLER we drink pottle-deep
> Of Learning's fount from thy translated tap![7]

For obvious reasons, the 'translated tap' was at full flow in the 'Classical Library', which was almost entirely composed of renderings from Latin and Greek. But even in the 'Standard Library', just over half the works were translations – an eloquent testimony to the importance of translation in British literary culture at this time. Of these, a little more than half were from German, slightly more than a third came from French, and a handful from Italian and Spanish. Bohn's recognition of the intrinsic value of translations in his series doubtless stemmed from his own work as a translator. He played a key role, for example, in producing a four-volume English edition (1858–60) of works by Schiller, of whom he was a particular devotee. With a good command of German, he was also well placed to aid Elise Otté in 1850 as she was struggling to complete her translation of Humboldt's *Ansichten der Natur*, which she was also translating at the same time as the first parts of the multivolume *Kosmos*.

It is not surprising that Bohn gave his translators the prominence they deserved in the works they translated for him. Their surnames and initials were, almost without exception, printed on the title page. For scholars of women's studies this is invaluable in understanding the role of female translators – otherwise largely 'invisible' figures in the world of publishing – in his undertaking. Around ninety-five translators worked for Bohn between 1846 and 1864, after which he sold his business to Bell and Daldy. Of the names still identifiable today, eleven were women, who between them were responsible for translating or co-translating just over a tenth of the output of his libraries. While they had no hand in preparing titles for the extensive Classical or Antiquarian Libraries – Latin and Greek remained male bastions of learning, and plenty of clergymen and schoolmasters came forward to translate for these series – women most frequently translated works for the 'Standard Library', which ranged broadly across history, literature, theology and philosophy. Leading female translators of the Victorian period worked for Bohn: the Quaker Mary Howitt (best known for her translations of Hans Christian Andersen and the Swedish novelist Fredrika Bremer),

Leonora and Joanna Horner (Charles Lyell's sisters-in-law, who specialised in non-fictional travel writing), and Anna Swanwick, whose translation of Goethe's *Faust* sold around 1,000 copies each year for almost half a century after its initial publication (Cordasco 1951: 16).

What is surprising, though, is that women translators were also well represented in the 'Scientific Library'. Of the eleven translations among the fifty-four titles making up this series, four were translated or co-translated by women. Although the term 'scientific' initially covered a rather heterogeneous range of titles, including Howard Staunton's *Chess Player's Handbook* (1847) and Ralph Wornum's *Lectures on Painting* (1848), the series gradually gained momentum and with it a sharper profile, as Bohn added titles more obviously related to the natural sciences. Humboldt's *Cosmos* (1849–58), *Views of Nature* (1850) and *Personal Narrative* (1852–3) were positioned alongside works such as Julius Stöckhardt's *Principles of Chemistry* (1851) and Gideon Mantell's *Wonders of Geology* (1857–8). The Danish physicist Hans Christian Ørsted's *Soul in Nature* (1852), translated from the German by the Horner sisters, and William Whewell's *Astronomy and General Physics* (1852) both appeared the same year as the first two volumes of Ross's translation of the *Relation historique* (Figure 4.1).

Was there a house style governing the character of Bohn's translations? How did Bohn, a translator himself, approach issues of translation? As Carol O'Sullivan observes, the endpapers to books in the 'Standard Library' series cast these translations as 'accurately printed in an elegant form, without abridgement', while fidelity was even more the guiding principle of books in the 'Classical Library', which bore 'literally translated' on their spine (O'Sullivan 2009: 111). In the 'Scientific Library', by contrast, works were not necessarily published in full. Thomas Wright justified his editing of a condensed version of the geologist George Fleming Richardson's *An Introduction to Geology, and its Associate Sciences Mineralogy, Fossil Botany, and Palæontology* (1855) by claiming to keep 'the Author's original object in view' yet also 'render the text still more concise by excluding all irrelevant matter' (Richardson 1855: n.p.). The abridgements that Ross made to the *Personal Narrative* were therefore in keeping with the broader agenda of enabling Bohn's series to convey relevant information in a compact form, and ensure these works remained affordable to a wide reading public.

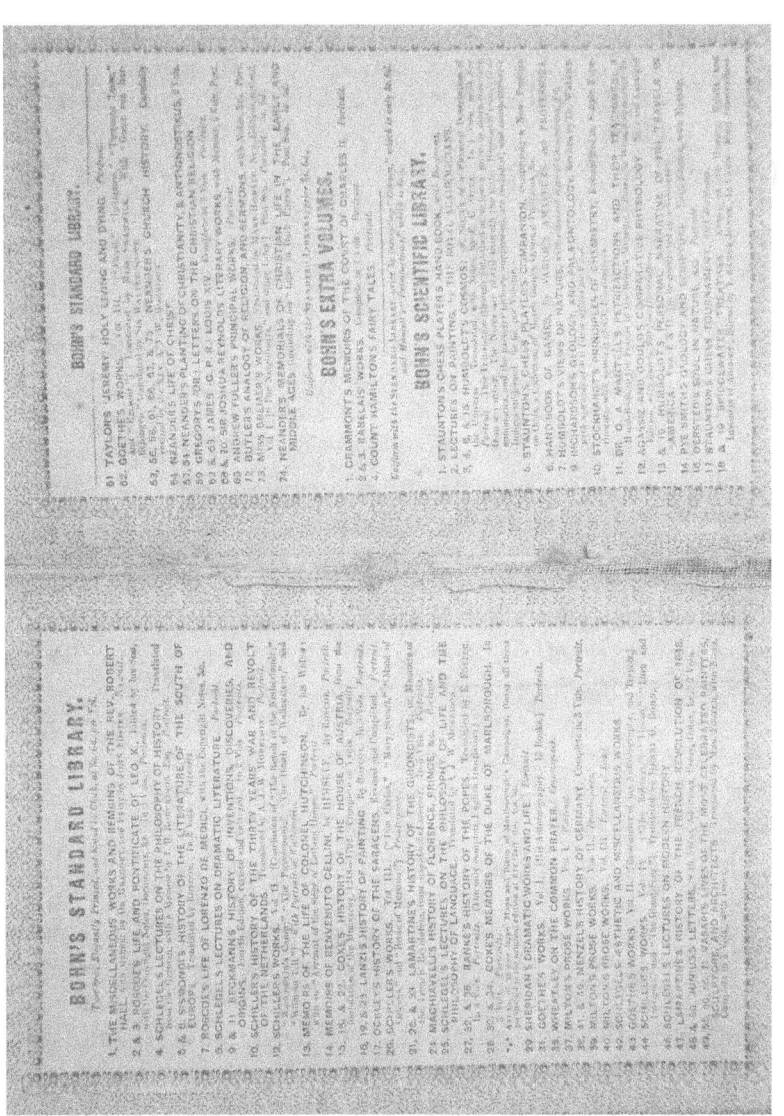

Figure 4.1 Front matter to Alexander von Humboldt, *Personal Narrative of Travels to the Equinoctial Regions of America, During the Years 1799–1804*, trans. Thomasina Ross, 3 vols (London: Bohn, 1852–3).

'Most Deserving and Reliable': Thomasina Ross as a Translator

'There is no doubt whatever, of her ability as a translatress,' wrote Charles Dickens of Ross in a letter to his publishers Chapman and Hall in 1842 (Dickens 1974: 353). 'If you can avail yourselves of her services, I shall be particularly glad,' he continued, '[f]or she is most deserving and reliable (though not beautiful); and it will be a high satisfaction to me, to have been instrumental in serving her' (Dickens 1974: 353). Related to Dickens by marriage through her elder sister, Thomasina came from a family of formidably productive journalists and translators. Her Edinburgh-born father, William Ross, settled in London in about 1795 – just after Thomasina's birth – and embarked on a career in journalism, first as editor of *The Albion*, then on *The Times*, where he worked as a reporter and translator from around 1816 until shortly before his death in 1852 (Carlton 1955: 59).[8] Described on the census return for 1851 as 'Formerly Employee at the *Times* Journal in the Department of German intelligence', his knowledge of foreign languages was probably what encouraged Thomasina to take up translation herself.[9] He also took a lively interest in politics, renouncing his family's Tory leanings on the outbreak of the French Revolution and joining 'The Friends of the People', a London association formed in 1792 to bring about universal suffrage and annual parliaments (Carlton 1955: 58). A member of the Corresponding Society, which pushed for parliamentary reform, William Ross was described by a fellow journalist as 'a tremendous democrat [. . .] and withal a very worthy man', adding that around 1818 he 'was a red-hot republican, though, like most others, he too, cooled down in later years' (Jerdan 1852: II, 178).

Thomasina's four brothers sustained their father's interest in politics and languages. William Ross junior, elected Professor of the Principles of Sculpture and Painting at the Andersonian University in Glasgow in 1829, produced the first English translation of the whole of Lessing's *Laocoon* in 1836, to wide critical acclaim. His concern to make the translation 'interesting and instructive' and to capture the 'spirit and energy of the author's style' (Lessing 1836: vi–xii), suggests that readability and accessibility by educated non-specialists informed his translational agenda. Dedicated to Lady Harriet Leveson-Gower, an energetic supporter of the anti-slavery movement, the translation also suggested that William Ross junior allied himself with Whig–liberal politics. The three other Ross brothers followed in their father's footsteps by pursuing careers as journalists with *The Times*. Charles Ross, like the young

Dickens, used his skills at shorthand to report on parliamentary debate in Westminster, and, as editor of the *Statistical Journal and Record of Useful Knowledge*, he supplied him with statistical data on juvenile offenders – valuable background material for *Oliver Twist* (Dickens 1965: I, 315).

While Dickens's friendship with Thomasina's brothers was cordial, his interest in her younger sister Georgina (1803–86) was, according to Ross family legend, warmer still. Dickens was reportedly even engaged to her for a while, until her formidable father refused his consent to a marriage with such a 'dissolute young man' (Slater 1983: 59). Georgina, like Thomasina, never married. The two lived together as spinsters with their father in Howard Street – just a stone's throw from Fleet Street – until his death, after which they moved to Chelsea and then, in the 1860s, northwards to Durham. Although Ross continued to describe herself as a 'Writer for Periodicals' in the 1860s, no major works are attributed to her as either a translator or a writer after her father's death in 1852, the same year that her translation of Humboldt's *Personal Narrative* appeared.[10] Given that she and her sister were named as the sole inheritors in his will, this suggests that translation and journalism were activities that supplied her with a welcome form of income, until her inheritance allowed her to enjoy a more leisured existence.[11]

Ross was an early example of the many freelance women writers and translators who emerged in the first half of the nineteenth century. Norbert Bachleitner has persuasively argued that such women, of whom many more have yet to be discovered, are probably 'more representative of the translation industry as a whole' at that time than the handful of prominent women translators on whom scholarly studies have hitherto focused (Bachleitner 2013: 173). The emergence of a mass market, combined with the rapid commercialisation of the book trade, meant that translation moved swiftly from being an activity carried out by scholars (or their wives and daughters) to one encouraging people who could not succeed in earning their living from journalism or other professions to start translating for money (Bachleitner 2013: 175). This was particularly favourable to women in need of an income, even if many translations were not attributed to their female translators, since translation seemed a 'rather mechanical and negligible activity' (Bachleitner 2013: 175). Bohn's handling of his female translators' 'visibility' was a notable exception, presumably because he well understood the skill and dedication that went into putting a text into a different language. Ross's marital status is also relevant in determining the career she could carve out for herself. As Bachleitner succinctly puts it, since husbands could forbid professional activity, it is 'not surprising that many female authors and translators were unmarried or divorced' (Bachleitner 2013: 176).

Ross's English rendering of Humboldt's *Relation historique* represents the culmination of her translation career, begun over three decades earlier. Her first translation appeared, in collaboration with Hannibal Lloyd, in the *Literary Gazette* in 1818. This journal – the first weekly review with a specifically literary focus – had been established the year before under the joint editorship of Ross and Lloyd (Haney 1970: xxxvii). William Jerdan, who later took over as sole editor, recalled some thirty years later that one of the successful pieces that the fledgling journal had published was 'an original translation of the "Remarks of the Austrian Arch-Dukes" (John and Lewis) tour in England by Mr. Lloyd [...] and [...] Miss Ross [...], whose talents were of a sound order, especially for a youthful female' (Jerdan 1852: II, 236–7). Lloyd was an experienced translator of non-fictional (scientific) travel writing, and could well have been influential in encouraging Ross to specialise in this genre as she gradually established a career for herself. Working alongside a man twenty years her senior, who had been a respected 'translator of continental intelligence' at the Foreign Office (Jerdan 1852: II, 190), Ross would certainly have gained an early taste of the realities of life as a translator from one of the ablest in the profession.

Through the 1820s and 1830s, Ross tackled various book-length projects. These included the *History of Spanish and Portuguese Literature* (1823), compiled from Friedrich Bouterwek's *Geschichte der spanischen Poesie und Beredsamkeit* (1804) and *Geschichte der portugiesischen Poesie und Beredsamkeit* (1805), and part, if not all, of the text for *The Memoirs of General Lafayette and the French Revolution of 1830* (1832). She also made substantial contributions to the eight volumes that comprised Laure Junot's *Memoirs of the Duchess d'Abrantes* (1833–5) and the three-volume *Pilgrimage to the Holy Land* (1835) by Alphonse de Lamartine.[12] In the second half of the 1840s she translated Johann Jakob von Tschudi's *Peru: Reiseskizzen aus den Jahren 1838–42* [*Peru: Sketches of Travel from the Years 1838–42*] (1845–6) for the London publisher Bogue and also supplied six- or seven-page articles or translations for the English literary magazine *Bentley's Miscellany*, which Dickens had initially edited. Taking either Spanish or German texts as her source material, Ross used these as an opportunity to consolidate her profile as an author and translator of biography and non-fictional travel writing. She also penned an obituary of Felix Mendelssohn Bartholdy in 1847, and, in 1848, a 'Memoir of Beethoven' and a condensed translation of parts of the *Ausflug von Lissabon nach Andalusien und in den norden von Marokko im Frühjahr 1845* [*Excursion from Lisbon to*

Andalusia and Northern Morocco in Spring 1845] (1846), written by Prinz Wilhelm zu Löwenstein, First Secretary of the Legation to the Prussian Embassy in London. Ross selectively translated the opening five chapters with a British audience in mind, and jettisoned Löwenstein's digressions on Portuguese colonial expansion, cockfighting in Cadiz and the insalubrious roads around the port of Lisbon. Concentrating instead on architectural sights, cultural events and Spanish customs, she sharpened up the narrative to enhance its linearity, pace and vibrancy.

In 1848, Ross also published two eight-page articles in *Bentley's Miscellany* entitled 'El Buscapié: The Long-Lost Work of Cervantes', which were by-products of a much longer monograph translation that would appear as Miguel de Cervantes's *El Buscapié, with the Illustrative Notes of Don Adolfo de Castro* with Bentley in 1849.[13] The manuscript of this work had just been discovered in a Cadiz library and was generating immense excitement among scholars of Spanish literature. But even the great Cervantes did not escape abridgement at Ross's hand. 'In the following English version of the *Buscapié*, care has been taken', she noted in her Translator's Preface, 'to adhere with all possible fidelity to the spirit of the original; some occasional redundancy of expression has been compressed, and here and there passages have been abridged, which, if literally rendered, would in our language appear prolix and tedious' (Cervantes 1849: vii). However, her 'careful' treatment of Cervantes's work was not reflected in the pay that her publisher offered for the work. As she tersely noted to Edward Morgan, who worked for the London publisher Richard Bentley,

> I regret to find that the remuneration proposal by Mr Bentley for the additional matter to make up the volume of Buscapié, will by no means adequately remunerate me for the trouble & labour I shall have to bestow on it.[14]

How influential was Dickens in Ross's career? The letter of recommendation he wrote in 1842 appears to have fallen on deaf ears with Messrs Chapman and Hall, since no records indicate that she completed any translation work for them. But by the late 1840s, Dickens had begun to take an active interest in her writing and translation work. In a letter from early April 1847, he commented on her translation of the Swiss scientific traveller Johann Jakob Tschudi's *Travels in Peru, During the Years 1838–42, on the Coast, in the Sierra, across the Cordilleras and the Andes, into the Primeval Forests*, which was just out:

> The book of travel in Peru has given me great pleasure in the perusal; and I assure you, honestly, that I have been very much struck by the ease and clearness of the style. It is as if the book were not a translation, but had been

written originally in English, and in thorough good English too. (Dickens 1981: V, 49–50)

In January 1849, Ross must also have sent him a copy of *El Buscapié*, fresh off the press. 'I look forward to putting my feet on the fender tonight', he replied, 'and thoroughly enjoying a prodigious cold in the head, and El Buscapie, together' (Dickens 1981: V, 893–4). But Dickens was not merely a passive consumer of her translation work. As he started to solicit manuscripts from his literary friends in 1850 for his periodical *Household Words* (1850–9), he actively sought material from Ross and took up her offer to translate and abridge passages for him from recently published Spanish or German works. Her contributions, he cheerfully remarked, could be 'to our mutual advantage', but advised that the subject matter 'should be interesting, of course; if somewhat romantic, so much the better; we can't be too wise, but we must be very agreeable' (Dickens 1988: VI, 13).

Household Words, a weekly journal that was a self-styled miscellany of 'Instruction and Entertainment', was aimed at a middle-class audience. Nation and national identity were key themes of the more 'instructive' contributions included in the journal, and travel writing cultivated an interest in foreign places that underpinned an ever-present preoccupation with what it meant to be British (Clemm 2009: 3, 49). Ross's specialism in translating non-fictional travel writing made her an ideal occasional contributor to this journal, as did her ability to turn foreign texts into good English. Of the eight pieces translated by Ross that appeared between June 1850 and October 1851, only one drew on a source text that was more than a year old: the first four were based on material from Johann Georg Kohl's *Reisen in den Niederlanden* [*Travels to the Netherlands*] (1850), 'A Lunatic Asylum in Palermo' was from the much older *Impressions de voyage: Le Capitaine Arena* [*Travel Impressions: Captain Arena*] (1842) by Alexandre Dumas *père*, 'Tahiti' and 'The "Mouth" of China' reproduced parts of Ida Pfeiffer's *Eine Frauenfahrt um die Welt* [*A Woman's Voyage around the World*] (1850), and the final 'Adventures of a Diamond' was from Eduard Jerrmann's *Unpolitische Bilder aus St. Petersburg* [*Apolitical Pictures from St Petersburg*] (1851). What these selective translations all had in common – apart from their snappy titles, designed to lure the reader into the article[15] – was the removal of prejudiced comments, the employment of a conversational tone and register, and the deliberate reorientation of the subject matter to appeal to a British reading public.

Humboldt's *Personal Narrative* was probably the last monograph that Ross translated. It was the culmination of thirty successful years of work as an editor and as a translator from the French, German and Spanish. She was awarded a Civil List pension of £50 by the Queen in March 1856 'in consideration of her literary merits' (Colles 1889: 33). When Ross came to work on the *Personal Narrative*, she was therefore an experienced translator, who, like Williams, was heading towards the end of her career. Although translation was undoubtedly central to Ross's world, it is unlikely that her livelihood turned wholly upon the success of Bohn's edition of the *Relation historique*, as had been the case with Williams. Most importantly, she had the approved Longmans translation readily at hand and could approach the process of retranslating the *Relation historique* with confidence, drawing if necessary on her connections in contemporary literary and journalist circles for advice on issues of style, content and form.

From the *Personal Narrative* to Humboldt's *Travels*

Ross stopped short of calling her work a 'translation' in her tellingly named Editor's Preface, appended to the Bohn edition. Instead, she talked of a new 'edition' characterised by the removal of those parts of Williams's translation that were stylistically deficient and by a reorganisation of Humboldt's own material to make her version of the *Travels* more linear and thus more readable. She moved the essay on 'Cuba and the Slave Trade' and the section on the 'Geognostic Description of South America' to the end of her translation, and used these to bring her account to a close. In so doing she made a clearer distinction between Humboldt's narrative of his travels and the more data-laden sections of his prose. The thirty-four-page index she appended to the work also helped to give it a new form in translation. This did not go unnoticed by reviewers. The critic in *Notes and Queries* considered it a 'copious' addition that had 'enriched' the *Travels*.[16] It certainly made the work less impenetrable by allowing readers to navigate within it more easily and enabled them to read it either as one long narrative, or dip into it and access specific pieces of information.

Although these two versions of the *Relation historique* were distinct from each other in terms of length and paratextual features, Ross's rendering was also, to some degree, a revised translation or retranslation that built on her predecessor's achievements. Antoine Berman's 'retranslation hypothesis', which suggests that each successive translation brings the target text closer to the essence of the original (Berman 1995: 57), has been criticised for over-eagerly embracing a theory of literary improvement. Siobhan Brownlie's more nuanced suggestion that later translators see new and different creative possibilities in a text, with which they refine their successive translations (Brownlie 2006: 148), is helpful in thinking about how Ross approached the task of re-visioning the *Relation historique*. More recently, Sharon Deane-Cox has used Bourdieu's notions of 'fields' and 'trajectories' to argue for a closer understanding of 'the specific contextual dynamics that have acted on the decisions to (re)translate, the physical appearances of the (re)translations, the relative values accorded to those (re)translations and the nature and extent of any interactions between these multiples of one' (Deane-Cox 2014: 23).

Ross's assertion in her Preface that she had cleansed the *Personal Narrative* of Williams's 'foreign turns of expression' and style 'unsatisfactory to the English reader' was largely true (*PNR* I, v). Her translation not only read more fluently and employed more common English terms, but also it drew on more modern vocabulary that had emerged to cater for the new processes, instruments and products that shaped scientific progress. Moreover, practitioners of science had come to be described differently in the period between the publication of the Longman and the Bohn editions of the *Personal Narrative*. Ross could profitably employ Whewell's term 'scientist', which had been in use for a couple of decades, to replace the 'natural philosophers' and 'naturalists' that Williams had used. Williams's spelling 'oxyds' – going out of use at the end of the eighteenth century – became 'oxides' in Ross's version, and the 'feldspath' of the early chapters of Williams's translation was 'feldspar' in the new edition. Ross also changed the 'radiant caloric' of Williams's translation (still used by Mary Somerville in *On the Connexion of the Physical Sciences* of 1834) into the term 'radiant heat', commonly found in non-specialist journals from the 1820s on. Similarly, 'New Holland' had been replaced by 'Australia' in 1824, in an official name-change sanctioned by the British government. The 'Stony Mountains', as they were still called in the Anglo-American Convention of 1818, had since become the 'Rocky Mountains'. Most strikingly, where the title of the Longman edition of the *Personal Narrative* carried travels to the 'New Continent', the Bohn edition set its readers on a journey to 'America'.

Ross's subtle reworking of other turns of phrase indicate a proto-feminist manipulation of the source text, in line with female translators' potential to engage in what Barbara Godard has termed 'womanhandling' of a text (Godard 1990: 91). Where Williams had opted for 'scientific men', Ross repeatedly altered this to a less openly sexist wording, presumably because she assumed a readership of men and women. Elsewhere in the text, Ross sustained this political agenda by making of the 'hommes imprudens' (*RH* II, 640) who fell prey to crocodiles in the swollen waters of the Orinoco not the 'imprudent men' of the Williams edition (*PNW* V.2, 702) but 'imprudent persons' (*PNR* III, 3).

Despite these divergences, the language of sensibility that had typified Williams's translation was very much in evidence in the Bohn edition. From 'lofty' summits to 'gloomy' curtains of mountains and the mind soothed by 'consolatory remembrances', stylistic choices made by Williams resonated through Ross's translation. Critical acclaim for Williams's literary achievements may well have caused Ross to be wary of rejecting such solutions. A comparison of the two translations shows that Williams's voice pervaded the text most strongly in those phrases borrowed directly or indirectly from literary sources. Ross's text similarly uses the expressions 'hushed into peace' and 'harrowed up by the sight of slavery', and retains the 'lapse of ages' and Williams's characteristic construction 'destitute of' to describe lack, as in 'destitute of vegetation' for 'dépourvues de végétaux' (*PNR* I, xxii, 53, 358, 359). Ross also appreciated the poetic qualities of Williams's neat constructions 'compact short-swarded turf' and 'keen-edged weapons' (*PNR* I, 86, 100).

Curtailing the 'Cumbrous'

If paratextual additions were the hallmark of Black's translation of the *Essai politique*, radical omissions were Ross's trademark in the *Personal Narrative*. To boil Humboldt's travelogue down, Ross jettisoned around 27 per cent of the original material (maps not included), partly by cutting large sections wholesale, partly by making shorter incisions.[17] While she dropped a mere 18 per cent or so from Williams's first volume and a similar amount from the third, almost half of the second volume disappeared, particularly those passages where Humboldt had discussed air and water temperature on the passage from Tenerife to Cumana and the colour of the water and sky. From the fourth volume only a minute amount of material was lost, and two sections of volume five were reduced by around a fifth. It was in the two parts of volume six and the final, seventh, volume where the largest amount of text was sacrificed, since Humboldt's lengthy 'Explanations' on the population and surface area of Continental America, sections on cocoa and tobacco production, lists of longitudes and sea temperatures, and discussions of sugar exports went completely by the board.

Ross's large-scale omissions offered a striking corrective to the difficulties of narrative organisation, which Humboldt had addressed in his own Introduction:

> Had I adopted a mode of composition which would have included in one and the same chapter all that has been observed on one particular point of the globe, I should have prepared a work of cumbrous length, and devoid of that clearness which arises in a great measure from the methodical distribution of matter. Notwithstanding the efforts I have made to avoid, in this narrative, the errors I had to dread, I feel conscious that I have not always succeeded in separating the observations of detail from those general results which interest every enlightened mind. (*PNR* I, xiv)

These more swingeing cuts tried to resolve the inherent tension in the *Relation historique* between immense 'cumbrous' scientific detail, which would interest only specialists, and 'general results' comprehensible to the general reader. Although these changes largely correspond to what Ross had intimated in her Editor's Preface, subtle omissions of quite a different type recurred, which would ultimately recast how Humboldt was presented in her translation as a scientific traveller, practitioner and member of the international scientific community.

The nineteenth century saw a marked shift in the ways that scientific authors positioned themselves in their texts. Dwight Atkinson has investigated how, between around 1825 and 1875, the strong authorial persona previously used in scientific writing was replaced by a more impersonal one, as the focus turned towards describing methods and recording results (Atkinson 1999: 77). As 'author-centred' rhetoric gave way to 'object-centred' rhetoric, four features disappeared: the use of witnessing (the naming of important figures present at the scientific event reported), indices of modesty and humility, a tendency towards digression, and 'elaborate politeness' (which involved the naming of fellow researchers as 'ingenious' or 'intelligent') (Atkinson 1999: 77). The demise of 'politeness' essentially marked the decline of a mode of scholarly communication and interaction dating back at least 200 years. Steven Shapin's 1985 study of civility and credibility in the work of the Anglo-Irish natural philosopher Robert Boyle demonstrates that the notion of the 'gentleman' was defined economically and socially, but also culturally. It was intimately related to notions of integrity, scientific sociability, and a modesty articulated through expressions of doubt and the recognition of experimental difficulties or failures. Humboldt, the embodiment of Enlightened 'polite' science, sustained these values well into the nineteenth century. Ellen Valle discusses how 'gentlemanly' behaviour remained a cornerstone of scientific exchange in the Victorian period, citing as proof Darwin's irritation at the 'mean-spiritedness' of members of the Zoological Society, where speakers snarled at each other 'in a manner anything but like that of gentlemen' (Valle 1999: 144). But by around 1860, the shift towards professionalisation and specialisation meant that this ethos had begun to suffer lasting erosion (Valle 1999: 132).[18]

Engagement with other members of the scientific community in a 'polite' fashion, through reference to their work and their standing in the scientific community, has been essential to scientific discourse. Bruno Latour contends that community consensus plays an essential role in ensuring that an individual's statement becomes an accepted fact (Latour 1987: 155–76). Dialogue has always been key to scholarly exchange, and articles were often written in response to previous publications. Valle suggests that the dialogic quality of scientific writing waned in the nineteenth century as a split developed between 'old' and 'new' knowledge, between a shared community-specific body of knowledge and the innovative, singular contribution by the author (Valle 1999: 291). It was in three areas – Humboldt's description of himself as a scientific experimenter and observer, his deployment of a 'modest' rhetoric to describe data collection and evaluation, and his 'polite' references to the wider scientific community in which he positioned himself – that Ross repeatedly made small-scale omissions that cumulatively gave her *Personal Narrative* a rather different feel.

Constructing Knowledge: Instrumentation, Observation and Experiment

One of Ross's early incisions in the *Relation historique* saw the loss from the first chapter of a 'List of the Physical and Astronomical Instruments' that Humboldt and Bonpland had taken on their journey, which had run to a good seven pages in the translation published by Longman and Murray. The line-up of chronometers and sextants, magnetometers, barometers and hygrometers, electrometers and theodolites, pendulums, compasses and microscopes was certainly impressive – not forgetting the cyanometer, that 'airy instrument', as Byron ironically termed it (*Don Juan*, Canto IV, stanza 122), shaded to measure the colour intensity of blue sky. But this was more than just an inventory of the instruments Humboldt and Bonpland had marshalled together for their journey. It offered a thumbnail description of the appearance of each – '*a snuffbox sextant by Troughton*, [...] very useful for travellers when forced in a boat to lay down the sinuosities of a river, or take angles on horseback without dismounting' (*PNW* I, 35) – to aid the reader's understanding of how these specialist instruments were used to generate findings.

Michael T. Bravo's account of the history of precision and travel in the eighteenth century highlights the importance of judgement and reliability in relation to scientific observation (Bravo 1999: 164). Although 'precision' might seem an intangible notion, he argues that it could be gauged in quite straightforward ways. Adopting working routines, holding a sextant correctly to read the sun's altitude while on board ship, winding a chronometer evenly, regularly writing up a traveller's log book or using standardised language to describe what had been seen, all invoked the notion of 'accuracy' (Bravo 1999: 164–5). Humboldt invested two-and-a-half years in learning how to use the instruments he intended to take on board (Weigl 1990: 206). They were fundamental to his encyclopaedic survey of heights and pressures, temperatures and climates, flora and fauna, which fed into his systematic mapping of the Americas geographically, geologically and botanically.

Humboldt's aim was not just to count and measure, though. It was also to find connections between facts that would enable him to understand the laws governing the natural world. As he recalled in his Introduction,

> whilst the number of accurate instruments was daily increasing, we were still ignorant of the height of so many mountains and elevated plains; of the periodical oscillations of the aerial ocean; the limit of perpetual snows under the polar circle, and on the borders of the torrid zone; the variable intensity of the magnetic forces, and so many other phænomena, equally important. (*PNW* I, vi)

Reflecting on his life's achievements in 1854, Humboldt recalled three publications he considered to be the most original – his essay on the geographical distribution of plants, the *Essai sur la géographie des plantes* (1807), his work on isothermal lines and his observations on terrestrial magnetism – all of which drew heavily on precise scientific measurement (Humboldt 2009: 227). Precision was a key component of his own scientific self-fashioning, and Marie-Noëlle Bourguet illustrates how it framed his expectations of the international network of local scientists who gathered accurate observations and measurements on his behalf (Bourguet 2002: 117).

In the *Relation historique*, Humboldt emphasised repeatedly just how difficult it was to obtain scientific readings. He remarked (in a paragraph that Ross cut from her translation):

> As I employed successively instruments of different constructions, I fixed my choice on those which appeared to me the most exact, and the least subject to break in the carriage. I had an opportunity of repeating measurements, which had been taken according to the most rigorous methods; and I learnt from experience the extent of the errors, to which I might be exposed. (*PNW* I, 4–5)

Making accurate measurements depended on the instruments having survived the journey in one piece. Humboldt's frequent reminders to the reader of the problems associated with travelling with instruments were always cut from the Ross translation, yet they gave the reader of the Longman edition a much clearer idea of the practical problems that scientific travellers faced as they attempted to make observations. 'In times of maritime warfare', advised Humboldt, 'it is highly prudent never to lose sight either of instruments, manuscripts, or collections' (*PNW* I, 33). The centrality of scientific apparatus to the success of Humboldt's undertaking was reflected in his constant references to the well-being of his thermometers, hygrometers and barometers. The barometers were particularly difficult to transport, and for the whole five years of the journey one was placed in the sole care of a guide who followed the party on foot, an expensive precaution that was not always failsafe. In the humid Amazon rainforest at La Esmeralda, the glass tube was cracked by the warping of the wood, an incident Humboldt 'regretted' all the more for the barometer having provided three years of sturdy companionship, braving journeys through the Styrian mountains, France and Spain, and from Cumana to the Upper Orinoco (*PNW* III, 6; V.2, 551).

Humboldt illustrated just how important it was to have the right instruments to hand as he described attempts to take measurements during a mosquito-infested night under stars near the island of Tomo on the Orinoco. In Williams's translation he observes:

> The night was beautiful and serene, but the stratum of moschettoes [*sic*] was so thick near the ground, that I could not succeed in levelling the *artificial horizon*; consequently I lost the opportunity of observing the stars. Had I been furnished with an *horizon of mercury* on this voyage, it would have been of great use to me. (*PNW* V, 127–8)

While measuring altitudes and angles without a clear horizon was no longer problematic by the end of the eighteenth century, Humboldt clearly knew that with the more sophisticated 'horizon of mercury', developed by the noted London instrument-maker George Adams the elder, more accurate measurements would have been possible, based on measuring the angle of reflection between the heavenly bodies and their image on the surface of the mercury. Ross's translation conveys well how the thick swarms of mosquitoes hindered the scientist's vision of the horizon:

> The night was beautiful and serene, but the torment of the mosquitos was so great near the ground, that I could not succeed in levelling the artificial horizon; consequently I lost the opportunity of making an observation. (*PNR* II, 294)

But in her rendering the final sentence has been cut, losing Humboldt's frustration at not having the right instrument to hand as he is taking his measurements.

Observation, measurement and the recording of scientific fact formed a permanent backdrop to Humboldt's activities. Waiting impatiently for permission to land on Tenerife, Humboldt idled away the hours by determining the longitude of the mole of Santa Cruz:

> I employed this time in making the necessary observations for determining the longitude of the mole of Santa Cruz, and the dip of the needle. Berthoud's chronometer gave, for the first 18°33'10". This differs three or four minutes from the result of former observations by Fleurieu, Pingré, Borda, Vancouver, and La Peyrouse. M. Quenot nevertheless obtained 18°33'36", and the unfortunate Captain Bligh 18°34'20". The precision of my result was confirmed three years after, the voyage of the chevalier Krusenstern, who found Santa Cruz 16°12'45" west of Greenwich, and consequently 18°33'0" west of Paris. (*PNW* I, 114–15)

These apparently minimal differences in reckonings of longitude do not make for captivating reading – which explains why Ross omitted this section entirely – even if inaccuracies Humboldt uncovered elsewhere meant that major mountain ranges were two-and-a-half degrees out on some maps, equating to 270 km on the ground (Weigl 1990: 220). By the 1850s, reliable marine chronometers were on the market, so that passages of this kind were obviously dated when Ross came to rework the *Relation historique*, but their inclusion would have given her readers a feel for how scientific knowledge had been advanced and honed through the gradual and collective accumulation of findings.

Lorraine Daston has charted how 'observation' as a scientific practice and epistemic category shifted over time. Closely connected to the practice of 'experimentation', the relationship between the two shifted 'from rough synonyms [...] to complementary and interlocking parts of a single method of inquiry throughout much of the eighteenth and early nineteenth centuries, to distinct procedures opposed as "passive observation" and "active experiment" by the mid-nineteenth century' (Daston 2011: 82). This last shift is relevant as we examine further what Ross cut and why. Central to Humboldt's narration of his observations was the emphasis on personal witness. Referring to the engravings used to accompany his travel account, Humboldt demonstrated the importance of first-person narrative perspective:

> I have drawn on the same plate, the profiles of the Peak of Teneriffe, Cotopaxi, and Vesuvius. I could have wished to have substituted Etna for

this last mountain, because its form is more analogous to that of the two volcanoes of America and Africa; but I chose to trace only the outlines of mountains that I had visited and measured myself; and with respect to Etna I should have wanted data for the intermediary heights. (*PNW* I, 204–5)

Ross, under pressure to cut material, removed these sentences entirely, perhaps because they seemed obvious reflections in what was, after all, a 'Personal Narrative'. In so doing, though, she sundered the link between observation and first-hand witness that gave Humboldt's remarks authority and scientific integrity.

Witnessing an event once did not suffice to give Enlightenment men of science the satisfaction of scientific certainty. By the mid-eighteenth century, repetition and observation were closely allied, and a cumulative observational tradition – based on findings recorded across several generations – was slowly giving way to a repetitive one, in which the existence of a phenomenon was confirmed not by single but by repeated observation (Daston 2011: 94–5). As Humboldt weighed up the advantage of a new route from Cadiz to Cumana that reduced sailing time by one-twentieth and involved steering a diagonal course from Cape St Vincent to America, rather than heading further south to pick up the trade winds, he observed, in the translation by Williams:

At the time of my abode in the Spanish colonies, I witnessed the arrival of several merchant ships, which the fear of privateers had determined to choose the oblique course, and that had a very short passage; it is only after repeated trials, that we can decide with certainty on an object, at least as important as the choice of the meridian, at which the equator should be cut in the navigation from Europe to Buenos Ayres or Cape Horn. (*PNW* II, 7–8)

Ross simply rendered this as, 'At the time of my abode in the Spanish colonies, I witnessed the arrival of several merchant-ships, which from the fear of privateers had chosen the oblique course, and had had a very short passage,' thereby completely removing the clause after the semicolon (*PNR* I, 128). A few pages later Humboldt again called for findings to be repeated in the interests of improving maritime knowledge, which Ross likewise cut:

This current bears the tropic grape into the high latitudes, toward the coast of Norway and France; and it is not the Gulf Stream, as some mariners think, which accumulates the fucus to the south of the Azores. It were to be wished, that navigators heaved the lead more frequently in these latitudes covered with weeds: for it is asserted, that Dutch pilots have found a series of shoals from the banks of Newfoundland as far as the coasts of Scotland, by using lines composed of silk thread. (*PNW* II, 11)

Ross's translation stopped sharply after 'Azores', leaving unuttered Humboldt's suggestion that more research could usefully clarify how far these banks of seaweed extended (*PNR* I, 130). Perhaps Ross simply failed to see the point of Humboldt's recommendations, or perhaps she did not grasp their wider relevance for research into ocean currents. But they had given the reader greater insight into how Humboldt processed the information he recorded himself or received from others. He used observation as a tool of conjecture – a way, as Daston puts it, of 'excluding some explanatory hypotheses and hatching new ones, which could in turn be submitted to a new round of observation and often experiment as well' (Daston 2011: 104–5). Observation was therefore a springboard from which Humboldt launched new questions and tried out new ideas. These would in turn ensure more information was collected, processed and analysed. Unlike Ross's version, the translation by Williams showed explicitly how scientific knowledge was advanced, and how, by collating and comparing the observations of others with one's own, new ideas could be generated.

Humboldt's observations were woven into a narrative that repeatedly emphasised intellectual gains – what he and fellow members of the scientific community had achieved or could yet accomplish. But the *Personal Narrative* also frequently registered losses, shortcomings and failures. Even with the right instrument to hand, Humboldt was sometimes forced to admit that his findings were probably inaccurate. In a paragraph specifically omitted from a section on the Orinoco otherwise given in full by Ross, Humboldt had noted:

> I have given with hesitation my opinion of the perpendicular height of the *raudales* [rapids] of the Oroonoko, limiting it to one extreme quantity. I carried the barometer to the little plain, that surrounds the mission of Atures, and to the cataracts, but I could not obtain any constant differences. Every one [*sic*] knows how delicate a business it is, to measure small heights by the barometer. (*PNW* V, 61)

Rather than hiding any reservations he had about these findings, Humboldt laid his results open to public scrutiny. By presenting inconclusive data rather than suppressing it, he tackled head on the problem of taking measurements during fieldwork and of accounting for inconvenient evidence. Williams's rendering therefore constructed a more sympathetic image of Humboldt as a working scientist whose objective reporting of both achievements and failures revealed that he was not prepared to distort his findings in the pursuit of scientific recognition.

Presenting Knowledge: Modesty and the Scientific Community

Humboldt did not feel responsible for all the scientific shortcomings that he mentioned in the *Relation historique*. Some could be laid at the door of tight-fisted politicians or lethargic academies that had not matched Humboldt's own investment of energy and money in supporting scientific projects. 'It is to be regretted', he lamented, as he closed his discussion of the submarine volcanoes near the island of St Michael in the Azores, 'that, notwithstanding the proximity of the spot, no European government, or learned society, has sent natural philosophers, to investigate a phenomenon, which would throw so much light on the history of volcanoes, and on that of the globe in general' (*PNW* IV, 7). Given the sheer enormity of Humboldt's encyclopaedic task and his repeated failure to meet all his goals, it is unsurprising that his writing was rarely triumphalist or openly self-promotional.

A rhetoric of modesty therefore pervades the *Relation historique*, yet so-called 'hedging devices', conveying uncertainty, were frequently removed by Ross in her translation, despite their importance in showing the limits of scientific knowledge. On leaving the Canary Isles, Humboldt noted:

> Je viens d'esquisser le tableau physique de l'île de Ténériffe; j'ai tâché de donner des notions précises sur la constitution géologique des Canaries [. . .]. Quoique je me flatte d'avoir répandu quelque lumière sur des objets [. . .] je pense pourtant que l'histoire physique de cet archipel offre encore un vaste champ à exploiter. (*RH* I, 188)

William translated this as:

> I have now given a physical sketch of the island of Teneriffe; I have endeavoured to lay down precise notions respecting the geological constitution of the Canaries [. . .]. Though I flatter myself with having thrown some light on objects, [. . .] I think nevertheless, that the natural history of this archipelago still offers a vast field to enquiry. (*PNW* I, 272–3)

Ross omitted this section completely, probably because it highlighted all that Humboldt had not achieved. He claimed only to have cast 'some' light on objects, and a 'vast field' was still left open to enquiry. A brief moment of confidence ('I flatter myself') stands in contrast with the rhetoric of the rest of the passage, in which he has only 'endeavoured to lay down' his findings.

Humboldt sometimes never even set out to resolve some of the problems he addressed. As he discussed the geology of the Canary Isles, he posed the question: 'Were these granites and these mica-slates of Gomera anciently united to the chain of Atlas, as the primitive mountains of Corsica appear to be the central nucleus of Bochetta and the Apennines?' His immediate response was that this 'can never be solved, till mineralogists shall have visited the islands that surround the Peak' (*PNW* I, 236). He therefore directly confronted his readers with this problem, and showed that by constantly querying how separate observations could be connected with other, some sense could be made of the laws underpinning the natural world. This was, once again, lost in the Bohn edition, as Ross removed passages of a speculative rather than informative nature.

Exchanging Knowledge: Community and Polite Discourse

For Humboldt, the advancement of scientific knowledge was a communal, not an individual, undertaking. His narrative construction of himself as a scientific traveller drew unremittingly on eighteenth-century rhetorical traditions that saw science as a cooperative activity, the process of intellectual dialogue and debate. The *Relation historique* abounded with references to Humboldt's place within the scientific community, not least in the fifty-page Preface, where he warmly thanked many fellow scientists by name for their cooperation and signalled the networks and groups in which he operated. Yet in the Ross edition this kind of information was frequently omitted. As Humboldt enumerated the works on which he had frequently drawn in the footnotes, he noted of the *Collections of Observations on Zoology and Comparative Anatomy* that this had benefited greatly from research by Georges Cuvier, who had done groundbreaking work on the axolotl of the 'lake of Mexico'. Humboldt considered him '[u]n savant illustre dont la constante amitié m'a été si honorable et si utile depuis un grand nombre d'années' (*RH* I, 20). Williams translated this as:

> A distinguished man of science, whose constant friendship has been highly honourable and advantageous to me during a great number of years, M. Cuvier, has enriched the collection with a very extensive treatise on the axolotl of the lake of Mexico.... (*PNW* I, xxvi)

In her translation Williams even heightened the esteem in which Humboldt held Cuvier by rendering 'si honorable' [so honourable] as 'highly honourable' and making of 'utile' [useful] 'advantageous'. Ross removed the evaluative predicates before Cuvier's name to pare the sentence right down to: 'M. Cuvier has enriched this work with a very comprehensive treatise on the axolotl of the lake of Mexico' (*PNR* I, xvi). Cuvier's relevance for Humboldt's work now appears to derive less from his position as a long-standing acquaintance and colleague than from his scientific work on the Mexican axolotl. And a few pages further on, the Ross edition loses references to the 'great geologist' von Buch, Joseph Gay-Lussac and Arago, 'who are endowed with that elevation of character, which is so congenial to an ardent love of the sciences' (*PNW* I, xxxvi–xlvii).

Humboldt not only showed his esteem for other scientists through warm testimony and a 'gentlemanly' scientific attitude. He also indicated his respect for their achievements as he discussed their findings within the context of his own research. Comparing fossil remains in the sandstone banks of Punta de Araya in Venezuela to those in the basin of Paris, Humboldt noted, in a passage neatly excised from Ross's translation:

> This mixture is in fact found in the lime-stone of very recent formation, that covers the chalk in the basin of Paris; but in order to verify a fact so important, we should have under our eyes the fossile [*sic*] shells of Araya, and examine them anew with that scrupulous exactness, which has been recently followed in this kind of investigation by Messrs. Lamarck, Cuvier, and Brongniart. (*PNW* II, 264–5)

These sentences explicitly signal Humboldt's awareness of recently published investigations on this subject and his public respect – another feature of 'gentlemanly' eighteenth-century scientific discourse – for the 'scrupulous exactness' of these scientists, which demonstrated that they shared a disposition towards the scientific precision that Humboldt valued so highly. By removing passages of this type from her translation, Ross constructed a narrative that wrote previous research and researchers out of Humboldt's account, portraying him as something of a lone hero in science.

Foreign Witness

In the *Relation historique*, Humboldt painted a vibrant picture of the international scientific scene, crowded with figures from different disciplines, countries and generations who were discussing, debating and collectively driving science forward. Humboldt was a highly 'present' figure in his own *Personal Narrative*, which abounded with uses of the first-person pronouns 'I' and 'we', and adhered to the author-centred rhetoric characteristic of late eighteenth-century scientific writing. While this was sustained – heightened, even – in Williams's translation, there were marked shifts in Ross's version towards the much less personal, more neutral, rhetoric that shaped nineteenth-century scientific writing. Where Humboldt had noted in the original French:

> Nous avons rappelé, au commencement de ce chapitre, que le lac de Valencia, comme les lacs de la vallée de Mexico, forme le centre d'un petit système de rivières dont aucune n'a de communication avec l'Océan. (*RH* II, 71)

Williams had translated this literally as:

> We observed at the beginning of this chapter, that the lake of Valencia, like the lakes of the valley of Mexico forms the centre of a little system of rivers, none of which have any communication with the ocean. (*PNW* IV, 141)

Ross opted for the impersonal:

> It has already been observed that the lake of Valencia, like the lakes of the valley of Mexico, forms the centre of a little system of rivers, none of which have any communication with the ocean. (*PNR* II, 8)

The use of the word 'we' to construct a sense of companionship between author and reader, upholding the notion that they were 'travelling' through the narrative together, is lost in the cooler 'it', followed by the passive construction in the Bohn edition.

Humboldt's personal investment in his narrative appeared less pronounced in the edition by Bohn; this in turn had repercussions for the way in which he was cast as a travelling scientist and scientific travel writer. It also changed how readers were asked to engage imaginatively with his text. Overwhelmed by the impression made on him by the rapids of Atures and Maypures on the Orinoco – even more stunning than the magnificent scenes of the *cordilleras*, he marvelled, Humboldt had remarked:

> Lorsqu'on se trouve placé de manière à embrasser d'un coup d'œil cette suite continue de cataracts, cette nappe immense d'écume et de vapeurs, éclairée par les rayons du soleil couchant, on croit voir le fleuve entier suspendu au-dessus de son lit. (*RH* II, 291)

Williams, twice turning the impersonal French 'on' into 'you', had rendered this as:

> When you are so stationed, that the eye can at once take in the long succession of cataracts, the immense sheet of foam and vapours, illumined by the rays of the setting Sun, it seems as if you saw the whole river suspended over its bed. (*PNR* V, 2)

Ross replaced the first 'on' with 'the spectator' and in the second made the river the subject of the clause:

> When the spectator is so stationed that the eye can at once take in the long succession of cataracts, the immense sheet of foam and vapours illumined by the rays of the setting sun, the whole river seems as if it were suspended over its bed. (*PNR* II, 234)

By classifying the narrator (and reader) as 'the spectator', Ross drew attention to them as viewing subjects and lost the spontaneity and directness that characterised the previous translation. Williams also used the personal pronoun 'you' as an inclusive device to involve readers as fellow travellers in the act of viewing this spectacular scene, which they were invited to reproduce in their imagination.

Colonialism and Race

Although the majority of Ross's omissions were related to how Humboldt was cast as a travelling scientist, some also presented in a slightly different light his approach to racial Otherness. Leask describes how Humboldt was able, on his journey through the Americas, to sustain the spatial freedom and cosmopolitan consciousness that he did because his travels were privately funded and not directly connected to any European agenda of colonial expansion (Leask 2002: 250). Nevertheless, while Humboldt openly campaigned against slavery, he still subscribed to the prevalent nineteenth-century notion that different parts of the globe had enjoyed different rates of development, with the advancement of Western, 'civilised', cultures held up as the model to which peoples elsewhere in the world should aspire.

Occasionally, Humboldt's descriptions of the indigenous tribes of Latin America conveyed a sense of Western cultural superiority in the translation by Williams. Ross considered this required urgent revision and excision to make the *Relation historique* a palatable text for a British readership in the 1850s. Her translation is characterised by extreme caution about casting the native peoples of the Americas in a negative light, in judging them from a Western perspective and in considering them intellectually, socially or culturally inferior. As Humboldt had noted in Chapter 5 of the second book, given here in Williams's literal translation:

> It was on a Sunday night, and the slaves were dancing to the noisy and monotonous music of the guitar. The people of Africa, of negro race, have an inexhaustible store of activity and gayety in their character. After having

passed through the painful labors of the week, the slaves, on days of festival, prefer the sounds of music, and the dance, to listless sleep. Let us not blame this mixture of carelessness and levity, which softens the bitterness of a life full of pains and sorrows! (*PNW* II, 243)

Although Humboldt essentially praised the African slaves for their mental and physical stamina, their 'noisy and monotonous' music suggested an uncultured, rather primitive, people. Ross's rendering was diplomatic in its abridgement:

It was on a Sunday night, and the slaves were dancing to the music of the guitar. The people of Africa, of negro race, are endowed with an inexhaustible store of activity and gaiety. After having ended the labours of the week, the slaves, on festival days, prefer to listless sleep the recreations of music and dancing. (*PNR* I, 177)

Gone are the negative judgements that Humboldt passed on the Africans' music, and while Williams simply described the negro race as 'having' an inexhaustible store of activity and gaiety, Ross considers them 'endowed' with it, implying it is one of their innate qualities. Humboldt's final sentence of exclamation is also missing from Ross's account, perhaps because it was a personal reflection rather than a factual statement, perhaps also because she felt its sentimentality did little to alleviate the plight of the people being described.

Ross was also sensitive to the power of the adjectives that Humboldt (and Williams) had used to portray non-Western cultures in more general terms. On his journey into the mountains of New Andalusia and the missions of the Chayma Indians, Humboldt had noted of the native peoples in the translation by Williams, 'we find beyond these mountains a people so lately nomade [*sic*], and still nearly in a state of nature, savage without being barbarous, and stupid rather from ignorance than long rudeness' (*PNW* III, 2). Evidently troubled by the string of negative associations – savage, barbarous, stupid – Ross reduced this to: 'we find beyond these mountains a people lately nomade, and still nearly in a state of nature, wild without being barbarous' (*PNR* I, 200). She thus remodelled the sentence to cast the native Indians in older, Rousseauvian, terms, which implied that they lived a simple existence 'in a state of nature'. By choosing the word 'wild' rather than the cruder terms 'savage' and 'barbarous', with their connotations of lawlessness and cruelty, Ross subtly presented them in a more favourable light.

The Bohn edition cast the rather stark differences that Humboldt posited between the Western travelling observer and the indigenous peoples in less absolute terms. Ross not only cut unsavoury passages from the text wholesale, but also translated Humboldt's comments more freely. Where Humboldt had observed that '[t]he Caribbee women are less robust, and uglier than the men' in Williams's translation (*PNW* VI, 12), Ross modified the sentence to convey the same content but in less pejorative terms, noting: '[t]he Carib women are less robust and good-looking than the men' (*PNR* III, 74–5). She also took issue with Humboldt's description of the different racial groups across the Americas as 'cet essaim de peuples répandus dans les deux Amériques' (*RH* III, 7), which Williams had translated literally as 'that swarm of nations spread over both Americas' (*PNW* VI, 13). Ross, sensing the potential offence carried by the word 'swarm', with its associations of infestation and destruction, talked instead of 'that multitude of nations spread over North and South America' (*PNR* III, 75). Finally, while Williams used the word 'mulattoes' (*PNW* VI, 91) to describe the racial identity of some of the inhabitants of Cumuna, Ross opted more cautiously for 'coloured people' (*PNR* III, 111) to indicate that their skin colour was different from that of the European visitors, without entering into more complex (and politically charged) discussions of racial difference.

Contextualising Curtailment

Bohn and Ross were by no means the first to see the sense in producing an abridged edition of the *Relation historique*. Writing to Pictet in August 1805, Humboldt had himself contemplated publishing the content of this account in two different ways: a shorter version that would offer the edited highlights such as barometrical measurements or astronomical positions, and a comprehensive edition targeted at readers seeking greater detail (Humboldt 1869: 43–4). He had also hatched similar plans for the *Voyage aux regions équinoxiales*, and confidently asserted in the same letter that either three volumes of an abridged version of the *Travels* or volume one of the 'Grand Voyage' could be out in less than twelve months. By December of that year, though, these plans were on hold: as Humboldt remarked to Cotta, acquaintances in England had advised him 'daß man die Idee der kleinen Reise für verderblich u. sehr unmerkantilisch halte' [that the idea of the smaller [i.e. abridged] Travels is considered detrimental and commercially most unviable] (Humboldt 2009: 71).

Ross's three-volume translation was not the first condensed edition of the *Relation historique* to appear on the Anglophone book market or on the Continent. The abridged four-volume German translation edited by Herman Hauff, which appeared with Cotta as the *Reise in die Aequinoctial-Gegenden des neuen Continents in den Jahren 1799, 1800, 1801, 1802, 1803 und 1804* [*Voyage to the Equinoctial Regions of the New Continent in the Years 1799, 1800, 1801, 1802, 1803 and 1804*] (1859–60), repays comparison with the Bohn edition of the *Personal Narrative*. Hauff shared with Ross a background in journalism and had worked as the editor of the *Morgenblatt* [*Morning Paper*] from 1827 onwards. His edition, like Bohn's, was immensely popular. But there the comparison ends. Where Ross's English version was competing with a translation that Humboldt had sanctioned, Hauff was in the much more comfortable position of working against an anonymous translation, published with Cotta between 1815 and 1832, which Humboldt had publicly denigrated. This first translation, the work of the Swiss publisher and politician Paulus Usteri and the German medical professor Ferdinand Gottlob Gmelin, was considered by Humboldt to be an inaccurate rendering and ugly deformation of the *Relation historique* (Fiedler and Leitner 2000: 83). Humboldt's openly derogatory remarks on this translation meant that it quickly fell into oblivion and constituted no threat to Hauff's translation when that appeared on the German-speaking book market in the late 1850s.

Ette's authoritative account of the various forms and faces that the *Relation historique* acquired in German translation reminds us that Hauff's translation is pre-dated by the first German-language adaptation (Ette 1996: 104). This appeared in four volumes in Vienna in 1830 in a scientific travel literature series aimed at adolescent readers, the 'Bibliothek naturhistorischer Reisen für die reifere Jugend'. Adopting much the same approach as MacGillivray a couple of years later, Gottlob August Wimmer, editor of this Austrian abridgement, changed the narrative perspective from first to third person and used this new narrative standpoint to subject the *Relation historique* to strict moral censure, while extolling the achievements of the Catholic Church in the former Spanish colonies. As Ette illustrates, the moralising character of the Viennese edition, together with its ideological shifts and radical narratological restructuring, meant that it constituted a rather different work from the original. Indeed, he argues, it stood at one remove to the *Relation historique* not only by dint of it being a translation, but also because of its 'parasitic' character, in that Wimmer drew on Humboldt's work at will, manipulating and recontextualising the original text as he saw fit (Ette 1996: 106).

Hauff was merciless in the amount of material he omitted. The footnotes and explanatory notes largely disappeared, as did whole swathes of text listing Humboldt's instruments, detailing his geological and physical observations of the Canary Isles or describing the population and surface area of South America. Of the third French volume alone, around 88 per cent of the text was missing.[19] Yet despite these drastic cuts, which required the editor to add bridging sentences to restore a sense of linearity otherwise completely lost, Humboldt declared himself satisfied. As Humboldt noted in his own Foreword, the material that only confirmed scientific results had all been compressed, in a bid to ensure that the vivid representation of events would be less frequently interrupted (Humboldt 1859–60: I, iii). In short, those findings that had given his work the scientific precision he most highly valued had all been cut. But it was time, Humboldt realised, to make the *Relation historique* more accessible to a greater circle of educated readers, and just as nature was characterised by creation and change, so too was his œuvre. Hauff's abridged translation received Humboldt's blessing and critics likewise lent it their approval. The *Illustrirte Zeitung* [*Illustrated Gazette*] considered it a laudable example of the 'German art' of translation.[20] The *Blätter für literarische Unterhaltung* [*Pages for Literary Entertainment*] judged the excision of strictly scientific digressions appropriate to the general readership that was Humboldt's declared public, and even noted that other parts of Humboldt's œuvre would gain from being 'recast' in a similar manner.[21] Abridgement was no longer seen as distorting or disfiguring: it represented a valuable means of keeping Humboldt's writing contemporary, vivid and accessible.

Conclusions

In a postscript to a letter written to the Prussian diplomat Christian Carl Josias Bunsen in November 1852, Humboldt briefly noted that he had just thanked Mr Bohn for sending him seven most elegant volumes comprising new translations of *Kosmos*, the *Ansichten der Natur* and the *Relation historique* (Humboldt 2006: 155). His acknowledgement of this new, abridged, edition of the *Relation historique* was at best cursory. But he would have done well to pay it more attention. Behind the 'plain and unassuming' front that Ross cultivated in her Preface, she produced a new edition of the *Personal Narrative* through selective erasure that recast it in three different ways. Firstly, it condensed Humboldt's account into a text neither extraordinarily long nor pricey, which more easily met the norms current in British publishing. With its greater sense of political correctness it also countered the chauvinistic and imperialistic inferences in Williams's translation of the *Relation historique*. Finally, Ross's edition brought Humboldt's narrative into line with developments that had occurred in British scientific writing over the past three decades. It drew attention away from the collaborative activities by which knowledge was generated, side-lining the importance of observation and experimentation, and making scientific fact became increasingly self-evident. Rhetorical devices that articulated modesty and uncertainty, characteristic of late eighteenth-century and early nineteenth-century scientific writing, and essential in shaping Humboldt's speculative, evaluative approach, were also removed to foreground his actual findings over the workings of his mind. By distancing the language of her *Personal Narrative* from the style adopted by both Humboldt and Williams, Ross opted for a more dispassionate form of communication that aligned Humboldt's text with the narrative modes current in scientific writing around the 1850s.

To what extent did the Bohn edition really constitute a new translation? Or was it in fact merely parasitic, feeding off Williams's version? Given that the Longman and Murray edition was largely a close rendering of the original, Ross would have struggled to produce a new version that was radically different. Mindful of critics' complaints, she removed the worst excrescences of Williams's Parisian English, but she also retained the literary turns of phrase that represented the more successful parts of Williams's text. Ross, who acknowledged her debt to Williams in the Preface, may never have even sought to expunge the voice of her predecessor from the original. By excising such large quantities of material, as well as making repeated stylistic alterations on a smaller level, she gave the text a radical overhaul that recast it (and its author) in quite a different light. It is easy, on this basis, to see Ross's translation as a reductive or even destructive reworking of the *Personal Narrative*. However, this does not entirely do her version justice. Ross's judicious editing and Bohn's savvy marketing of this second translation of the *Relation historique* enabled Humboldt to reach a far greater audience than the Longman edition had ever attracted. For all its excisions, Ross's version should more properly be understood as a modernising account of the *Relation historique* that revivified it for a new generation of readers, and brought it back on to the Anglophone market just as English translations of Humboldt's *Ansichten der Natur* and *Kosmos* were beginning to enjoy the peak of their success.

Notes

1. *Athenæum*, 1346 (1853), p. 966.
2. *Examiner*, 2295 (1852), p. 53.

3. *Daily News*, 1774 (1852), p. 2.
4. *Literary Gazette*, 1833 (1852), p. 229.
5. *Examiner*, 2295 (1852), p. 53.
6. *Daily News*, 2215 (1853), p. 2.
7. *Punch*, 87 (1884), p. 110.
8. Thomasina's year of birth is unclear. On the census return for 1851, she is recorded as being fifty-six (that is, born in either 1795 or 1796), yet on the return for 1841 she is recorded as forty, and on the return for 1861 sixty-one years of age (that is, born in either 1800 or 1801). See census returns for St Clement Danes, Borough of Westminster, Middlesex HO 107/731/2 (1841) and HO 107/1512 (1851), as well as for Chelsea St Luke, Chiswick Ward, Middlesex, RG9/33 (1861) and for Bishop Wearmouth, Sunderland, RG10/5005 (1871). On all these returns she is recorded as having been born in Edinburgh. Thanks to Patricia and Ian Martin for investigating the census returns made by Ross.
9. Census return for St Clement Danes, Borough of Westminster, Middlesex, 1851, HO 107/1512.
10. Census return for Chelsea St Luke, Chiswick Ward, Middlesex, RG9/33 (1861); St Clement Danes, Borough of Westminster, Middlesex, 1851, HO 107/1512; for 1861, RG9/33.
11. London Metropolitan Archives, Diocese of London Consistory Court Wills, September 1852, Strand 1b 236.
12. Richard Bentley Papers, vol. XC, British Library Manuscripts, ADD Ms. 46649ff.12, 13, 36, 57, 79.
13. *Bentley's Miscellany*, 24 (1848), pp. 199–206, pp. 295–302.
14. Bentley Papers, vol. LVI, British Library Manuscripts, ADD Ms. 46615.f 113, letter of 15 September 1848 from Ross to Edward Morgan.
15. This was a common selling feature of articles in *Household Words*. See Lohrli 1973: 9.
16. *Notes and Queries*, 190 (1853), p. 610.
17. I base my calculations about the amount of text missing from Ross's translation on the Longman edition of the *Personal Narrative*, which I take as representing 100 per cent of the text. My rough estimates are as follows: Volume I, 18 per cent; II, 47 per cent; III, 15 per cent, IV, 0.1 per cent; V:1, 22 per cent, V:2, 17 per cent; VI:1, 59 per cent; VI:2, 39 per cent; VII, 40 per cent.
18. This is not to say that 'politeness' has completely disappeared from scientific discourse. As Greg Myers (1989) convincingly argues in his analysis of Francis Crick and James Watson's 1953 *Nature* article on the structure of DNA, the demonstration of intellectual solidarity and the avoidance of outright criticism continue to be key features of scientific writing.
19. Ulrike Leitner's detailed list of the pages missing from the Hauff edition, based on the French original, is extremely useful in illustrating quite how much material was lost (Fiedler and Leitner 2000: 87).
20. *Illustrirte Zeitung*, 855 (1859), p. 341.
21. *Blätter für literarische Unterhaltung*, 4 (1860), p. 731.

Chapter 5

The Poetry of Geography: The *Ansichten der Natur* in English Translation

In 1849, the year Humboldt turned eighty, a considerably updated edition of his essay collection, the *Ansichten der Natur*, came out with Cotta in Stuttgart. Initially published in 1808, then in a second edition in 1826, it was now appearing in its third and final version. One of the earliest pieces he had written on his return from the Americas, it gained in length as he successively added material of scientific interest and with aesthetic appeal that sustained the hybrid character of his prose. The first edition had comprised just three short texts: 'Über die Steppen und Wüsten' [On Steppes and Deserts], 'Über die Wasserfälle des Orinoco' [On the Cataracts of the Orinoco] and 'Ideen zu einer Physiognomik der Gewächse' [Ideas for a Physiognomy of Plants]. Twenty years later, the collection had been extended to include an essay on volcanoes and on 'Die Lebenskraft oder das rhodische Genius' [The Vital Force or the Rhodian Genius]. The final, two-volume third edition, covering almost 800 pages in the German original, also included a couple of new pieces, 'Das Hochland von Caxamarca, der alten Residenzstadt des Inka Atahualpa' [The Plateau of Caxamarca, the ancient capital of the Inca Atahuallpa] and 'Das nächtliche Tierleben im Urwalde' [The Nocturnal Life of Animals in Primeval Forests]. While the shifting character of the *Ansichten der Natur* charted Humboldt's development as a natural historian and writer over more than forty years, these editions also reflected the political climate of their time. The Foreword to the 1808 edition deliberately spoke to 'minds oppressed with care' (*AdN* 8), gesturing to the Prussian defeat of 1806. Revolutionaries were stoning the building in which Humboldt was writing as he updated the third edition in 1848 (Humboldt 1908: 299).[1]

Unlike their French counterparts, English publishers had seen no value in translating the first two editions.[2] However, the preparation of a third German edition in 1849 galvanised Longman and associates, working in collaboration with John Murray III, and their rival Bohn into action. The translation by Elizabeth Sabine, titled *Aspects of Nature, in Different Lands and Different Climates*, appeared in autumn 1849 with Longman and Murray. The publishing house of John Murray had already begun to specialise in narratives of travel and exploration in the late eighteenth century, and by the mid-nineteenth century it was a leading publisher of accounts of non-European exploration. In their significant study of Murray's prominence in the field of travel writing, Innes M. Keighren, Charles W. J. Withers and Bill Bell demonstrate how this publisher harnessed the considerable momentum that travel literature had gained from 'the combination of scientific discovery, public interest and political competition' (Keighren, Withers and Bell 2015: 31). Sabine's translation of the *Ansichten der Natur* appeared at the tail end of the 1840s, the decade in which the firm, under the aegis of John Murray III, issued over fifty new travelogues, which constituted almost a quarter of its total output of travel writing in the period from 1773 to 1859 (Keighren, Withers and Bell 2015: 25).

The *Views of Nature: or Contemplations on the Sublime Phenomena of Creation*, on which Elise C. Otté worked with Bohn, came out at the very start of 1850. Since Sabine's translation was already on the market, this compelled those working on the edition for Bohn to ensure it distanced itself categorically from its rival. Bohn again used price to tempt his audience. The two-volume work published by Longman and Murray retailed for six shillings, Bohn's one-volume edition for five. Other differences made Bohn's account more attractive to a reading public. His presentably bound red cloth edition of the *Ansichten der Natur* was issued as part of the 'Scientific Library' series and was therefore a 'collectable' item. Moreover, readers were lured into the Bohn edition by a vibrant oil colour printing, facing the title page, of llamas grazing leisurely among fearsome cacti at the foot of Chimborazo (Figure 5.1).

The smattering of reviews that appeared in British journals in autumn 1849 focused less on the scientific detail of the *Ansichten der Natur* than its descriptive vibrancy. The crocodiles and electric eels, boa constrictors and huge 'vampyre-like' bats that crawled, writhed and fluttered across its pages were, *The Athenæum* remarked, what made this an exciting work for children.[3] The work's sublime qualities appealed more to the *Literary Gazette*, which enthused about the 'gigantic wonders of eternal mountains, interminable plains, and vast rivers laving thousands of miles of land', while the *Examiner* admired its literary merits and reminded its readers that the great Goethe had compared Humboldt 'to a source of ever-gushing sweet waters'.[4] Others considered that Humboldt's literary achievements typified less the man himself than his nation. 'The Germans do contrive to make poetry of very matter of fact

Figure 5.1 Oil colour printing of Chimborazo, facing the title page of the *Views of Nature* (London: Bohn, 1850).

materials,' mused a critic in the *Daily News*, adding that there were 'no modern poems so imaginative, yet so true, and so full of all those elements, which ought to constitute verse, than some of those prose essays of Humboldt'.[5] This ability to ally the scientific with the literary was considered Humboldt's greatest achievement, and the *Ansichten der Natur* was even heralded as the very 'poetry of geography'.[6]

But critics were equally quick to condemn the work's structural flaws. David Knight has leniently described this as a collection of 'lyrical essays dwarfed by more scientific commentary' (Knight 2004: viii). Humboldt's contemporary British reviewers were less generous. 'We have', calculated the critic in *The Athenæum*, '185 pages constituting the original matter, and 422 pages of "annotations and additions" [. . .] the additional portions being dry, uninteresting for the most part, and often unimportant,' and presented a neat tabular summary of the work's shortcomings:

	Text.	Annotations.
Steppes and Deserts	26 pages.	177 pages.
Cataracts of the Orinoco	24 "	22 "
Nocturnal Life of Animals	33 "	12 "
Physiognomy of Plants	31 "	179 "
Structure of Volcanoes	28 "	6 "
The Vital Force	8 "	5 "
The Plateau of Caxamarca	35 "	21 "[7]

Even Brewster, in his forty-page discussion of the translation by Sabine for the *North British Review*, had difficulty finding praise for the structure of the work. It defied proper critical review, he lamented, because 'the great length of the "annotations and additions", which extend to more than twice the length of the original chapters which form the text' meant that he 'found it impossible to give such copious and continuous extracts as the reader might have desired'.[8] The language of the work further complicated matters. It was, Brewster complained, an 'admixture of scientific with popular details', which used technical terms that were bound to fox the general reader.[9]

Since landscape description underpinned at least four of the seven essays in the *Ansichten der Natur*, this chapter starts by considering how Humboldt's narrative presentation of scenery engaged with late eighteenth- and early nineteenth-century theories of the sublime and the picturesque. It then examines the background history to the appearance of the two rival translations and the difficulties that the text and its (re)translation presented to Sabine and Otté. In the close textual analysis of the differences in style and presentation between Longman's *Aspects of Nature* and Bohn's *Views of Nature* that forms the central focus of this chapter, a discussion of how these two translations addressed aesthetic, literary and spiritual concerns throws further light on the ways in which they were recast and reframed for a British audience. It also asks, though, what the divergences between the two texts reveal about the environment in which the translations were prepared. How free were Otté and Bohn to follow their own aesthetic, political and spiritual agenda, and to what extent did Sabine's translation limit the choices they made? Should we understand these different editions as a translation and a retranslation of the *Ansichten der Natur* or rather as two near-contemporaneous distinct versions deriving from the same German source text?

'An impression like Nature herself': The *Ansichten der Natur*

The *Ansichten der Natur* essentially started life as the written version of lectures given at the Prussian Academy of Sciences in Berlin in 1806 and 1807. The three essays at the core of the first edition – on steppes and deserts, on the waterfalls of the Orinoco and on the physiognomy of plants – were pieces that melded rich, intensely visual, accounts of the natural world with detailed scientific analyses of the phenomena at work in them. This holistic, unifying vision of landscape, which brought images of great richness and diversity together with physical, geological and meteorological data, was, Malcolm Nicolson notes, 'a very typical Humboldtian production' (Nicolson 1990: 178). But it also sprang from Humboldt's own engagement with German *Naturphilosophen* in Schelling's circle, as well as with Goethe, to whom Humboldt had dedicated the German edition of the *Essai sur la géographie des plantes* [*Essay on the Geography of Plants*] (1807).

Humboldt's 'Author's Preface' to the first edition of the *Ansichten der Natur* was instrumental in clarifying his aims as author and narrator. Each was 'designed to be complete in itself' yet underpinned by 'one and the same tendency', which wrought of these individual parts a whole (*VN* ix). However, as Humboldt freely admitted, his aesthetic approach to treating objects in the natural world was 'fraught with great difficulties in the execution, notwithstanding the marvellous vigour and flexibility of my native language' (*VN* ix). How could the fecundity of nature be represented narratively, given that it 'presents an accumulation of separate images, and accumulation disturbs the harmony and effect of a picture' (*VN* ix)? Humboldt's 'survey of nature at large' aimed to bring all the sense impressions together in the mind's eye of the reader to recreate 'the enjoyment which the immediate aspect of the tropical countries affords to the susceptible beholder' (*VN* ix). The different angles of perspective that he adopted, the changes of pace that he built into his prose and the magnitude of the scenes he described were all important visual components in his narrative experiment.

In a letter to Cotta in early 1807, Humboldt related how he had just completed a small work on deserts, in the style of his essay on the physiognomy of plants, which Goethe had just reviewed favourably in the *Jenaische Allgemeine Literatur-Zeitung* [*Jena General Literary Gazette*]. These two pieces formed the basis of a new book – the first edition of the *Ansichten der Natur* – which would simply contain aesthetic representations of natural objects and general overviews (Humboldt 2009: 78). Bettina Hey'l singles out these contributions to the first edition of the *Ansichten der Natur* as particularly essayistic, since they present a synthesis of ideas yet to be proven and are characterised by a provisional and open endeavour to gain some sense of the whole (Hey'l 2007: 216). As Hey'l suggests, this essayistic style of writing, which echoes the oral nature of Humboldt's original enterprise, was an essential ingredient that made this work the most successful literary piece of Humboldt's entire œuvre (Hey'l 2007: 215–16).

Humboldt's chapter 'On Steppes and Deserts' exemplifies how he made the traveller the focal point of his narration. It also demonstrates how his modes of description could oscillate between the objective scientific descriptions of geobotanical features and the expressive articulation of emotional responses to the landscape. Humboldt singled out this essay, in a letter to Cotta, as one that was specifically designed to have an imaginative effect on its readers (Humboldt 2009: 82). The piece opens as the traveller is leaving behind him the 'Alpine valleys of the Caracas' and the verdant shores of Lake Tacarigua, shaded with banana trees, sugar cane, and cocoa groves, to head into the 'vast and boundless' plain of the steppes, 'whose seeming elevations disappear into the distant horizon' (*VN* 1). The treeless wastes of this expanse stand in immediate contrast to the lush, luxuriant scenes just described. This creates a moment of narrative 'surprise', designed to set the picturesque exoticism of the previous landscape against the sublime starkness of the steppes. As the observer at the centre of this contrastive panorama, Humboldt was able to orchestrate an immediate transition from one scene to another. Annette Graczyk perceptively notes that this frees him from using the traditional pictorial categories of fore-, middle- and background, which would otherwise need to be described in separate stages (Graczyk 2004: 336). Standing on the edge of the steppes, looking out over the landscape, Humboldt registers the difference between landscapes previously encountered and the present 'shore-less ocean [. . .] spread before us' (*VN* 2):

> Aus der üppigen Fülle des organischen Lebens tritt der Wanderer betroffen an den öden Rand einer baumlosen, pflanzenarmen Wüste. Kein Hügel, keine Klippe erhebt sich inselförmig in dem unermeßlichen Raume. Nur hier und dort liegen gebrochene Flözschichten von zweihundert Quadratmeilen Oberfläche, bemerkbar höher als die angrenzenden Theile. (*AdN* 15)
> [Leaving behind the luxuriant richness of organic life, the astonished traveller finds himself on the dreary margin of a treeless desert, devoid of plants. No hill, no cliff rises, as an island in the ocean, out of the boundless plain: only here and there lie broken strata of floetz, extending across two hundred square miles, appearing distinctly higher than the surrounding areas.]

The cumulative acquisition of visual impressions is reinforced through repetition ('kein Hügel, keine Klippe') and the sweep of the observing eye ('hier und dort') across a landscape that provides no other visual interest than limestone outcrops to break the surface monotony. In the *Ansichten der Natur*, we perceive the landscape either in abstract and schematic ways or through highly detailed description (Graczyk 2004: 337). We are also shown it from a variety of different perspectives: as the geographer and cartographer, the botanist, the geologist and as the landscape artist.

Movement was central to the perspectival shifts in landscape viewing that underpinned Humboldt's presentation of the life that crossed the steppes. From the 'swift-footed ostriches and herds of gazelles' grazing across this boundless space to the itinerant tribes of the Huns and

Mongols, Humboldt cast this region as a locus of movement past and present, human and animal, that charted cultural changes, as well as shifts of habitat (*VN* 3). As Graczyk notes, it emphasises how nature is constantly in motion, a dynamic play of rhythms and forces operating independently of the travelling observer (Graczyk 2004: 340). Humboldt's essay 'On the Cataracts of the Orinoco' displayed a similar wealth of movement deriving from the landscape itself, found in the stream 'foaming down' the eastern declivity of the Cunavami mountain range, the enormous whirlpools or the immense mass of water flowing along the Orinoco: all were constituent elements in the effect that surroundings had on the 'inner susceptible world of the mind' (*VN* 154).

Humboldt likewise opened his 'Ideas for a Physiognomy of Plants' by painting a vibrant scene of life in universal profusion, which appealed as much to the aural as the visual – with birdsong and humming insects as background accompaniments to images of the air and the sea that teemed with microscopic life. The energy of these pieces derives from the accretion of visual impressions and the denseness of factual detail. This essay, Graczyk argues, exemplifies Humboldt's goal of achieving a total impression or 'Totaleindruck', in which the spatial continuum of the earth and sky is divided up into individual areas to which discrete intrinsic aesthetic values can be ascribed (Graczyk 2004: 299). In this piece on plant life, Humboldt used even more shifts of perspective and orders of magnitude than in his essay on the steppes, since here he focused on the very smallest visible elements of life. Once again, the observer's susceptibility to the aesthetic appeal of nature was essential in powerfully conveying the scene to Humboldt's readers and encouraging them to visualise it in their mind's eye. Reverting briefly to more familiar objects after a long discussion of exotic nature, Humboldt asked of his audience, 'who is there that does not feel himself differently affected beneath the embowering shade of the beechen grove [...] where the breeze murmurs through the trembling foliage of the birch?' (*VN* 219). This 'total impression' relied on an appeal to the senses, but also to something of a higher order:

> This influence of the physical on the moral world – this mysterious reaction of the sensuous on the ideal, gives to the study of nature, when considered from a higher point of view, a peculiar charm which has not hitherto been sufficiently recognised. (*VN* 219)

Writing to Varnhagen von Ense in October 1834, after the first and second German editions had been published, Humboldt observed of the *Ansichten der Natur*:

A book on Nature ought to produce an impression like Nature herself. The point, however, to which I have especially, as in my 'Aspects of Nature', paid attention, [. . .] is this, that I have endeavoured in description to be truthful, distinct, nay even scientifically accurate, without getting into the dry atmosphere of abstract science. (Humboldt 1860: 19)

This essay collection was Humboldt's stylistic exemplar of how narrative could achieve factual accuracy yet be engaging. By showing how his explicitly physical descriptions could be brought to bear on the development of scientific knowledge, he avoided the pitfall of generating 'dry' narrative that dealt only in abstraction. To some extent, this held true for the style of the two additional essays that formed part of the second edition, 'Über den Bau und die Wirkungsart der Vulkane in den verschiedenen Erdstrichen' [On the Structure and Mode of Action of Volcanoes, in Different Parts of the Globe] and 'Die Lebenskraft oder das rhodische Genius' [The Vital Force or the Rhodian Genius]. Originally read at a public meeting of the Academy in January 1823, the piece on volcanoes was fiery and fearsome, treating readers to showers of cinders and eruptions of subterranean water, fish and mud, and leaving them, like the terrified peasants of Humboldt's narrative, in awe of nature's violence (*VN* 366–7). No clearer evidence was needed of Humboldt's debt to the Vulcanists (or Plutonists), who endorsed the notion that the earth had been shaped by fire, in contrast to the Neptunists' belief that it was water that had moulded the earth's surface. The main text of this essay was more heavily laden with scientific detail than those in the previous edition, but Humboldt narrated the events at Pompeii and Herculaneum in such a way as to sustain their human interest, while also allowing for discussions of volcanic storms, ash strata and the periodic recurrence of tremors. His essay on volcanic action not only was characterised by a wealth of factual details, but it was also extremely up to date, including reports of an eruption of Vesuvius in 1822.

Although Humboldt tried to sustain a contemporary feel in the *Ansichten der Natur*, his short essay on the Rhodian Genius from 1795 was long obsolete by the time it appeared in the second German edition of this essay collection in 1827. As his brother Wilhelm had observed, such 'semi-poetical clothings of grave truths were more in vogue at the time this was written than they are at present' (*VN* xii). At the core of this essay were two paintings, both allegorical representations of the phenomenon of 'Lebenskraft' or 'vital force'. This force had been the subject of Humboldt's early experiments on nerve and muscle tissue. His research on how muscles were excited by electricity had fascinated intellectuals in Jena and Weimar, notably Schiller, who had invited Humboldt to contribute a piece on these enquiries into the secret of life itself for the 1795 issue of the short-lived literary journal *Die Horen* [*The Horae*]. But by the time Humboldt came to republish his essay, he had retracted the earlier ideas he had entertained about any 'vital principle' governing life. Indeed, in his *Versuche über die gereizte Muskel- und Nervenfaser* [*Experiments on Excited Muscle and Nerve Fibre*] of 1797 he had concluded that the existence of such vital forces had not been formally established. Nevertheless, Humboldt continued to include this essay on the Rhodian Genius in later editions of the *Ansichten der Natur* under the illusion that Schiller had approved of it (Humboldt 1860: 2). Only in July 1849 would he discover that the poet had in fact cast him, in a letter to the German writer and lawyer Christian Gottfried Körner from 1797, as a person without an iota of imagination, a narrow-thinking man of reason (Humboldt 2009: 394).[10]

The third and final edition of the *Ansichten der Natur* gave Humboldt ample opportunity to revise, rework and update. Two-thirds of the text of the final German edition was new, he noted to Arago in November 1849, additions that he believed would enhance the work's lyrical value (Humboldt 1908: 299). The two final pieces – one on the plateau at Caxamarca in Peru, and the other on the nocturnal activities of animals in primeval forests – which now extended the collection to its full seven essays, saw Humboldt return to narrative modes that emphasised the arduous travel that he had undertaken with Bonpland through the marshlands skirting the mountain passes of the Andes, their fascination at the remains of Inca roads and their frustration at the impenetrability of the primeval forests in the torrid zone of South America. Rhetorically, too, it was a dense, multivocal piece in which a host of earlier commentators on South America were brought into dialogue with each other (*VN* 196). As Ette reminds us, though, this was not the first time that Humboldt had used this material. Some information had also been published elsewhere and in different versions and contexts that complement each other in highly productive ways (Ette 2002: 184–91). The description of the caves at Ataruipe in the essay 'On the Cataracts of the Orinoco' had already been recorded once in Humboldt's diary, once in the fifth volume of Williams's translation of the *Relation historique* and once again in earlier editions of the *Ansichten*, now with a different emphasis placed on the narrative significance of the mummified remains found within the caves.

Humboldt set great store by the literary value of his essays, which itself posed challenges for his translators. The scientific character of his writing proved still more problematic for them, though. Since the text had evolved over forty years, Humboldt was not always able to provide a swift answer to the queries they had about terminology and scientific content. In response to three points raised by Elizabeth Sabine as she worked on the chapter 'On Steppes and Deserts', which Humboldt had written forty years earlier, he was obliged to consult the geographer Carl Ritter (Humboldt 2010: 106). Could he please confirm the existence of the Indian River Goschop, a tributary of the White Nile? Did snow really lie on the Mountains of the Moon – today's Rwenzori Mountains, a range that forms the source of the White Nile – in a southern latitude of 2° to 3°, as the German missionary Johannes Rebmann had reported? And how far west of Zanzibar were the Lupata Mountains? Humboldt's French translator of the third edition, Charles Galusky, similarly 'tormented' him with questions that he could not readily answer. 'I no longer know what I wrote in 1807' – 'Ich verstehe nicht mehr was ich 1807 geschrieben [habe],' complained a harried Humboldt to Ritter in September 1850 (Humboldt 2010: 114).

Correspondence with his associates also brought home to Humboldt why his stylistic enterprise might not 'translate' well across different cultures. Writing to Bunsen in November 1849, Humboldt observed that the *Ansichten der Natur*, a piece calculated to appeal so specifically to German sensibility – 'ein so rein auf deutsche Gefühlsweise berechnetes Buch' – would be unlikely to meet with approval in Britain (Humboldt 2006: 118). He reiterated this with greater charm, irony and self-deprecation to the translator Sarah Austin, who had suggested that she translate this book, once she had finished her three-volume translation of the historian Leopold von Ranke's *History of the Popes* for John Murray. Elegantly deflecting her proposal, Humboldt suggested playfully:

> After Ranke should come 'Ansichten der Natur'. You will never find a book offering the same advantages, notes longer than the text, information how in the tropics rain is followed by fine weather, and sentiment induced by the sight of sand, rocks, river-foam, palm-trees, and wild sheep. This Teutonic sentimentality, which has stood me in such good stead in my own country, would be ridiculous in the land of Positive Philosophy. I cannot believe that you wish to translate me, glad and pleased as it would make me. Luckily you can find no title for my pre-Adamite work. (Ross 1893: 206–7)

His humorous enumeration of the elements he sensed British critics would lambaste indicates a sensitivity towards the problems of cultural transfer inherent in the *Ansichten der Natur*: more notes than text, information neither novel nor remarkable, and overtly articulated sentimentality, all of which was combined with the potential for religious controversy that a 'pre-Adamite' work might ignite. Humboldt did not consider Austin an ideal translator for this work and therefore went out of his way to dissuade her; he also turned down her offer to translate *Kosmos*, noting to Cotta that the 'Queen of English Translations' would be capable of translating only the literary part and not the rest (Humboldt 2009: 273). While the *Ansichten der Natur* did not present translational difficulties of quite the same order as those in *Kosmos*, Humboldt's summary of its shortcomings did forewarn of the problems both Elizabeth Sabine and Otté would encounter some five years later as they sought to convince British readers (and reviewers) of its inherent qualities.

In the previous chapter, we explored why retranslation takes place and what shapes the relationship between the first translation and those that follow. The justifications put forward by Ross for reworking the *Relation historique* several decades after the publication of Williams's version have little in common with those motivating Otté, who would produce a rival translation of the *Ansichten der Natur* that would appear almost contemporaneously with that by Sabine. Signalling dissociation from this other translation would have been important to Otté and Bohn. Dirk Delabastita's discussion of 'translational intertextuality' is significant in understanding how translators build distance into their retranslations, namely:

> when translations display avoidance strategies by *not* using certain solutions that in every other respect would have seemed to be the logical choice, except that the solution in question has been adopted and observed in earlier translations and that having recourse to it might create suspicions of undue dependence or even plagiarism. (Delabastita 2008: 240)

We cannot know for sure when Otté or Bohn came to see the translation for Longman and Murray, but its presence cannot be ignored. While it is interesting to chart the similarities between Otté's translation and its rival, just as intriguing are the disparities between the two. As Delabastita persuasively argues, any 'anxiety of influence' is not marked by the presence of one text in the other but by the very opposite – a phenomenon that operates '*in absentia* by definition' and is 'much harder to discern and pin down than the cases where patterns of influence are positively on textual record' (Delabastita 2008: 240).

Mrs Sabine's *Aspects of Nature*

Scrawled in an addendum to a letter from Humboldt to Cotta dated 20 January 1849 are his initial concerns about the appearance of an unauthorised English translation (Humboldt 2009: 347). The endpapers of books published in Bohn's 'Scientific Library' in 1849 had already alerted readers to the forthcoming seventh volume in the series,

'Humboldt's Views of Nature, *with Coloured View of Chimborazo, &c.*'. In a letter to Cotta, Humboldt suggested how the publication of this rival version – what he darkly prophesied as a 'bad translation' – could be averted (Humboldt 2009: 347). Edward Sabine and his wife Elizabeth, to whom Longman had assigned the translation, were to outmanœuvre the competition by ensuring that their translation appeared first. The timing was crucial. Print proofs of the first volume were to be sent to the Sabines to facilitate their head start with the translation, while nothing should appear in England that had not already been published in Stuttgart (Humboldt 2009: 347).

With Elizabeth Juliana Sabine (1807–79) as its translator, Humboldt's *Ansichten* was in safe hands. She had previously completed the *Narrative of an Expedition to the Polar Sea, in the Years 1820, 1821, 1822 and 1823* (1840), a translation of Ferdinand von Wrangell's 1839 account of the navigable parts of the Arctic seas and the inhabitants of the circumpolar regions. Yet there is no mention of Elizabeth's name on the title page of the *Narrative*, described as 'Edited by Major Edward Sabine, R.A., F.R.S.', and her 'presence' in the translation seems literally marginal, since she speaks to us only through translator's footnotes that merely serve to update the names of places and geographical features. However, Wrangell's *Narrative* stands as an early example both of Elizabeth's prodigious (if hidden) linguistic talents and of her ability to work at speed on scientifically complex texts.

With the notable exception of the *Ansichten der Natur*, all her translations of Humboldt's works would be ascribed to her husband, under whose 'superintendence' they had been completed. The Dublin-born Edward Sabine (1788–1883) had been recommended by the Royal Society of London to serve as the astronomer on board John Ross's expedition of 1818 in search of the Northwest Passage. Sabine's broad scientific remit extended to measuring the direction and intensity of the earth's magnetism, observing atmospheric refraction or noting the tides, currents, temperature and salinity of the seawater. A year later, he again sailed to the Arctic under the command of William Edward Parry and continued his magnetic observations, earning him the prestigious Copley Medal of the Royal Society in 1821 for his findings. By the mid-1820s, Sabine was involved, together with Herschel, in measuring differences of longitude between the Greenwich and Paris observatories. In the 1830s he became involved in the 'magnetic crusade', an extensive cooperative project to understand the laws governing the true causes of geomagnetism. At Sabines' Woolwich home in southeast London he collected and processed masses of data coming in from the furthest corners of the Empire. His excellence as a scientist in his own right and as a scientific administrator culminated in his election as President of the Royal Society, an office he held for a decade from 1861.

The works that Elizabeth Sabine translated reflected her husband's interests. The mathematical prodigy Carl Friedrich Gauss had published his *Allgemeine Theorie des Erdmagnetismus* [*General Theory of Terrestrial Magnetism*] in 1838.[11] In it, he postulated that the causes of normal terrestrial magnetism were to be found in the interior of the earth and calculated the position of the earth's magnetic axis. Elizabeth Sabine, who accompanied her husband to the 'Magnetic Congress' at Göttingen in mid-October 1838, produced a translation that appeared in the 1841 edition of Richard Taylor's *Scientific Memoirs* and would primarily be used by army officers embarking on the Antarctic magnetic expedition. The brief recognition '[Translated by Mrs. Sabine, and revised by Sir John Herschel, Bart.]' immediately below the title of Gauss's article cannot be overlooked, and it was in this smaller project that she began to gain some public acknowledgement. The next translation that the Sabines would tackle after the first two volumes of Humboldt's *Kosmos* and the *Ansichten der Natur* was the Berlin physicist Heinrich Wilhelm Dove's *Die Verbreitung der Wärme auf der Oberfläche der Erde* (1852). Published in 1853 as the *Distribution of Heat over the Surface of the Globe* with Taylor and Francis for the British Association for the Advancement of Science, it was awarded the Royal Society's Copley Medal in 1853. This achievement was undoubtedly due to Elizabeth Sabine's excellent translation into English, even if she again went unnamed on the title page and was given the barest of mentions in her husband's Preface. The final piece on which the Sabines concentrated, aside from their undertakings for Humboldt, was Arago's *Meteorological Essays: With an Introduction by Alexander von Humboldt* (Longman, 1855), a compilation that included essays on the characteristics of thunder clouds, the relative duration of lightning and the luminosity of storm clouds, the geography and seasonality of storms, and injuries caused by lightning strikes.

It is difficult to assess whether Longman and Edward Sabine were complicit in deliberately stifling public acknowledgement of Elizabeth's input into these long, complex translations or whether she herself shunned the public eye. Either way, Healy is right to choose Elizabeth Sabine as one of the central figures in her study of invisibility in relation to English women, translation and science. Details on Sabine 'remain obscure in inverse proportion to the number, range, and importance of the texts she translated', Healy succinctly observes, and points up an important distinction between the visibility of the translator and the visibility of the texts they put into a foreign language (Healy 2004: 282). It is all too easy to cast Elizabeth as little more than an exploited amanuensis to her husband. Rather, the glimpses we do get of their relationship suggest it was warm and mutually fulfilling, '[s]uch is the sympathy between these married magnetists', declared Caroline Fox (daughter of Robert Were Fox, inventor of the 'Deflector Dipping Needle') when she visited them in August 1841 (Pym 1882: I, 227).

Despite the competence of the Sabines as translators and editors, Humboldt's involvement in the translation of the *Ansichten der Natur* was not limited to basic practical issues, such as ensuring that they had access to the German source text. He became quite intensely involved in discussions about how the work should be translated English – not least its title. On 13 March 1849 Edward Sabine had written to John Murray, assuring him that the title of *Ansichten der Natur* would be 'in conformity with Mr de Humboldt's desire [. . .] knowing that he is particular on that point', since Humboldt had already turned down Sarah Austin's suggestion of calling it 'Pictures of Nature'.[12] Early on, Humboldt had signalled his desire to call the work 'Views into Nature' and voiced his disapproval of 'Views of Nature'. He also disliked 'Scenes of Nature' – just one of a series of possible alternatives proposed by Elizabeth Sabine.[13] The German title of the *Ansichten der Natur* echoed how Forster had titled the work he published on his return from travels with the young Humboldt through the Low Countries and Great Britain in 1790. Forster's *Ansichten vom Niederrhein* [*Views from the Lower Rhine*] (1791–4) openly drew on contemporary meanings of the word 'Ansicht', which referred to a picture or a drawing of a landscape, like a *veduta* or panorama (Hey'l 2007: 218–19). Humboldt may have discarded Austin's suggestion of 'picture' because it implied a flat, two-dimensional representation of nature that failed to convey the imaginative vibrancy and depth he sought. The term 'scene' was near-impossible to divorce from theatrical associations, which would highlight mediated forms of representation rather than emphasising narrative immediacy. Of the three terms debated in early discussions, 'view' was the one most directly associated with landscape appreciation, which encompassed the notion of the observing eye scanning a landscape, examining, inspecting and surveying. It could stand both for a single scene and for the whole range the eye could cover, from the 'prospect' view of the observer looking on high across a landscape to the perspective of someone standing in the scene described. Humboldt would have been familiar with it as a term used in pictorial art – for example, from coloured engravings such as the lavish prints in William Hodges's *Select Views of India* (1793) – so it is unsurprising that he swiftly alighted upon it and considered it appropriate as a translation for the title of his own work.

With his mind firmly set on 'Views', Humboldt then sought an English preposition to convey movement as the reader was drawn through imaginative 'transport' into the landscapes so vividly described. The title 'Views into Nature' seemed to him ideal. But it was difficult to persuade Humboldt that 'view into' would sound odd to an Anglophone ear. 'I wish Mr de H. had been contented to propose Views of Nature. I dare say however that deep meditation on the subjects would elaborate Views into Nature as the more appropriate,' sighed Edward Sabine in a resigned postscript on the matter to Murray.[14] Sabine's unease at Humboldt's suggestion of 'into' did not abate. The next day he announced to Murray: 'We will retain M. de Humboldt's title therefore, unless Mr Longman or You have objections to it; and Mrs Sabine will state in her preface that the English Title is given by Mr de Humboldt himself.'[15] By signalling that the title was Humboldt's own, the Sabines would be distancing themselves from criticism of the awkward choice of preposition and strengthening their claims to have produced the authorised, definitive translation. Two weeks later, though, things were still not settled. Sabine noted to Murray that 'Mr Longman does not dislike "Views into Nature" on the contrary he thinks it [...] a good selling title,' but Sabine's misgivings still compelled him to ask a further three or four people for their opinion.[16] By the start of April, opinions had shifted again. It was to be 'Aspects of Nature in different Lands, and different climates', Sabine wrote to Murray, hoping 'that Mr de Humboldt might be brought to like it as well, and possibly even better than his own'.[17] Humboldt remained unconvinced, remarking to Bunsen in November 1849 that he had not yet received a copy of the English translation of the *Ansichten der Natur* and that he would have preferred something other than the 'curious Aspects-title', such as 'Fragments' or 'Views into Nature' (Humboldt 2006: 118). Later that month, Humboldt boasted to Arago that the *Aspects of Nature* had sold out within the week, but was still grousing that the title displeased him immensely (Humboldt 1908: 308).

The subtitle of the German original, 'mit wissenschaftlichen Erläuterungen' [with scientific explanations], was also an important descriptor characterising the book to its readers. 'This work consists of Text and *Illustrations and Additions*,' Edward Sabine tried to hammer home to Humboldt's British publishers, 'not text and *notes* like Cosmos.'[18] They were to be called 'illustrations and additions', he emphasised, because they represented new material that supplemented the narrative of the essays: moreover, Sabine insisted, their style was not that of notes 'but rather an enlargement of the subjects'.[19] In the end, Longman and Murray settled for 'Scientific Elucidations', which in fact gestured in the direction of explanatory notes, where the Bohn edition would, ironically, carry the 'Scientific Illustrations' Sabine had originally favoured.

Finally, and most intriguingly, the name that appeared on the title page just below Humboldt's was not the 'Lieut.-Col. Sabine' readers would come to expect in the Longman and Murray edition of *Cosmos*. Presented as 'Translated by Mrs. Sabine', this edition of the *Ansichten der Natur* at last brought Elizabeth into the limelight and won her the public acclaim that her husband's greater prominence had hitherto eclipsed. It also earned her the considerable sum of £300 for translating the piece at breakneck speed (Humboldt 2009: 394). Longman and Murray even made her role in the Englishing of Humboldt's text an important selling point, advertising in papers such as the *Daily News*, 'SABINE'S AUTHORISED ENGLISH TRANSLATION OF BARON HUMBOLDT'S NEW WORK', adding that it was translated 'with the Author's sanction and co-operation, and at his express desire, by Mrs. Sabine'.[20]

That Sabine's wife should have been presented so conspicuously as the translator of the *Ansichten der Natur* was no coincidence. Indeed, her husband seems to have deliberately orchestrated it. Early on in the proceedings, he noted rather enigmatically to Murray:

> Altho' I fully purpose to give just the same pains in the superintendence of the printing of this work as in the case of Cosmos, the nature of the work does not require that it should be formally announced as under my superintendence; Its annunciation simply as translated by Mrs Sabine will fully answer the purpose quite as much as if My name were to appear also – and such being the case, I have other reasons (quite unconnected with this work) for preferring that my name should be omitted.[21]

Sabine's sudden reticence at being associated with this work – a sea change from *Cosmos*, in which Mrs Sabine was mentioned only in passing as the author of the translation – may best be explained by the type of work the *Ansichten der Natur* was. With its overt appeals to the imagination and its strong aesthetic and literary qualities, it may not have dovetailed neatly into the soberly scientific profile that Sabine, recent winner of the Royal Society's Royal Medal for his geomagnetic writings, was earnestly cultivating for himself.

'Late into the Field': The Bohn Translation

> I am at present engaged on Humboldt's 'Ansichten der Natur' – a graceful outpouring of his love of nature; and a picturesque reflection of his own impressions of natural scenery amid the deserts, llanos and savannahs – the alpine valleys, mountain ranges and volcanic regions – and the tropical lands and the glacial seas of our chequered planet

enthused Elise Otté to the American historian William Hickling Prescott on 15 April 1849.[22] In a reply one month later, Prescott thanked her most warmly for sending him a copy of Humboldt's *Ansichten der Natur*. It was, he affirmed, a collection of essays 'which display his peculiarities in quite as marked a degree as any other of his works', and enthused that the 'splendour of philosophy and poetry all combine to produce an approach which results from no other works of a scientific character that I am acquainted with'.[23]

Given her scientific interests and expertise, Elise Charlotte Otté (1818–1903) was a particularly appropriate figure to undertake the translation of the *Ansichten der Natur*. Edmund Gosse's obituary in *The Athenæum* of 1904 described her as 'unquestionably one of the most learned women of her time' and as 'modern, audacious and liberal'.[24] Otté, the daughter of a Danish father and English mother, received her education from her stepfather, the English philologist Benjamin Thorpe, who gave Elise a thorough grounding in Anglo-Saxon and Icelandic. But Thorpe was a difficult taskmaster and Otté left for the States in 1840 to teach in a Boston family. In America she developed an interest in the New England Transcendentalists and studied geology, physiology and anatomy at Harvard. She may also have met Prescott there, whose *History of the Conquest of Mexico* (1839) and *History of the Conquest of Peru* (1847) were informed by Humboldt's work on South America. Gosse relates how she then lived in Frankfurt for a while, translating scientific texts for the professors there, but later returned to Britain to help Thorpe with his translation project, the *Edda of Saemund* (1856). Finding the relationship with him difficult, she moved on, and by 1851 had joined the household of George Edward Day, Chandos Professor of Anatomy and Medicine at the University of St Andrews, in Scotland. In 1863, she accompanied Day and his wife on their move down to Torquay on the south coast of England and, following his death in 1872, made London her new home.

These two places – St Andrews and Torquay – were particularly influential in her intellectual development. The lively and progressive scientific community at St Andrews enabled Otté to develop a range of scientific interests, including marine biology and botany. In December 1855, she was elected Honorary Member of the St Andrews Literary and Philosophical Society, during the time that Sir David Brewster was serving as vice-president and Day as joint secretary. The Society was clearly open to women's involvement in its various scientific projects. In 1859, Otté was appointed to a committee considering how herbarium specimens could best be mounted, and later rearranged a case of moths and beetles that had been donated for exhibition.[25] St Andrews, situated on the Fife coast, enabled her to establish a reputation as one of the most renowned amateur marine biologists in the region. As the first edition of the *Journal of the Marine Biological Association of the United Kingdom* of 1887 noted, 'Prof. Gr. E. Day [. . .] and Miss Otte [sic], lost no opportunity of interesting the students in marine zoology.'[26] Day may also have been the driving force behind her readiness to take on *Kosmos*, much of which she translated while living in the Days' house at the end of St Andrews' bustling South Street. A physician by training, Day had used reviewing and translation to supplement his early income as a lecturer at Middlesex Hospital and had himself produced the English editions of Johann Franz Simon's *Medizinisch-analytische Chemie oder Chemie der näheren Bestandtheile des thierischen Körpers* (1840) [*Animal Chemistry* (trans. 1845–6)], Julius Vogel's *Pathologische Anatomie des menschlichen Körpers* (1845) [*Pathological Anatomy of the Human Body* (trans. 1847)], the fourth volume of Karl Rokitansky's *Handbuch der allgemeinen pathologischen Anatomie* (1842–6) [*Pathological Anatomy of the Abdominal Viscera* (trans. 1852)], and Karl Gotthelf Lehmann's *Lehrbuch der physiologischen Chemie* (1850) [*Physiological Chemistry* (trans. 1851)].[27] Beyond assisting Otté with technical terminology, Day might also have alleviated the disability that had already begun to develop during her stay in America, a 'spinal curvature of a very pronounced and painful character'.[28]

The Days' move to Torquay was fortuitous for Otté, since she met the eminent zoologist Philip Henry Gosse and his son Edmund there, and became acquainted with a number of renowned geologists hammering away at the chalk and Devonian limestone of the cliffs on the south coast of England, including William Pengelly.[29] As Pengelly's wife Mary noted shortly after the Days' move down to the south coast, Otté was a 'clever, interesting woman' (Pengelly 1897: 175). Philip Gosse, a member of the Plymouth Brethren and author of a number of seashore studies including *A Naturalist's Ramble on the Devonshire Coast* (1853) and the *Handbook to the Marine Aquarium* (1855), is now best remembered for *Omphalos: An Attempt to Untie the Geological Knot* (1857), a work that attempted to show that the fossils found in the rocks were not the remains of life-forms that had once inhabited prehistoric shores and seas, but had been divinely created *in situ*. Otté may well have been more interested in Philip Gosse's coastal writing than *Omphalos*. She translated the two-volume *Souvenirs d'un naturaliste* (1854) by the Parisian professor of ethnology and anthropology Jean Louis Armand de Quatrefages de Bréau, as the *Rambles of a Naturalist on the Coasts of France, Spain and Sicily* for Longman and partners in 1858. The translation by 'E. C. Otté, Honorary Member of the Literary and Philosophical Society of St. Andrews' was dedicated to Brewster, whom she asserted was, like Quatrefages, endowed with the 'gift of popularising science in a spirit at once earnest and genial' (Quatrefages de Bréau 1857: I, n.p.). A work that was essentially right up Otté's street, with its combined interests in botany, zoology and geology, it was favourably received by critics as a piece well translated for the British public. The *Athenæum* declared that Otté 'has now rendered them more popularly available in an admirable translation, which preserves more of the freshness of the original than is usually to be met with'.[30] Brewster, in an equally indulgent account for the *North British Review*, noted that the volumes were 'written with much perspicuity and elegance; and they have been so admirably translated by Miss Otté, the well-known translator of Humboldt's Cosmos, that we never doubt, in the perusal of the work, that it was originally written in our own tongue'.[31]

She certainly shared Edmund Gosse's interest in the marine biology of the south coast. He nostalgically remembered beachcombing in summer 1874:

> The collecting mania seized us both, I rushed back to the inn, borrowed a hammer and a chisel, begged a pickle-bottle, and raced back. I found Miss Otte [sic] already on her knees working away with a hairpin and her fingers. [. . .] The abundant forms proved to be bellis and gemmacea, these two occurred in thousands. [. . .] I turned great stones, being rewarded by the strangest echinodermata – queer things [. . .] that wriggled and heaved. (Gosse 1931: I, 63)

It is difficult to imagine the elegantly reclining figure in a photo dating from Otté's period in Torquay (Figure 5.2) clambering among the rock-pools for wriggling, heaving specimens. It is only a little easier – given her advancing disability – to envisage her botanising on the chalk downs, finding several very rare plants, including 'some strange trefoils'.[32] Despite the onset of her illness, she was a reasonably frequent visitor to the Gosses' home, joining guests at dinner parties in the late 1870s and early 1880s that included aesthetes such as the Rossettis and Alma Tadema, and the translators Helen Zimmern and Anna Swanwick.

Figure 5.2 Photograph of Elise C. Otté, AR1438A, Hester Pengelly Collection, Torquay Museum.

Edmund Gosse ensured that Otté was awarded £100 from the Royal Literary Fund towards the end of her life on account of her being 'one of the most learned women I have ever known' (Thwaite 1984: 452).

Otté's infirmity had already begun to impinge upon her work by the time she was translating for Bohn. As his own 'Preface by the Publisher' to the *Views of Nature* indicates, much had gone awry between the cheerful, busy letter Otté had written to Hickling Prescott in April 1849 and the appearance of the translation in early 1850. While the genesis of the Sabines' *Aspects of Nature* can now be reconstructed only through private correspondence, Bohn ensured that the background to the *Views of Nature* was a public affair. He had originally aimed to produce a translation that would be quick on to the market and appear either before or contemporaneously with a rival Longman and Murray edition, thus automatically disarming critics' assertions that his was a derivative piece, a copy, a 'retranslation'. He now wanted his readers to understand why his edition had been completed only with some delay. By summer of 1849, Bohn reported, Otté had fallen ill and could not continue the translation scheduled for completion by early autumn. 'I had every reason to expect that I should fulfil my engagement to publish it in October last, or at latest in November,' Bohn confessed in his preface (*VN* vi). But, he continued, 'after much of the manuscript was prepared, the translator's indisposition [. . .] occasioned a serious suspension' (*VN* vi). The only solution was to draw on the expertise of other German-to-English translators to complete the work quickly. Bohn even knew how to market successfully this piecemeal approach to finalising the translation. Every sheet, he asserted, had been 'at least trebly revised' by a series of scholars to get the translation on to the market swiftly. Furthermore, the translation had received the sanction of the author himself:

> when I wrote to Baron Humboldt, more than a year and a-half ago, presenting him with my then unpublished edition of Cosmos, I announced my intention of proceeding with his other works, and consulted him on the subject. He replied in the kindest spirit, without intimating any previous engagement, and honoured me with several valuable suggestions. [. . .] In consequence of what I then presumed to be his recommendation, I determined to make the *Ansichten* my next volume, and announced it, long before any one [sic] else, though not at first by its English name. (*VN* vi–vii)

By closing the Preface with a facsimile of Humboldt's letter (Figure 5.3), probably illegible to most English readers (and, in any case, in French), Bohn submitted to public inspection convincing evidence that the right to translate the *Ansichten der Natur* lay with him. He could thus forestall any legal attack that Longman and Murray might wish to launch on the grounds of copyright infringement. Although Bohn conceded that his *Views of Nature* had indeed come 'late into the field' (*VN* vii), this did not detract in the long term from its popularity. Longman could boast to Humboldt by mid-November 1853 that 4,428 copies of the *Aspects of Nature* had sold (Humboldt 2009: 521), but it was the edition by Bohn's team of translators that was still being bought at the end of the century, following its reissue with George Bell & Sons in 1896.

The title and subtitle of the Bohn edition (Figure 5.4) suggest that the

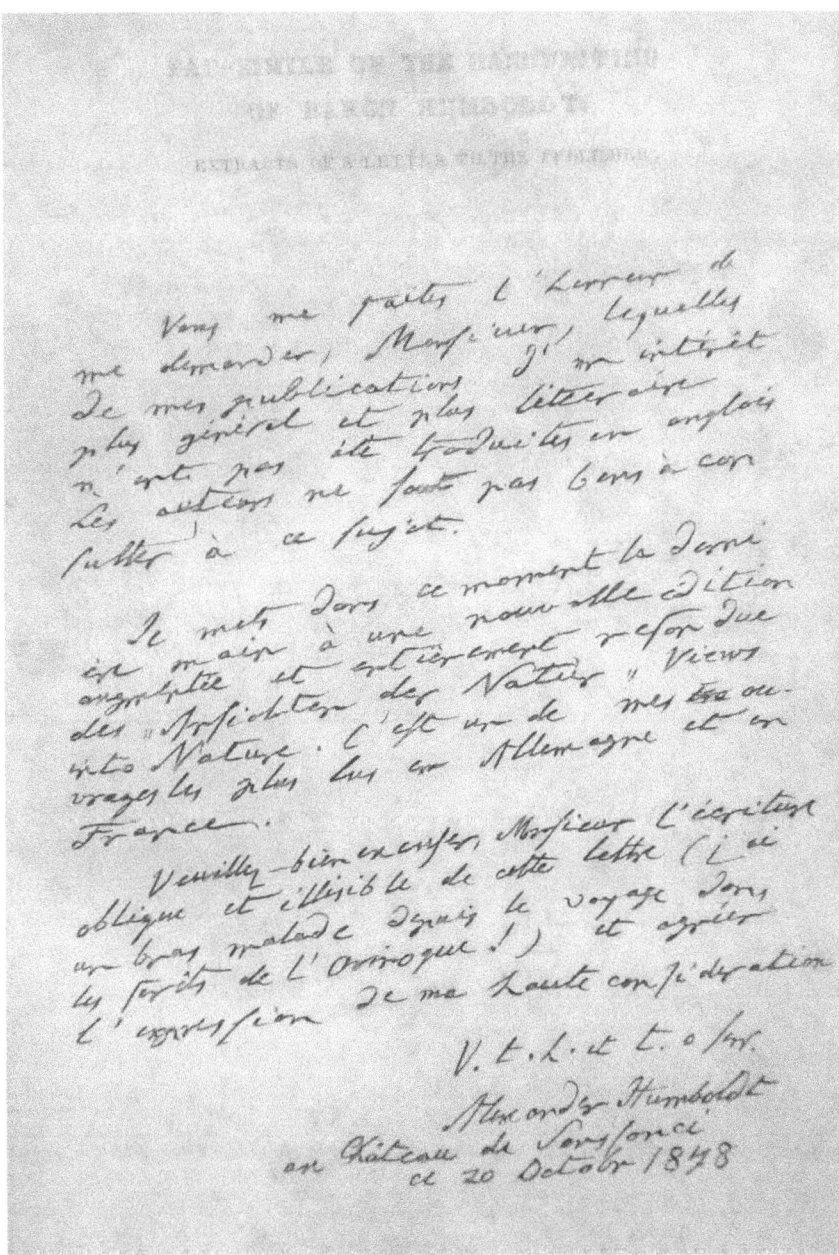

Figure 5.3 Facsimile of a letter of 20 October 1848 from Alexander von Humboldt to Henry George Bohn, printed at the end of the front matter, *Views of Nature* (London: Bohn, 1850).

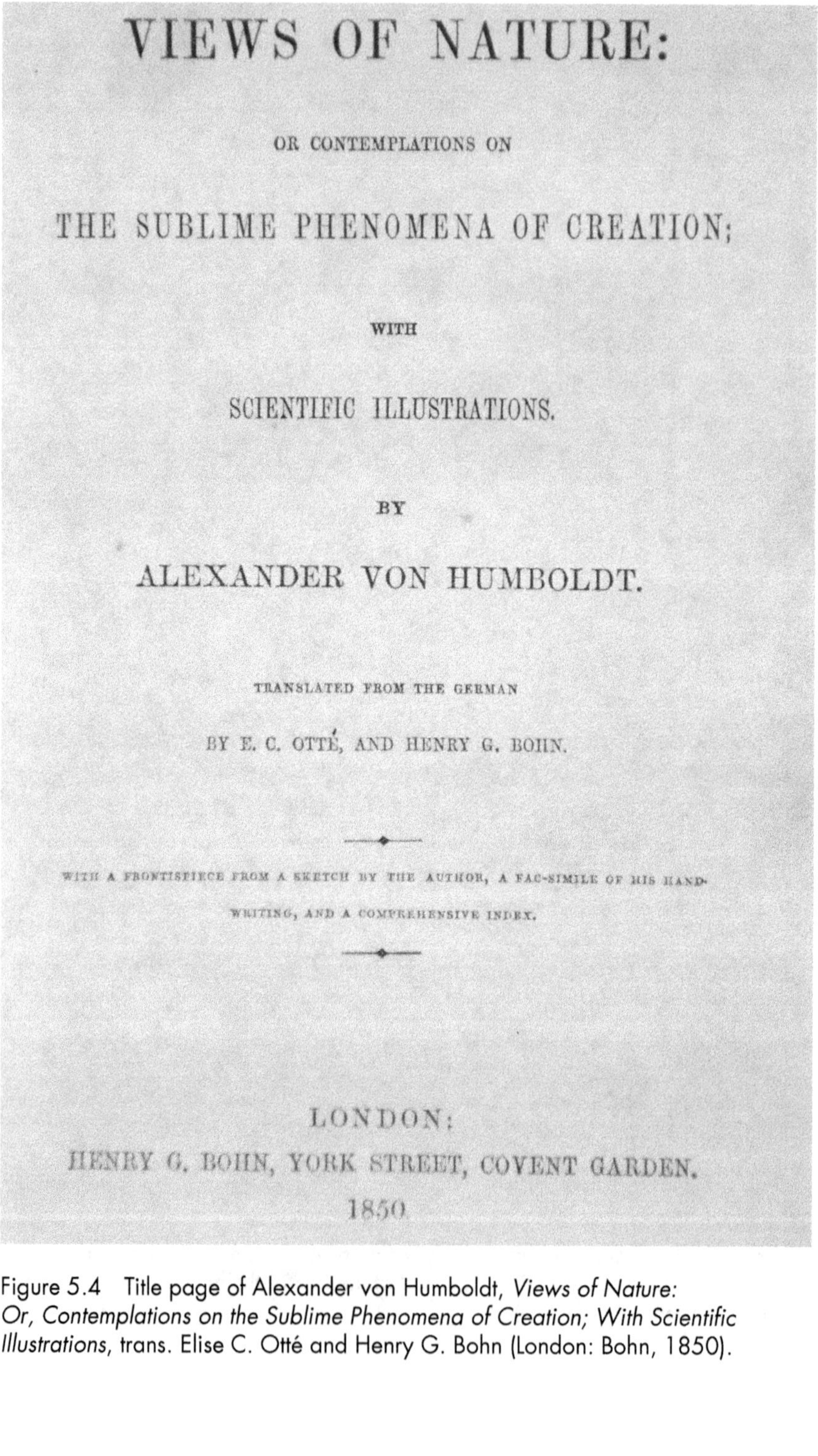

VIEWS OF NATURE:

OR CONTEMPLATIONS ON

THE SUBLIME PHENOMENA OF CREATION;

WITH

SCIENTIFIC ILLUSTRATIONS.

BY

ALEXANDER VON HUMBOLDT.

TRANSLATED FROM THE GERMAN

BY E. C. OTTÉ, AND HENRY G. BOHN.

───◆───

WITH A FRONTISPIECE FROM A SKETCH BY THE AUTHOR, A FAC-SIMILE OF HIS HAND-
WRITING, AND A COMPREHENSIVE INDEX.

───◆───

LONDON:
HENRY G. BOHN, YORK STREET, COVENT GARDEN.
1850.

Figure 5.4   Title page of Alexander von Humboldt, *Views of Nature: Or, Contemplations on the Sublime Phenomena of Creation; With Scientific Illustrations*, trans. Elise C. Otté and Henry G. Bohn (London: Bohn, 1850).

Views of Nature was oriented towards a slightly different public from the one that Longman and Murray had envisaged. Sabine's version presented itself as *Aspects of Nature, in Different Lands and Different Climates; with Scientific Elucidations*: this addition of 'Different Lands and Different Climates', a phrase nowhere present in the German *Ansichten der Natur, mit wissenschaftlichen Erläuterungen*, had stressed a geographical diversity implicit in Humboldt's work, while the focus on 'climate', followed by 'Scientific Elucidations', made Sabine's edition a book oriented towards those with (amateur) scientific interests. Otté and Bohn, by contrast, exploited the creative potential gained from translating Humboldt's work unfettered by the demands of the author himself, to inject an overt religiosity into their work by titling it *Views of Nature: Or Contemplations on the Sublime Phenomena of Creation; With Scientific Illustrations*. The use of the words 'view' and 'sublime' gave the title aesthetic associations that encouraged the reader to consider this less a primarily scientific piece than a collection that would appeal to specialists and the general reader alike.

Presentation and Paratext: *Aspects of Nature* and *Views of Nature*

The two English translations of the *Ansichten der Natur* needed clear points of divergence to cast them as individual pieces in their own right. The layout of the texts and the translators' handling of footnotes and endnotes already gave them a different overall appearance. Variations in the translations themselves also lent Sabine's *Aspects of Nature* and Otté and Bohn's *Views of Nature* their own specific characters. Both essentially adopted the same translation strategy, 'domesticating' Humboldt's writing for a British audience. They omitted the dedication to Humboldt's brother Wilhelm, who had been a diplomat in Rome in May 1807 as the first edition came off the press – a reference to a person and a time long past (he had died in 1835), which only served to highlight how dated parts of the *Ansichten der Natur* really were.

Where Bohn used footnoting for brief textual references and kept the remainder of the notes in the same sections of the 'Scientific Illustrations and Additions' as in the original, Longman and Murray, almost without exception, used endnotes. This slightly different organisation of the paratextual material, coupled with narrower line spacing, was not inconsequential. Bohn's edition went some way to righting the huge imbalance between text and annotations that had been so viciously attacked by Humboldt's British critics. In his edition, the notes on the 'Plateau of Caxamarca' were reduced to just half the length of the chapter (they were six-tenths in the Longman and Murray edition), in the essay on the structure of volcanoes to a mere three pages (where Sabine's translation had six) and in the 'On Steppes and Deserts' piece to a slimmer 130 pages in the Bohn edition (where his rivals had 177 sides of supplementary material).

Sabine's translation left a number of titles and the opening quotation from Schiller in German. By contrast, the rendering by Otté and Bohn offered a less foreignising version by putting these in English. Additional footnotes marked as 'By the Editor' also gave readers of *Views of Nature* a greater sense that they were being steered through Humboldt's writing by a commentator familiar with different facets of cultural and scientific life in Britain. In stark contrast to Elizabeth Sabine's near-'silence' and 'invisibility' in the text for Longman and Murray – her 'Note by the Translator' extended to a mere paragraph on the conversion of units of length, temperature and longitude – the editor–translator Bohn was present in ways that extended beyond the four-page Preface. His notes cast him as widely read, yet not excessively learned, and he used them to demonstrate his awareness that the science addressed in Humboldt's *Ansichten der Natur* catered to both specialist and more amateur interests. The mere handful of footnoted comments that he appended to his 452-page edition were hardly intrusive. But they showed familiarity with Edward Gibbon's *History of the Decline and Fall of the Roman Empire*, Arnold Ludwig Heeren's work on the common origins of different nations, Johann Friedrich Blumenbach's Latin account of craniology, and 'Grindlay's *Scenery of the Western Side of India*' (VN 5, 15, 172). Cynics might have contended that this was just another marketing ploy cleverly used by Bohn to sell his own books, since Gibbon's seven-volume *History* appeared with Bohn between 1853 and 1855, as did a whole raft of works by Heeren in the 1840s. But more general comments that modern naturalists 'affirm that *all* bats are insectivorous', or that fossil remains of the *Colossochelys atlas* (a giant prehistoric tortoise) could now be found in the British Museum, confirmed that this work was aimed at a more general readership interested in natural history (*VN* 15, 222).

Differences of Perspective: *Ansichten der Natur*, *Aspects of Nature* and *Views of Nature*

British reviews of the two English translations indicated that the appeal of Humboldt's *Ansichten der Natur* lay less in its scientific detail than in its ability to portray the natural world imaginatively. The aesthetic categories of the sublime and the picturesque on which Humboldt repeatedly drew were not, however, always accorded the same prominence by his two sets of British translators. In his essay on the plateau of Caxamarca, Humboldt had, for example, looked down into the valley and described the view from above. It was, he wrote, a charming prospect because it was 'von einem Flüsschen durchschlängelt' (*AdN* 454): in the words of Sabine, it was a plain through which 'a small river winds' (*AN* II, 287), while in the Bohn edition it was one through which 'winds a serpentine rivulet' (*VN* 407). The word 'serpentine' emphasised the sinuous, curving course of the stream that the plainer 'winds' only more briefly evoked. The Bohn edition explicitly allied Humboldt's work with a picturesque idiom. The serpentine line, asymmetric and emblematic of nature's liberty – Hogarth's 'line of beauty' – was recognised well into the nineteenth century by devotees of William Gilpin's aesthetics, like Thomas Fosbrooke, who described the River Wye as consisting of 'rude and broken elevations, and rough valleys, irregularly serpentine, adorned with purling streams' (Fosbrooke 1822: 16). Otté and Bohn therefore deliberately employed a language familiar to British readers acquainted with the literature of landscape aesthetics, to ally their translation with other accounts that exploited the possibilities for imaginative 'transport' that such descriptions offered.

The vibrancy with which Humboldt had imbued his landscapes also received rather different treatment at the hands of his British translators. From the outset, Otté and Bohn were keen to reinforce nature's vitality, personifying it by attributing human verbs of action to natural phenomena. For example, in the essay 'On Steppes and Deserts', Humboldt had described how vegetation could proliferate in the forests of the Orinoco, where it would never suffer destruction by humans and could be halted in its growth only by tightly winding creepers. Humboldt's 'üppiger Andrang schlingender Gewächse' (*AdN* 27) was translated by Sabine as the 'pressure of the luxuriant climbers' (*AN* I, 15), whereas Otté and Bohn opted for the freer 'luxuriant embrace of the plants' (*VN* 12). They therefore personified the climbing plants by enabling them to 'embrace' the host vegetation around which they twisted, and presented nature as a more powerful force than was implicit in Sabine's 'pressure'.

Humboldt was particularly fascinated by the fecundity of plant life in extreme conditions. Describing the arid climate around the Haruj Mountains in Libya, he had wondered at the richness of the Oasis of Siwa with its many date trees, surrounded by a sea of burning-hot sand. In the desert encircling the oasis, he noted that: 'Kein Thau, kein Regen benetzt diese öden Flächen und entwickelt im glühenden Schooß der

Erde den Keim des Pflanzenlebens' (*AdN* 17). Otté and Bohn translated this as: 'Neither dew nor rain refreshes these barren wastes, or unfolds the germs of vegetation within the glowing depths of the earth' (*VN* 2). This was subtly different from Sabine's version: 'Neither dew nor rain bathe [*sic*] these desolate plains, or develope [*sic*] on their glowing surface the germs of vegetable life' (*AN* I, 3). While neither translation had taken up the image of seeds germinating in the glowing lap – 'im glühenden Schooß' – of Mother Nature, the Bohn edition had cast rain as a force that could penetrate both soil and seeds, unpacking the latter and facilitating their sprouting. Although 'unfolds' seems at first sight a rather free translation of Humboldt's original 'entwickelt' (normally taken to mean 'develops', as in Sabine's version), reading this verb as 'ent-wickeln' – in the sense of 'un-wrap' or 'un-wind' – enabled Otté and Bohn to convey more successfully how rain causes seeds to swell and burst from their casing. They could therefore forge a more direct link between rain and germination, which cast nature as a dynamic and life-giving force.

Otté and Bohn also employed to good effect the multiple associations a word could carry to sustain their personification of nature in their English translation. As Humboldt described how the rainy season brought dramatic changes to the steppes, he focused on its effect upon 'sensitive' plants such as the mimosa:

> Vom Lichte gereizt, entfalten krautartige Mimosen ihre gesenkt schlummernden Blätter, und begrüßen die aufgehende Sonne, wie der Frühgesang der Vögel und die sich öffnenden Blüthen der Wasserpflanzen. (*AdN* 31–2)

Sabine translated it as:

> The herbaceous mimosas, with renewed sensibility to the influence of light, unfold their drooping slumbering leaves to greet the rising sun; and the early song of birds, and the opening blossoms of the water plants, join to salute the morning. (*AN* I, 20)

Otté and Bohn, by contrast, put this:

> Excited by the power of light, the herbaceous Mimosa unfolds its dormant, drooping leaves, hailing, as it were, the rising sun in chorus with the matin song of the birds and the opening flowers of aquatics. (*VN* 16)

The word 'excited' in the translation by Otté and Bohn, with its dual meaning of stimulated activity and aroused emotions, sustained their personification of the plant as an active force 'hailing' the sun 'in chorus' with other life forms as the new day unfolded.

Modes of viewing were also cast rather differently in the editions by

Bohn and by Longman and Murray. The traveller's gaze tended to be more penetrating and wide-ranging in the Bohn edition, encompassing a greater visual field than that of the observer in the translation for Longman and Murray. Humboldt had opened his essay on plant physiognomy by expressing sublime awe at the seeming boundlessness of natural diversity and its ability to generate life in the most inhospitable of climes. In Sabine's translation, Humboldt had observed:

> When the active curiosity of man is engaged in interrogating Nature, or when his imagination dwells on the wide fields of organic creation, among the multifarious impressions which his mind receives, perhaps none is so strong and profound as that of the universal profusion with which life is everywhere distributed. (*AN* II, 3)

Otté and Bohn opted for:

> When the active spirit of man is directed to the investigation of nature, or when in imagination he scans the vast fields of organic creation, among the varied emotions excited in his mind there is none more profound or vivid than that awakened by the universal profusion of life. (*VN* 210)

The 'wide spaces' – 'die weiten Räume' of the original German – were scaled up in the later translation to the sublime 'vast fields', while the observer in the edition by Otté and Bohn was portrayed 'scanning' the scene, rather than simply 'dwelling' on it, which gave it a greater sense of breadth (*AdN* 237). In their reading, the curious observer was also 'excited' by 'varied emotions', conveying something of Humboldt's own fascination at the natural world, which was formulated by Otté and Bohn in a much more active fashion than Sabine's rather impersonal 'multifarious impressions which his mind receives'.

This persistent appeal to the emotions more generally characterised the Bohn edition of the *Ansichten der Natur*. In his essay on steppes and deserts, Humboldt had commented that as whirling currents of sand-laden air bear down on the traveller, the horizon seems to draw closer, constraining the spirit of the observer: 'Der Horizont tritt plötzlich näher. Er verengt die Steppe, wie das Gemüth des Wanderers' (*AdN* 30). Opting for a close translation of the original, Sabine gave: 'The horizon draws suddenly nearer; the Steppe seems to contract, and with it the heart of the wanderer' (*AN* I, 18). Otté and Bohn chose: 'The horizon suddenly contracts, and the heart of the traveller sinks with dismay as the wide Steppe seems to close upon him on all sides' (*VN* 14). By providing a freer rendering, Otté and Bohn could emphasise the claustrophobic feel of the scene, with the sand closing in 'on all sides', and by explicitly recording the traveller's sense of 'dismay' as the landscape seems to 'close upon him', they heightened the affective power of the description over the reader.

Similarly, in a passage where Humboldt stood gazing back to the lush forests of the valleys and on to the dry monotony of the steppes, he tried to emphasise the conspicuously different landscapes around him: 'Aus der üppigen Fülle des organischen Lebens tritt der Wanderer betroffen an den öden Rand einer baumlosen, pflanzenarmen Wüste' (*AdN* 15). Sabine rendered the sentence as: 'Fresh from the luxuriance of organic life, he treads at once the desolate margin of a treeless desert' (*AN* I, 1). She apparently overlooked – or considered unimportant – Humboldt's adverb 'betroffen', meaning 'affected' or 'moved', which had expressed the traveller's surprise at the sudden change in the landscape. Otté and Bohn made of this sentence: 'From the rich luxuriance of organic life, the astonished traveller suddenly finds himself on the dreary margin of a treeless waste' (*VN* 1). The placing of 'astonished' close to 'suddenly' in this sentence reinforced the sharp contrast that Humboldt had consciously built into his narrative to give it interest.

The deliberately contrastive nature of Humboldt's prose was a key element in his landscape descriptions, since it enabled him to inject structure into the scenes being described. While Sabine was alert to the distinctions between fore- and background in Humboldt's portrayal of the savannah, she did not play up the contrasts to quite the same degree as Otté and Bohn. In his description of a scene from his essay on the cataracts of the Orinoco, Humboldt immediately juxtaposed 'die Ferne' [the distance] and 'nah' [near] in the opening sentence:

> So die Ferne; nahe umher ist alles öde und eng. Im tief gefurchten Thale schweben einsam der Geier und die krächzenden Caprimulge. An der nackten Felswand schleicht ihr schwindender Schatten hin. (*AdN* 190)

Although Sabine did not ignore this contrast, she made it less stark, juxtaposing 'distance' with the comparative 'nearer':

> Such is the distance; the nearer prospect is desolate, and closely hemmed in by high and barren rocks. All is motionless save where the vulture or the hoarse goatsucker hover [*sic*] solitarily in mid-air, or as they wing their flight through the deep-sunk ravine, their silent shadows are seen gliding along the face of the bare rocky precipice until they vanish from the eye. (*AN* I, 228)

Otté and Bohn brought the foreground still closer to the reader by using 'immediately around':

> Such is the distant view; but immediately around all is desolate and contracted. In the deep ravines of the valley moves no living thing save where

the vulture and the whirring goat-sucker wing their lonely way, their heavy shadows gleaming fitfully past the barren rock. (*VN* 170)

Where they were more successful at distinguishing between the distant and the close-up, Sabine conveyed more effectively the oppressive sublimity of the 'deep-sunk ravine' with its 'high and barren rocks', and also attached greater importance to the dynamism of the birds and insects flying through it, even if she opted for the pleonastic 'winged their flight'.

Creative Solutions

The rich complexity of the scenes portrayed in the *Ansichten der Natur* often derived from the juxtaposition of different concepts, conveyed through compound nouns and metaphorical allusion. These were particularly difficult to convey neatly in English. In his essay on the plateau of Caxamarca, Humboldt discussed the relative levels of civilisation that the Peruvian empire had achieved. He reflected on how one particular step pyramid had essentially been copied from that built by an earlier culture, and one civilisation was founded both culturally and architecturally upon another: 'So dringt man durch jegliche Civilisationsschicht immer in eine frühere ein' (*AdN* 445). The Bohn edition simply had: 'Thus through every stage of civilization, we pass into an earlier one' (*VN* 398). Sabine, by contrast, took more trouble to convey the imagery conveyed by the compound noun 'Civilisationsschicht' [level of civilisation]. She gave this sentence as: 'Thus as we penetrate through each successive stratum of civilisation we arrive at an earlier one' (*AN* I, 276). This geologically connoted language well suited Humboldt's discussion of the various cultural 'strata' that were uncovered as the archaeologist and historical anthropologist worked chronologically backwards to reveal the different layers left by past civilisations, cultures and societies.

Humboldt's narrative also played creatively with language through the use of compound nouns such as 'Naturgefühl' [literally: 'feeling of nature'] and 'Naturwahrheit' [literally: 'truth of nature']. These had a brevity that was hard to capture in English. The essay on the 'Nocturnal Life of Animals in the Primeval Forest' was particularly challenging in this regard. In the opening sentence, Humboldt declared:

> Wenn die, stammweise so verschiedene Lebendigkeit des Naturgefühls, [. . .] die Sprachen mehr oder minder mit scharf bezeichnenden Wörtern für Berggestaltung, Zustand der Vegetation, Anblick des Luftkreises, Umriß und Gruppirung der Wolken bereichern; so werden durch langen Gebrauch und durch litterarische Willkühr viele dieser Bezeichnungen von ihrem ursprünglichen Sinne abgewendet. (*AdN* 215)

This passage itself problematised the nature of language, the uninhibited growth of terminology in the natural sciences and the literary deployment of such terms for different ends. Otté and Bohn described this word 'Naturgefühl' as 'the faculty of appreciating nature' (*VN* 191), while Sabine rose to the challenge differently by circumscribing the meaning of this term using word pairs:

> If the vivid appreciation and sentiment of nature which differ so greatly in nations of different descent, [. . .] have rendered different languages more or less rich in well defined and characteristic expressions denoting the forms of mountains, the state of vegetation, the appearance of the atmosphere, and the contour and grouping of the clouds, it is also true that long use, and perhaps their arbitrary employment by literary men, have diverted many such words from their original meaning. (*AN* I, 259)

The expression 'Lebendigkeit des Naturgefühls' became 'vivid appreciation and sentiment of nature', as she attempted to convey in a paraphrase all the facets of meaning she felt these terms to hold, both aesthetic (in 'appreciation') and emotional (in 'sentiment').

The compound noun 'Naturwahrheit' was also problematic. As Humboldt again reflected on how language could be used to convey what the traveller has observed and experienced, he argued that simplicity of narration, coupled with a close examination of the natural phenomenon being described, best enabled the object to be depicted well.

> Speech is enriched and animated [he observed (in Sabine's translation)] by everything that tends to and promotes truth to nature, whether in rendering the impressions received through the senses from the contemplation of the external world, or in expressing thoughts, emotions, or sentiments which have their sources in the inner depths of our being. (*AN* I, 261)

She made of 'alles [. . .], was auf *Naturwahrheit* hindeutet' (*AdN* 216) the phrase 'everything that tends to and promotes truth to nature'. Her rival translators, by contrast, would opt for 'everything which bears the true impress of nature', personifying nature as a figure who gave her stamp of authenticity to everything that truthfully conveyed the innermost sensations of the observer (*VN* 192).

By comparing and contrasting the different choices made by the translators, we come slightly closer to assessing whether the Bohn edition was really a retranslation based on Sabine's 'original' for Longman and Murray. Although geographical fact, mathematical calculations and experimental observations were unlikely to be translated in radically different ways, the more creative passages did offer room for linguistic manœuvre, which Otté and Bohn exploited to the full. Some solutions did not always fit as well as those that Sabine had already selected, but metaphorical expressions tended to be better rendered in the Bohn edition, perhaps because Otté and Bohn could use the Longman and Murray edition as a basis from which to devise a better solution. In the essay 'On Steppes and Deserts', Sabine had described the onslaught of the Mongol hordes in this way: 'Thus went forth from the Mongolian deserts a deadly blast, which withered on Cisalpine ground the tender long-cherished flower of art' (*AN* I, 6). Otté and Bohn sustained the metaphor better by writing: 'Thus swept a pestilential breath from the Mongolian deserts of the fair Cisalpine soil, stifling the tender, long-cherished blossoms of art!' (*VN* 5). Their choice of the words 'swept', 'breath' and 'stifling' sustained the imagery of suffocation more strongly. Similarly, in the chapter 'Ideas for a Physiognomy of Plants', where Humboldt had described how the quality of light changed in the heat, Sabine had talked of 'The azure of the sky, the lights and shadows, the haze resting on the distance' (*AN* II, 13). Otté and Bohn's version was better honed to 'The azure of the sky, the effect of light and shade, the haze floating on the distant horizon' (*VN* 217). They captured more vividly how the scene shimmered in the heat by using the line of the horizon as a fixed point above which the haze floated, rather than simply leaving it to 'rest' on a 'distance'.

Humboldt's awareness of the creative possibilities of the German language urged his translators to be ingenious in other ways in their translation decisions. Echoing the technique applied by Williams in her translation of the *Relation historique*, Sabine and Otté resourcefully borrowed turns of phrase and vocabulary from literary sources of the period. As in Williams's *Personal Narrative*, these translations were shot through with Romantic vocabulary. Otté and Bohn in particular used turns of phrase like the 'lofty summits of the Andes' (*VN* x). While Sabine clearly had a sense of the lyrical qualities of the German original, at times she would reach for rather uncommon pairings such as 'carbonized turfy' (*AN* I, 17), where Otté and Bohn offered the less jarring 'parched sward' (*VN* 14). The Longman and Murray edition also included expressions that echoed literary phrasing of the period, so that Sabine offered 'skyey canopy' (*AN* I, 23) – also to be found in Scott's *The Monastery* (1820) – which Otté and Bohn had suffused with greater religiosity as 'the wide canopy of heaven' (*VN* 19). Similarly, Sabine talked of the 'changeful destinies' of man (*AN* I, 25) – a word pair found in poetry and prose by Elizabeth Montagu, Edward Bulwer-Lytton and Charles Mackay – set against Otté and Bohn's 'changing destinies of nations' (*VN* 20). Elsewhere, however, Otté and Bohn showed themselves well able to rise to the stylistic challenge of weaving more obviously 'literary' solutions into their writing. Their 'embowering shade of the beechen grove' (*VN* 219) echoed a phrase from Scott's *Kenilworth* (1821) and offered a solution far more lyrical than Sabine's 'dark shade of the beech' (*AN* II, 16).

Nature and Spirituality

Brewster had ended his critical reading of the *Aspects of Nature* for the *North British Review* with the observation that 'only the irreligious man [...] can blindly gaze upon the loveliness of material nature', and argued that of all the sciences, it was geology that could best call upon the 'depths of the earth to praise the Lord'.[33] But how could Brewster consider that Sabine's English translation presented nature as the handiwork of the Almighty when references to God were palpably missing from the German original? Humboldt had hinted that his depiction of nature was underpinned by a religious world view, noting in his essay on the 'Nocturnal Life of Animals' that '[e]verything proclaims a world of active organic forces. [...] It is one of the many voices of nature revealed to the pious and susceptible spirit of man' (*VN* 201). The 'pious and susceptible spirit', as the edition by Bohn had it – or 'the sensitive and reverent ear of her true votaries', in the words of Sabine (*AN* I, 261) – was a faithful translation of Humboldt's original 'frommen, empfänglichen Gemüthe des Menschen' (*AdN* 226). But such subtle, implicit religious references were rather too few and far between for their presumed readership – a mid-Victorian audience that expected to be presented with the natural world as the product of a beneficent Creator.

On several occasions, Sabine and Otté took the issue into their own hands, supplying translation solutions that departed quite unambiguously from the original to steer this work away from controversy. In his essay on the steppes, Humboldt had described the movement and adaption of plants and animals to their climate thus: 'Ein solcher Anblick erinnert unwillkührlich den ernsten Beobachter an die Biegsamkeit, mit welcher die alles aneignende Natur gewisse Thiere und Pflanzen begabt hat' (*AdN* 33). Sabine formulated it as: 'Such a sight reminds the thoughtful observer involuntarily of the capability of conforming to the most varied circumstances, with which the all-providing Author of Nature has endowed certain animals and plants' (*AN* I, 21). The Bohn edition gives: 'This spectacle involuntarily reminds the contemplative observer of the adaptability granted by an all-provident nature to certain animals and plants' (*VN* 17). Although Otté and Bohn rendered 'die alles aneignende Natur' as 'an all-provident nature', and therefore echoed notions of divine providence, Sabine's translation sent stronger signals that nature was the product of divine creation. Sabine made an explicit shift from making 'nature' the power that brings about adaptability in plants and animals to the 'Author of Nature'. In a nineteenth-century context, her capitalisation would have pointed unequivocally to the Christian deity, even if she neatly sidestepped directly using the word 'God'.

While there was outspoken opposition to theology by scientific naturalists in the Victorian age, Fyfe reminds us that there were also plenty of devout (and prominent) scientific practitioners (Fyfe 2008: 121). Indeed, the 'inevitable conflict' between science and religion was not in fact a reality since, Fyfe argues, the contrast was not between 'science' on the one hand and 'religion' on the other, but between two competing visions of the sciences, one Christian and the other secular (Fyfe 2008: 123). Otté and Bohn were less explicit in their enhancement of the spiritual message of the source text. Approaching the problem rather differently – perhaps a case of *force majeure*, since Sabine had got there before them – they carefully drew on expressions that echoed the language of the King James Bible and silently imbued the *Ansichten der Natur* with a religious feel that would have been distinctly familiar to British readers. In Humboldt's chapter on the steppes and deserts of the globe, they exploited the range of meanings that could be attributed to the word 'geistig', which can signify 'intellectual', 'mental' or 'spiritual'. The original German had: 'erfüllt die Steppe das Gemüth mit dem Gefühl der Unendlichkeit, und durch dieses Gefühl, wie den sinnlichen Eindrücken des Raumes sich entwindend, mit geistigen Anregungen höherer Ordnung' (*AdN* 16). This was translated by Sabine as 'the Steppe fills the mind with the feeling of infinity; and thought, escaping from the visible impressions of space, rises to contemplations of a higher order' (*AN* I, 2). Otté and Bohn opted for 'the Steppe fills the mind with a sense of the infinite, and the soul, freed from the sensuous impressions of space, expands with spiritual emotions of a higher order' (*VN* 2). Their subtle injection of a religious feel into this sentence – the 'spiritual emotions' coupled with the use of 'soul' rather than 'mind' – nudged Otté and Bohn's *Views of Nature* away from seeming atheistic, pantheistic or downright irreligious.

In the opening to the chapter 'On the Cataracts of the Orinoco', Humboldt had remarked on the spiritual relationship between human beings and nature that 'alles steht in altem, geheimnißvollem Verkehr mit dem gemüthlichen Leben des Menschen' (*AdN* 172). This appeared in the Longman and Murray edition as '[a]ll that determines the character of a landscape [. . .] all are in antecedent mysterious communication with the inner feelings and life of man' (*AN* I, 208) and in the Bohn volume '[a]ll stand alike in an ancient and mysterious communion with the spiritual life of man' (*VN* 154). The words 'communion' and 'spiritual' again imbued Otté and Bohn's translation with a religiosity that was not always to be found in Sabine's rendering of this text.

Conclusions

Aaron Sachs remarks that purchasing an English translation of Humboldt's *Ansichten der Natur* in the nineteenth century did not always mean having to choose between the edition by Longman and Murray and that by Bohn: the American essayist and philosopher Henry David Thoreau purchased both translations when they appeared in the States in 1850 (Sachs 2006: 96). On reading them, though, Thoreau would have gained subtly different impressions of Humboldt's aesthetic and spiritual preoccupations. In Otté's translation, greater attention was paid to the picturesque element of Humboldt's impressions of natural scenery, and in her stirring evocations of landscape nature tended to be presented more vividly and with greater dynamism. Donald McCrory rightly emphasises that Humboldt was constantly thinking in terms of 'scenes, portraits, views, depictions and perspectives' (McCrory 2010: 165). Humboldt's aesthetic representation of nature was a play of painterly elements such as scale, perspective, light and shade. This was particularly well captured by Otté, in descriptions that demonstrated she had understood how these compositional principles informed Humboldt's narrative project.

Sabine, by contrast, was more concerned to convey the detail of Humboldt's prose. She recognised that his poetic descriptions of landscape pertained to places he had also explored as a travelling naturalist keen to make global connections. Her attempts to portray fully the figurative dimension of his prose meant that she sometimes sacrificed brevity to communicate all the aspects of meaning she perceived were present in the German original. By using pairs of adjectives or paraphrasing to ensure that nothing was lost in translation, she did not necessarily provide the most economical translation solutions, but they were always thorough and careful, and almost without exception succeeded in putting Humboldt's complex sentence structure into elegant English. Whether Otté and Bohn produced their translation with an eye on the rendering by Sabine, or completed their version first and then revised it with her edition to hand, is now impossible to tell. Either way, this could account for why some passages that required the figurative use of language were done better in the Bohn edition, simply because Sabine's translation supplied Otté with a first version that could be borrowed and honed.

If Bohn's Preface is to be believed, then Otté was well on her way to finishing the *Views of Nature* before the *Aspects of Nature* appeared. If so, she would have been presented with the doubly challenging situation of having to revise her translation to ensure it still conveyed the source text accurately yet avoided overlap with the edition published by Longman and Murray. That readers understood these as two different renderings of the same work is testament to the skill of Otté and Bohn at exploiting the creative possibilities of Humboldt's more poetical descriptions and the lexical range of the English language to produce an alternative, competing version that still passed muster with British readers.

Although the translations by Sabine and by her rivals Otté and Bohn necessarily differed in terms of the lexical choices made, on a broader strategic level they had one key aspect in common. They shared a desire to reinforce the spiritual associations in the *Ansichten der Natur* so that it would meet the expectations of more conservative British readers at mid-century. They were both, in different ways, successful at edging the *Views of Nature* and the *Aspects of Nature* away from the fierce controversies that would arise following the publication of *Cosmos* in English translation.

Notes

1. In the Preface to its authorised French translation of 1851, the *Tableaux de la Nature*, Charles Galusky registered Humboldt's profound sadness at witnessing a similar degree of political turmoil some forty years after the *Ansichten* had first appeared (Humboldt 1851: I, vii).
2. Gide Fils in Paris brought out a translation by J. B. B. Eyriès in 1808 and an updated edition by the same translator in 1828.
3. *The Athenæum*, 1147 (1849), p. 1060.
4. *Literary Gazette*, 1711 (1849), p. 807; *Examiner*, 2175 (1849), p. 628.
5. *Daily News*, 1053 (1849), p. 2.
6. *Daily News*, 1053 (1849), p. 2.
7. *The Athenæum*, 1147 (1849), p. 1060.
8. *North British Review*, 12:23 (1849), p. 263.
9. *North British Review*, 12:23 (1849), p. 263.
10. For a detailed discussion of Schiller's lack of appreciation for Humboldt, see Schwarz 2003.
11. For a modern retranslation of this text by K.-H. Glassmeier and B. T. Tsurutani in the *History of Geo- and Space Sciences* (2014) 5: 11–62, see <https://www.hist-geo-space-sci.net/5/11/2014/hgss-5-11-2014.pdf> (last accessed 30 January 2018).

12. JMA, Ms.41052, letter of 13 March 1849 from Edward Sabine to John Murray.
13. JMA, Ms.41052, letter of 13 March 1849 from Edward Sabine to John Murray.
14. JMA, Ms.41052, letter of 13 March 1849 from Edward Sabine to John Murray.
15. JMA, Ms.41052, letter of 14 March 1849 from Edward Sabine to John Murray.
16. JMA, Ms.41052, letter of 30 March 1849 from Edward Sabine to John Murray.
17. JMA, Ms.41052, letter of 2 April 1849 from Edward Sabine to John Murray.
18. JMA, Ms.41052, letter of 14 March 1849 from Edward Sabine to John Murray.
19. JMA, Ms.41052, letter of 14 March 1849 from Edward Sabine to John Murray.
20. *Daily News*, 1139 (1850), p. 8.
21. JMA, Ms.41052, letter of 14 March 1849 from Edward Sabine to John Murray.
22. Massachusetts Historical Society, William Hickling Prescott Papers, letter of 15 April 1849 from E. C. Otté to William Hickling Prescott.
23. Torquay Museum, Hester Pengelly Collection, AR1521, letter of 16 May 1850 from William Hickling Prescott to Elise C. Otté.
24. *The Athenæum*, 3975 (1904), p. 15.
25. Minutes of the St Andrews Literary and Philosophical Society, 1838–61, University of St Andrews Library, MS UY8525/1, 160r, 160v, 161v.
26. *Journal of the Marine Biological Association of the United Kingdom*, 1 (1887), p. 202. See also Leighton (1901: 188): 'Years ago Mdme. Otté was well known for her work on marine zoology.'
27. *The British Medical Journal*, 1.581 (1872), p. 198.
28. *The Athenæum*, 3977 (1904), p. 83.
29. See the letters in Torquay Museum, Hester Pengelly Collection, AR3806.
30. *The Athenæum*, 1575 (1858), p. 14.
31. *North British Review*, 28 (1858), p. 160.
32. Cambridge University Library, Gosse Papers, Letters between Philip and Edmund Gosse, Add. 7018: 59: letter of 3 August 1874 from Edmund Gosse to his father.
33. *North British Review*, 12.23 (1849), p. 264.

Chapter 6

Cosmos: The Universe Translated

'We all know that it has become fashionable to talk of Cosmos. Not to have read it is to be a boor,' declared a reviewer for the *Literary Gazette* in March 1849.[1] But as this critic also recognised, Humboldt's *Cosmos* could hardly be referred to in the singular 'it'. Three different, competing, translations of the opening volumes of Humboldt's *Kosmos: Entwurf einer physischen Weltbeschreibung* were now out on the British book market, and their appearance heralded the start of a publishing war that would last the best part of a decade. The first volume of *ΚΟΣΜΟΣ: A General Survey of the Physical Phenomena of the Universe*, translated by the Bristol eye-surgeon Augustin Prichard, had already appeared in 1845 with the Regent Street publisher Hippolyte Baillière. The second volume came out in 1848 and corresponded to the second part of the German original, which had itself reached booksellers' shelves in autumn the previous year. Just behind Baillière in the race to publish *Cosmos* were Longman and associates and Murray, who had brought out their first volume of *Cosmos: Sketch of a Physical Description of the Universe* in 1846 and the second in 1848, both translated under Edward Sabine's superintendence, but primarily the work of his wife Elizabeth.[2] Although these two different editions occupied the market concurrently, the real threat came from the translation made by Otté for Bohn's 'Scientific Library' series, similarly titled *Cosmos: A Sketch of a Physical Description of the Universe*, which entered on to the market in 1849.

The reviewer for the *Literary Gazette* marshalled an extract from *Kosmos* and its three English translations into four columns covering two full pages of the journal (Figure 6.1).[3] Readers were explicitly invited to compare the different versions, identify their stylistic and linguistic variations and judge which best rendered Humboldt's original German. As Bohn and his rivals Longman and Murray jostled for a greater market share in the work, each laid claim to providing the most authoritative, comprehensive and accurate translation. In the Translator's Preface to

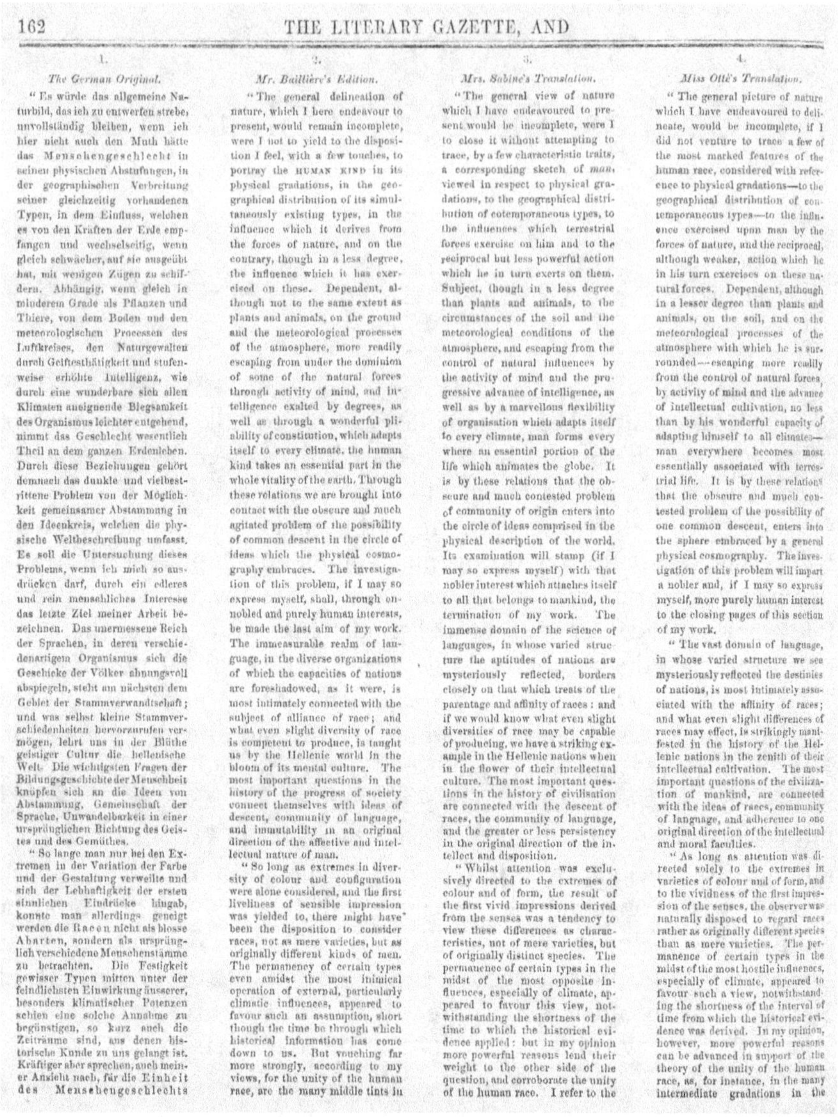

162 THE LITERARY GAZETTE, AND

1.

The German Original.

"Es würde das allgemeine Naturbild, das ich zu entwerfen strebe, unvollständig bleiben, wenn ich hier nicht auch den Muth hätte das Menschengeschlecht in seinen physischen Abstufungen, in der geographischen Verbreitung seiner gleichzeitig vorhandenen Typen, in dem Einfluss, welchen es von den Kräften der Erde empfangen und wechselseitig, wenn gleich schwächer, auf sie ausgeübt hat, mit wenigen Zügen zu schildern. Abhängig, wenn gleich in minderem Grade als Pflanzen und Thiere, von dem Boden und den meteorologischen Processen des Luftkreises, den Naturgewalten durch Geistesthätigkeit und stufenweise erhöhte Intelligenz, wie durch eine wunderbare sich allen Klimaten aneignende Biegsamkeit des Organismus leichter entgehend, nimmt das Geschlecht wesentlich Theil an dem ganzen Erdenleben. Durch diese Beziehungen gehört demnach das dunkle und vielbestrittene Problem von der Möglichkeit gemeinsamer Abstammung in den Ideenkreis, welchen die physische Weltbeschreibung umfasst. Es soll die Untersuchung dieses Problems, wenn ich mich so ausdrücken darf, durch ein edleres und rein menschliches Interesse das letzte Ziel meiner Arbeit bezeichnen. Das unermessene Reich der Sprachen, in deren verschiedenartigem Organismus sich die Geschicke der Völker ahnungsvoll abspiegeln, steht am nächsten dem Gebiet der Stammverwandtschaft; und was selbst kleine Stammverschiedenheiten hervorzurufen vermögen, lehrt uns in der Blüthe geistiger Cultur die hellenische Welt. Die wichtigsten Fragen der Bildungsgeschichte der Menschheit knüpfen sich an die Ideen von Abstammung, Gemeinschaft der Sprache, Unwandelbarkeit in einer ursprünglichen Richtung des Geistes und des Gemüthes.

"So lange man nur bei den Extremen in der Variation der Farbe und der Gestaltung verweilte und sich der Lebhaftigkeit der ersten sinnlichen Eindrücke hingab, konnte man allerdings geneigt werden die Racen nicht als blosse Abarten, sondern als ursprünglich verschiedene Menschenstämme zu betrachten. Die Festigkeit gewisser Typen mitten unter der feindlichsten Einwirkung äusserer, besonders klimatischer Potenzen schien eine solche Annahme zu begünstigen, so kurz auch die Zeiträume sind, aus denen historische Kunde zu uns gelangt ist. Kräftiger aber sprechen, auch meiner Ansicht nach, für die Einheit des Menschengeschlechts

2.

Mr. Baillière's Edition.

"The general delineation of nature, which I here endeavour to present, would remain incomplete, were I not to yield to the disposition I feel, with a few touches, to portray the HUMAN KIND in its physical gradations, in the geographical distribution of its simultaneously existing types, in the influence which it derives from the forces of nature, and on the contrary, though in a less degree, the influence which it has exercised on these. Dependent, although not to the same extent as plants and animals, on the ground and the meteorological processes of the atmosphere, more readily escaping from under the dominion of some of the natural forces through activity of mind, and intelligence exalted by degrees, as well as through a wonderful pliability of constitution, which adapts itself to every climate, the human kind takes an essential part in the whole vitality of the earth. Through these relations we are brought into contact with the obscure and much agitated problem of the possibility of common descent in the circle of ideas which the physical cosmography embraces. The investigation of this problem, if I may so express myself, shall, through ennobled and purely human interests, be made the last aim of my work. The immeasurable realm of language, in the diverse organizations of which the capacities of nations are foreshadowed, as it were, is most intimately connected with the subject of alliance of race; and what even slight diversity of race is competent to produce, is taught us by the Hellenic world in the bloom of its mental culture. The most important questions in the history of the progress of society connect themselves with ideas of descent, community of language, and immutability in an original direction of the affective and intellectual nature of man.

"So long as extremes in diversity of colour and configuration were alone considered, and the first liveliness of sensible impression was yielded to, there might have been the disposition to consider races, not as mere varieties, but as originally different kinds of men. The permanency of certain types even amidst the most inimical operation of external, particularly climatic influences, appeared to favour such an assumption, short though the time be through which historical information has come down to us. But vouching far more strongly, according to my views, for the unity of the human race, are the many middle tints in

3.

Mrs. Sabine's Translation.

"The general view of nature which I have endeavoured to present would be incomplete, were I to close it without attempting to trace, by a few characteristic traits, a corresponding sketch of *man*, viewed in respect to physical gradations, to the geographical distribution of cotemporaneous types, to the influences which terrestrial forces exercise on him and to the reciprocal but less powerful action which he in turn exerts on them. Subject, though in a less degree than plants and animals, to the circumstances of the soil and the meteorological conditions of the atmosphere, and escaping from the control of natural influences by the activity of mind and the progressive advance of intelligence, as well as by a marvellous flexibility of organisation which adapts itself to every climate, man forms every where an essential portion of the life which animates the globe. It is by these relations that the obscure and much contested problem of community of origin enters into the circle of ideas comprised in the physical description of the world. Its examination will stamp (if I may so express myself) with that nobler interest which attaches itself to all that belongs to mankind, the termination of my work. The immense domain of the science of languages, in whose varied structure the aptitudes of nations are mysteriously reflected, borders closely on that which treats of the parentage and affinity of races: and if we would know what even slight diversities of race may be capable of producing, we have a striking example in the Hellenic nations when in the flower of their intellectual culture. The most important questions in the history of civilisation are connected with the descent of races, the community of language, and the greater or less persistency in the original direction of the intellect and disposition.

"Whilst attention was exclusively directed to the extremes of colour and of form, the result of the first vivid impressions derived from the senses was a tendency to view these differences as characteristics, not of mere varieties, but of originally distinct species. The permanence of certain types in the midst of the most opposite influences, especially of climate, appeared to favour this view, notwithstanding the shortness of the time to which the historical evidence applied: but in my opinion more powerful reasons lend their weight to the other side of the question, and corroborate the unity of the human race. I refer to the

4.

Miss Otté's Translation.

"The general picture of nature which I have endeavoured to delineate, would be incomplete, if I did not venture to trace a few of the most marked features of the human race, considered with reference to physical gradations—to the geographical distribution of contemporaneous types—to the influence exercised upon man by the forces of nature, and the reciprocal, although weaker, action which he in his turn exercises on these natural forces. Dependent, although in a lesser degree than plants and animals, on the soil, and on the meteorological processes of the atmosphere with which he is surrounded—escaping more readily from the control of natural forces, by activity of mind and the advance of intellectual cultivation, no less than by his wonderful capacity of adapting himself to all climates—man everywhere becomes most essentially associated with terrestrial life. It is by these relations that the obscure and much contested problem of the possibility of one common descent, enters into the sphere embraced by a general physical cosmography. The investigation of this problem will impart a nobler and, if I may so express myself, more purely human interest to the closing pages of this section of my work.

"The vast domain of language, in whose varied structure we see mysteriously reflected the destinies of nations, is most intimately associated with the affinity of races; and what even slight differences of races may effect, is strikingly manifested in the history of the Hellenic nations in the zenith of their intellectual cultivation. The most important questions of the civilization of mankind, are connected with the ideas of races, community of language, and adherence to one original direction of the intellectual and moral faculties.

"As long as attention was directed solely to the extremes in varieties of colour and of form, and to the vividness of the first impression of the senses, the observer was naturally disposed to regard races rather as originally different species than as mere varieties. The permanence of certain types in the midst of the most hostile influences, especially of climate, appeared to favour such a view, notwithstanding the shortness of the interval of time from which the historical evidence was derived. In my opinion, however, more powerful reasons can be advanced in support of the theory of the unity of the human race, as, for instance, in the many intermediate gradations in the

Figure 6.1 *Literary Gazette*, 10 March 1849, 1677 (1849), p. 162.

the first volume of the Bohn edition, Otté asserted that her version was superior to that of the Sabines by dint of it being a full account of the original: 'I have not conceived myself justified in omitting passages, sometimes amounting to pages, simply because they might be deemed slightly obnoxious to our national prejudices' (*CO* I, viii). Sabine retaliated in the Preface to the 1849 edition of this volume that, on the two occasions when passages had been omitted, 'in both instances the sense and reasoning were complete without them' and that both omissions were known to Humboldt, one even in time to receive his sanction (*CS* I, n.p.).⁴ Bohn launched a counter-attack in the seven-page 'Bohn's Rejoinder' of February 1849, which listed the eight distinctive features of his own edition as well as the mistakes and omissions in the Sabines' translation, and, in the words of a contemporary reviewer, accused them of acting as 'literary laundresses to clearstarch the productions of thinkers like Humboldt'.⁵

While the critic at the *Literary Gazette* encouraged readers to compare the different editions and evaluate their relative merits, other critics were less tolerant. The *Examiner* forthrightly denounced Bohn for considering it 'expedient to speculate in rivalry to so good a translation as Mrs Sabine's authorized by Humboldt himself, and accepted as worthy of him by the most competent authorities'.⁶ Far from deeming such competition productive, the reviewer argued that so many different editions could only confuse the British reading public. Two years later, the same journal would reiterate this point: 'Of late, rival translations, such as those of Humboldt's *Cosmos*, have so injured each other and perplexed the public, that both booksellers and authors have been obliged to surrender as hopeless even the best works by the best translators.'⁷ In 1852, when Baillière had dropped out of the race and left Bohn and his rivals Longman and Murray to stay the course alone, reviewers were beginning to reach ultimate fatigue at having to report continually on two competing translations of the same work. As *The Athenæum* noted: 'We trust we shall ere long have the international question of literary property placed in a more satisfactory condition than that which at present prevails.'⁸ The battle became so entrenched because the publication of Humboldt's *Kosmos* itself spanned seventeen years, the first volume appearing in 1845 with Cotta in Stuttgart, and the last in 1862, three years after its author's death. As with Humboldt's previous works, reviewers were quick to criticise 'a character of diffuseness in his labours'.⁹ The Scottish physicist James D. Forbes observed sourly of *Kosmos* that life was 'too short and uncertain to encourage the undertaking of encyclopædiacal [*sic*] publications by individuals'.¹⁰ Prince Albert's physician Henry Holland even suggested that Humboldt was 'haunted by a spectre of his own creation'.¹¹

The differences between the Bohn and the Longman and Murray editions did not revolve solely around structure, authority or accuracy. Price was also crucial to their ultimate success. As a critic in the *Art Journal* observed on the appearance of the first volume of the Bohn edition, the translation by Mr. Otté [sic] now before us, is quite equal to either of those which have preceded it, and it comes recommended to the public, not only by additional notes, [. . .] but in being published at less than one third the price of either of those which have yet been published.[12]

The reviewer in *The Athenæum* spelt out the price war in clearer terms:

> Since the publication of Mr. Otté's [sic] translation by Mr. Bohn in two neat volumes at 7s., Messrs. Murray and Longman & Co. have been induced to issue Sabine's edition, in two volumes stitched, at 5s. instead of 24s. – the original price. Thus, by the rivalry of publishers, the public may now enjoy the benefit of a valuable work at an exceedingly cheap rate.[13]

Other reviewers were drawn less to price than to presentation. The extensive index at the end of the first volume of the Bohn edition and the portrait in the front matter of a youthful Humboldt were other features that made this an attractive book for the general reader.

The tension brewing in Britain between these publishers did not pass Humboldt by as he laboured over *Kosmos* in Berlin. By autumn 1848 he was well aware that the third translation of this work had appeared with Bohn in Covent Garden (Humboldt 2007: 203). The next spring, he informed Cotta that *Cosmos* was exciting pure madness in England – '[i]n England ist reine Tollheit mit dem Cosmos' – as he described how Longman was being forced to reduce the price of each volume to almost a fifth of the original to undercut Bohn (Humboldt 2009: 359). Humboldt had been horrified at the damage that Prichard's 'incomprehensibly wooden' translation – 'unverständlich hölzerne Uebertragung' – could do to his reputation among British scientists (Humboldt 2006: 76). He responded with relief and enthusiasm to Sabine's 'exalted style' and 'delicate handling of even the finest nuances of language'.[14] Casting Bohn by turns as a 'menace', a 'corsair' and the 'foe', Humboldt did his utmost to ensure that the Longman and Murray edition should appear first (Humboldt 2009: 372, 378). He considered this the authoritative English version of his work, even if he did not publicly denounce the rival editions, since he wanted to keep his hands out of the 'wasp's nest' of stinging recriminations fuelled by the competing translations from Baillière and Bohn (Humboldt 2009: 372).

This chapter begins by exploring what Humboldt had set out to achieve in writing *Kosmos* and how it was received by a German-speaking public. It then investigates the copyright situation in Britain, before considering how the Sabines and Otté handled the fact that neither party was, for the first two volumes at least, translating *Kosmos* from scratch and that Prichard's translation needed, in some way, to be taken into account. The main body of this chapter draws on a select group of examples to illustrate the key differences between the English renderings of *Kosmos*, placing particular emphasis on the translators' different strategies with regard to the style, scientific terminology and religious message of their translations as the first volumes of this work appeared. It goes on to trace how, while tackling the final volumes of *Cosmos*, Humboldt's British translators had to rethink their working practices to negotiate the increasing pressures upon them to complete their translation ahead of the competition. Finally, we consider what lasting impressions of Humboldt these translations left on a British public and how they continued to disseminate his scientific vision after his death.

Creating *Kosmos*

'In the late evening of an active life, I offer to the [. . .] public a work, whose undefined image has floated before my mind for almost half a century,' observed Humboldt in the opening sentence to *Cosmos* (CO I, ix). Over thirty years in the making, this work embodied his belief that it was possible to unite natural phenomena into a whole, based on what was known about their relationship to each other.[15] The first seeds of an idea for a physical geography of the world had germinated in Humboldt's mind as early as 1796 (Humboldt 1869: 17). They acquired more concrete form only as Humboldt presented the first lectures on which *Kosmos* would be based at the salon of the Countess of Montauban in Paris in 1825. By 1827 Humboldt was offering public lectures, of which sixty-one in total were held: sixteen at Berlin's musical academy, the Singakademie. Humboldt was swift to publish an advertisement in the *Berlinische Nachrichten* [*Berlin News*] of 1827, declaring that any third-party publications deriving from these lectures infringed his intellectual property rights. But he would be slow to honour the contract drawn up with Cotta in March 1828, which had set the extent of this work at two volumes of at least forty-five octavo sheets (Humboldt 2009: 643). Seventeen years passed between the signing of the contract and the appearance of the first volume, with Humboldt distracted by other projects, not least his journey to Russia in spring 1833.

Humboldt later expressly asked Cotta to remove from the subtitle a reference to his Berlin lectures as fundamental to the genesis of *Kosmos* (Humboldt 2006: 232). Yet the spoken quality of the narrative at the core of the work remained central to his authorial project. The oratorical, lively form was what characterised the text, Humboldt noted to Cotta as he described the different elements of the work, while the erudite part, which would interest only a few more learned readers, would be contained in the notes following each section (Humboldt 2006: 240). *Cosmos* was, as Humboldt observed repeatedly, the most important literary work of his œuvre. The form of the book was such that it floated above the level of scientific detail – 'so schwebend über alle Einzelnheiten gehalten', he assured Cotta, which ensured that it would not quickly date (Humboldt 2009: 232, 309).

It is little wonder, then, that he was delighted at the positive responses in the German press at a time when, he asserted, readers little appreciated clarity of expression and stylistic harmony (Humboldt 2009: 331). Equally understandable is his bitter dejection at British criticism of his style. In Germany, Humboldt sighed to Bunsen, his prose was considered too lyrical, while in England it was apparently too dull and lifeless; his home audience approved in particular of the introductory sections in *Kosmos*, whereas the English found them dull and redundant (Humboldt 2006: 83). 'Vivid description, close and convincing reasonings, and terse composition are not in general characteristic of Humboldt's writings,' sneered Forbes in the *Quarterly Review*.[16] The gentleman scientist and electrical experimenter Andrew Crosse concurred in the *Westminster Review* that 'his sentences are often too long and complicated' and wondered why Humboldt's translators did not edit his text more fiercely to ensure such long sentences were 'occasionally broken up, and at times reorganised'.[17] More worryingly, Crosse asserted that *Kosmos* was an utterly godless publication. Humboldt retorted that the man was a 'galvanic louse' (Humboldt 2006: 81).

Although *Kosmos* is considered the apogee of Humboldt's career, it was not the culmination of his efforts alone. Rather it represented the achievements of a figure at the heart of international scientific exchange. Dassow Walls rightly stresses the highly collaborative nature of the work, as she pictures Humboldt at the head of a large, international team of correspondents, cajoled into assisting him in his endeavours and receiving from him welcome praise and patronage (Dassow Walls 2009: 218). The collaborative nature of *Kosmos* makes it seem a rather different piece from the rest of his œuvre, marking a conspicuous shift away from his active involvement in natural science towards the more passive role of observer and historical documenter. But his repeated requests for the latest scientific data and his obsessive need to revise the manuscript of *Kosmos* – clause 4 of his 1849 contract with Cotta even permitted other scientists to update this work after his death – attest to his self-positioning at the cutting edge of scientific knowledge.[18] His network of fellow scientists not only provided him with information. Von Buch and

Herschel also gave advice and criticism as the manuscript took shape (Werner 2004: 115ff, 136ff).

As Dassow Walls stresses, Humboldt's considerable talent as a 'literary eco-critic' was ably supported in the writing of the second volume of *Kosmos* by the great German Shakespeare specialist, Tieck (Dassow Walls 2009: 243). This quotation from Shakespeare's *Love's Labour's Lost* was originally to have graced the opening pages of *Kosmos* in German translation:

> These earthly godfathers of Heaven's lights,
> That give a name to every fixed star,
> Have no more profit of their shining nights,
> Than those that walk, and wot not what they are;
> Too much to know, is to know nought but fame,
> And every godfather can give a name. (I, i, 88–93)[19]

It would have explicitly borne witness to the ease with which Humboldt straddled both the arts and the sciences, and the relevance of literary sources in being able to comment on scientific knowledge. It would also, however, have denied the centrality of the German literary heritage to his project, in a piece that Humboldt deliberately wrote in his native tongue to gain the approval of his home audience.

Humboldt's ultimate rejection of the Shakespearean quotation indicates the instability of the text in its pre-publication phase. The title also underwent a long gestation period. The cumbersome *Entwurf einer physischen Weltbeschreibung; Erinnerungen aus Vorlesungen von Alexander von Humboldt in zwei Bänden* [*Sketch of a Physical Description of the World: Recollections from Lectures by Alexander von Humboldt in Two Volumes*] would metamorphose into the neater *Kosmos*, after Humboldt had first rejected 'Essai sur la physique du monde' ['Essay on the Physics of the World'] and, subsequently, 'Buch von der Natur' ['Book of Nature'] in homage to the work of the German philosopher and theologian Albertus Magnus (Humboldt 1860: 17–18). 'I know', admitted Humboldt, 'that Kosmos is very grand, and not without a certain tinge of affectation; but the title contains a striking word, meaning both heaven and earth' – and since his brother Wilhelm had also approved of it, the matter was settled (Humboldt 1860: 18). Humboldt's British reviewers saw in the title the work's immense terrestrial and cosmic range, as well as the Greek term for 'beauty', doubtless prompted by the font used in the title to the Baillière edition, *ΚΟΣΜΟΣ*. 'Kosmos, the adornment, the orderly arrangement, the ideal beauty, harmony, and grace, of the universe!' exclaimed Herschel of the Sabines' translation of volume I in his critique for the *Edinburgh Review*.[20]

The five volumes of *Kosmos* adopt rather different angles of approach. The first contains a general view of nature, 'from the remotest nebulæ and revolving double stars to the terrestrial phænomena of the geographical distribution of plants, of animals, and of races of men' and includes 'some preliminary considerations on the different degrees of enjoyment offered by the study of nature and the knowledge of her laws' (CS I, n.p.). Subdivided into parts on celestial phenomena, terrestrial phenomena, physical geography and organic life, it approaches the earth gradually from space, working from the outside to the inside and journeying from discussions of surface terrestrial phenomena such as volcanoes, through to the density of the earth, its internal heat and core temperature (Werner 2004: 46). The second volume is more historical in focus and comprises two main sections: the 'Incitements to the Study of Nature' and the much longer 'History of the Physical Contemplation of the Universe'. It discusses literary and pictorial descriptions of nature down the ages, from classical literature to Titian's panoramas, followed by key moments in the history of human involvement with the natural world. It traces the progress of humankind's view towards nature from Sanskrit texts through works by the classical authors Plato, Strabo and Pliny, to Arab contributions to the mathematical sciences. Discovery was the motif of the next sections as Humboldt turned his focus to the terrestrial explorations conducted by Columbus, Vasco da Gama and Marco Polo, and the celestial discoveries of Galileo and Kepler, Newton and Halley. Finally, Humboldt dwelt upon the intellectual findings of his own age, singling out electromagnetism as a scientific phenomenon of 'cosmic' character (in the widest sense of the word), which held him utterly in thrall.

In the third volume of *Kosmos*, Humboldt revisited attempts by past scholars to reduce the phenomena of the universe to one principle of explanation and worked his way from the Ionic school of philosophy, the Pythagoreans, Aristotle and the Stoics, to Bacon and Magnus, Galileo and Copernicus, Kepler and Newton. The onward march of knowledge and the boundlessness of science were realities with which Humboldt was constantly grappling. As the third volume was beginning to take shape, Humboldt admitted to Bunsen that the sheer mass of material he had collated now compelled him to accept that he could not compress heaven and earth into one of his little volumes – 'daß ich Himmel und Erde [. . .] nicht in eines meiner Bändchen zusammenzwängen könnte' (Humboldt 2006: 138).

The 'little' fourth volume was so complex that it took eight years to write and publish. It discussed the size, form and density of the earth, its internal heat and magnetic activity, and movements in the earth's interior revealed by earthquakes, thermal springs and mud-volcanoes. Freighted with over 600 footnotes (occupying more than 170 pages in the Longman and Murray edition), this volume relied on a vast corpus of scientific papers, articles and monographs. Although the Sabines had put the text into English in a matter of months, a feat achieved only through having access to the German proofs, thanks to Humboldt and Cotta, the English translation still failed to present mid-nineteenth-century readers with the most up-to-date information. Edward Sabine, with Humboldt's sanction, made additional extensive corrections to the text, enlarging it to account for more recent advances in scientific knowledge.

Of the fifth volume of *Kosmos* Humboldt managed to complete little more than the Introduction and part of a section on telluric phenomena, which had been meant to conclude the investigations at the end of volume four. This final volume brought untimely closure to a book project begun over twelve years earlier, which had taken Humboldt so long to write because he had been hampered by the complexity and the mass of material he had originally set out to discuss (*K* 868).

In her excellent study of the making of *Kosmos*, Petra Werner suggests that Humboldt initially did not reckon with the international fame of this work, since he blithely left all questions regarding translations to his publisher and had no separate clause drawn up to address translated editions of the work (Werner 2004: 15). But the immediate success of his Berlin lectures must have alerted him to the wider appeal of *Kosmos*. Cotta's wholesaler was left near-speechless at the stampede of agents who descended on his bookshop to get their hands on the second German volume, and could only watch helplessly as parcels destined for St Petersburg and London were unceremoniously torn open and their contents redirected to Vienna or Hamburg (Humboldt 2009: 329). With brisk self-irony, Humboldt again observed that no one was less skilful at deriving profit or use from his work than himself (Humboldt 2009: 273). Cotta's frequent prods and pleas to Humboldt to consider his intellectual rights and the financial reward to be gained from his work in translation largely fell on deaf ears (Humboldt 2009: 323, 325).

'The lynx-eyed perfection of competition': *Cosmos*, Copyright and Retranslation

In the eighteenth century, it was near-impossible for authors to get legal recognition as the owners of the products of their intellectual labours (Woodmansee 1994). Piracy had grown to epidemic proportions by the start of the nineteenth century, and legal institutions could not keep pace with the book trade on a national, let alone an international, scale. While the period in which the competing translations of Humboldt's *Kosmos* appeared still falls under what some have termed the 'Great Age of Piracy', it also saw a series of settlements drawn up between Britain and other European countries that ensured increasing copyright protection for authors (Sonoda 2007: 4). The International Copyright Act of 1838 allowed reciprocal copyright agreements with other nations to be reached, as a result of which bilateral agreements were concluded with Prussia and other German states between 1846 and 1847, with France in 1852, and with Belgium and Spain in 1854. Translation was frequently a grey area within copyright law in the nineteenth century, even if it played a vital role in enriching emerging national literatures (Bassnett and France 2006: 57). Degrees of adaptation were difficult to assess and questions of originality and piracy hard to settle. As Catherine Seville observes, translations into English from other languages 'generated little discussion, because often there was no English copyright holder to protest', and although translations were sometimes considered to have the same status as original works, which meant that translations of copyrighted works could not be considered piracy, this was not always the case (Seville 1999: 245–6). While the Copyright Act of 1842 stated that 'the Copyright in every translation shall be deemed to be the property of the Translator thereof and his assignees as though it were an original work', this clause was subsequently overruled and deleted by the House of Lords, leaving the matter unresolved until the early 1850s (Seville 1999: 247).

Changes in international copyright law at mid-century were welcomed by the international scientific community. The German physicist Georg Adolf Erman wrote to Edward Sabine in December 1848, 'I have been told that by new contracts between our English and Prussian governments, writers have finally become, like other men, the proprietors of their own hands or minds [sic] work' and might even enjoy 'somewhat like a visible income from many years [sic] thoroughly unpaid work'.[21] Humboldt was more reticent about asserting his right to any income from translations of his books. For one thing, the first volume of the German edition had already appeared in 1845. For another, *Kosmos* had been published in Stuttgart, in the state of Württemberg, which had not participated in the first round of intellectual property agreements. The contracts drawn up between Prussia and England on 16 June 1846 had been ratified only by Saxony, Brunswick and Hanover, so it remained unclear whether rival translations and retranslations fell under such agreements (Humboldt 2009: 389). Despite reading the 1842 Copyright Act from cover to cover, Humboldt was none the wiser: you were more likely to lose money in legal battles than gain your rightful dues, he concluded cynically (Humboldt 2009: 383–4). Longman and Murray could only ensure that their English translation was not reprinted by another publisher in England. But as Humboldt was well aware, this would not hinder Bohn or other British publishers from exercising their right to translate the German edition again from scratch (Humboldt 2009: 384).

For the critic in the *Literary Gazette*, the Bohn edition did not really represent competition. 'It must, we think, strike every one that Miss Otté's translation could hardly have been made had not Mr. Baillière's and Mrs. Sabine's preceded it,' the reviewer observed, adding that the 'construction of the sentences is far too close to have been made from the inverted order of the German, and the earlier translation or translations must have been lying before the writer of the last'.[22] Edward Sabine had written in a similar vein to Murray:

> I have read some pages of Mr Bohn's translation, and although I can have no doubt of the influence which the previous translation has exercised on the subsequently published, and believe that it could not fail to be recognised by 99 out of 100 careful and impartial [?] readers, yet I should [. . .] doubt the expediency of attempting legal measures of restriction.[23]

While the Bohn edition ably circumvented some of the awkward formulations in the Sabines' translation and corrected a few errors on the way, Otté was also at times forced to go out of her way to use different turns of phrase that would distinguish her work from that of her predecessor. The critic in the *Literary Gazette* recognised the power of retranslation to improve and enhance the qualities of the first translation, yet also constrain the second translator's creative potential. In the case of Sabine versus Otté the distinctions were minimal, the reviewer asserted, and any shortcomings in either version could be put down only to the 'lynx-eyed perfection of competition'.[24]

Kosmos in British Hands: Openings

The first translator to tackle Humboldt's *Kosmos* has largely disappeared from scientific memory, yet Augustin Prichard (1818–98) might not, at first glance, have seemed an unpromising candidate to translate this multifaceted work. He was the son of the Bristol physician, anthropologist and ethnologist James Cowles Prichard (1786–1848). Cowles Prichard, a devout Quaker, had written the popular *Natural History of Man* (1843), which attempted to vindicate the truth of the anthropological precepts laid down in the book of Genesis, by arguing that all humankind belonged to one species. A fierce refutation of the polygenism to be found in theories promoted by figures such as the Scottish philosopher Henry Home, Lord Kames, Humboldt drew on it in the first volume of *Kosmos* as he compared racial difference with facial similarity (*K* 185–6). While Augustin Prichard may not have shared his father's spiritual inclinations, he did follow in his footsteps by studying medicine, specialising early on in ophthalmology. Augustin left for Berlin in 1840 to pursue his studies under the physiologists Johann Lucas Schönlein and Johann Friedrich Dieffenbach, and gained an MD from Berlin for a thesis on the inflammation of the iris. He then moved to Vienna and attended demonstrations by the Austrian pathologist Karl von Rokitansky and the cataract specialist Friedrich Jaeger. Having settled in Bristol in 1842, Prichard worked as a surgeon in the Eye Dispensary of the Bristol Royal Infirmary.[25] He therefore brought to his translation extensive knowledge of medicine, a good understanding of optics (he was a keen photographer) and proficiency in German. Despite these accomplishments, he still did not have the skills set that would be needed to translate *Kosmos* proficiently. As Humboldt outlined to Murray, the ideal translator needed to have expertise in other disciplines: 'The work is very difficult to translate. It above all requires knowledge of astronomy, magnetism, geology, botany.'[26]

The second British translator to work on *Kosmos* was Elizabeth Sabine. A good ten years older than Prichard, she brought to the undertaking experience in translating scientific (travel) writing into English, and, as with the *Ansichten der Natur*, could rely on her husband's scientific competence. While she was beavering away at *Kosmos*, she also tackled the Berlin physicist Heinrich Wilhelm Dove's *Die Verbreitung der Wärme auf der Oberfläche der Erde* (1852). Published in 1853 as the *Distribution of Heat over the Surface of the Globe* with Taylor and Francis for the BAAS, it was awarded the Royal Society's Copley Medal in 1853. This slim, but significant, volume built on Humboldt's famous 1817 paper 'Sur les lignes isothermes' ['On Isothermal Lines'], charting heat distribution across the northern hemisphere in terms of isotherms, lines demarcating zones of similar average temperature. Dove's work extended Humboldt's research by providing isothermal charts for the entire earth and discussing thermal anomalies, and enabled him to establish meteorology as an independent discipline. He was absolutely delighted that this piece would be rendered into English by none other than the translator of Humboldt's *Kosmos*.[27] In a letter to Edward Sabine in 1847, he noted that when he wanted to present his own spouse with an example of how the wife of a learned man should be, he took his example from the Sabines at Woolwich.[28]

Elizabeth did not live a cloistered existence. Since she was not burdened with extensive domestic duties – they had no children – she accompanied her husband on some of his expeditions and journeys through Europe, and on one such tour breakfasted with Humboldt in Berlin. He was always assiduous in his recognition of Elizabeth Sabine's achievements, and impressed on Edward the need to thank his worthy spouse for her translation.[29] In a letter to Cotta from July 1849 (after the second English volume of *Cosmos* had appeared), in which Humboldt was extolling her talents, he quoted correspondence from Longman in which the latter had emphasised that it could only be 'a disadvantage to omit Mrs Sabine's name' from the front (Humboldt 2009: 393).

Although Sabine seems initially to have suggested that his own name need not be mentioned in *Cosmos*, or at least only in an advisory role, later correspondence suggests a change of heart. For one thing, it would bolster his own scientific standing. It was now not to be issued as 'translated by Lt.-Col. S', he informed Murray, but rather as 'translated under the superintendence of Lt.-Col. Sabine R.A., F.R.S', which played up his scientific status and played down his direct involvement in the arguably derivative practice of translation.[30] For another, the translation became an important showcase for his own research. In a letter to Faraday in 1852 he noted:

> Humboldt wrote me some time since an account of Mr. Wolfe's discovery of the connexion between the Magnetic variations & the solar spots, & remarked that I had preceded him in publication by between 4 & 5 months: and I have reason to believe that he will notice my priority in his forthcoming vol. of Cosmos, which I am glad of, as the thing itself has excited but little interest in this country. (James 1999: IV, 444)

Cosmos therefore embodied various forms of collaboration – as a scientific narrative in which Humboldt emphasised the importance of Sabine's publications, and as an accomplished translation by the Sabines that highlighted the centrality of Humboldt's thought to mid-century European science.

Humboldt had relied heavily on his own international networks of scientists to supply him with the data and terminology in *Kosmos*. The Sabines likewise mobilised their friends in the scientific community to help with queries that arose during translation. A flurry of letters in summer and autumn 1846 between Sabine and the Arctic explorer James Clark Ross show how valuable such connections were as the Sabines struggled to make headway with the first volume of *Cosmos*. Could Ross please check whether Humboldt's quotation 'that you sounded with 25400 french [*sic*] feet of line without hitting [?] bottom' from the first volume of *Cosmos* was actually correct, enquired Edward

Sabine.[31] 'I return the sheet of Declinations with many thanks – it has served the purpose – I must have mislaid the one you sent me before,' Sabine noted to Ross a couple of months later.[32] And on one occasion Ross must have asked for prior knowledge of what would be written in *Cosmos* about the results collected from his own voyage. Sabine gave him a sneak preview and appears to have asked for verification of the details in the passage where Humboldt would describe marine life in the Gulf of the Erebus and Terror, in which 'eight siliceous-shelled polygastrica and phytolitharia, together with a single calcareous-shelled polythalamia, were brought up by the lead from depths of 1242 to 1620 English feet'.[33]

Not all the Sabines' friends and correspondents were men of science. Bunsen, who was Prussian ambassador to London between 1841 and 1854 (but no relation of the chemist Robert Bunsen, after whom the 'Bunsen burner' is named), was drawn into the *Cosmos* project in 1848. Bunsen's main preoccupations were political and spiritual. He moved in circles that envisioned Prussia as the historical incarnation of a German *Kulturnation* and believed it should adopt an English model of rule to see change brought about through the development of modern bourgeois political institutions (Toews 2004: 67–9, 99). The sheer range of Bunsen's publications – including an evangelical hymn and prayer book (1846), the seven-volume *Christianity and Mankind, Their Beginnings and Prospects* (1854), and a five-volume work on Egypt's role in world history (1844–56) – attest to an interest in exploring the role of religion in culture and history, as well as in examining how God could be accommodated in a society shaped by an increasingly mechanistic world view.

Bunsen's London residence was a cultural hub for the exchange of ideas. Susanne Stark's lively account of his congenial and popular gatherings stresses his crucial role 'in establishing social and intellectual links not only between various groups of people with an interest in reading German texts but also between their varied scholarly activities' (Stark 1999: 27). A figure gifted with 'encyclopaedic polymath knowledge' (Stark 1999: 26), he would have embraced the complexities of *Kosmos* wholeheartedly. He was also well anchored in British translation networks, and had approached Austin in 1848 to ask if she would put a biography of his mentor, Barthold Georg Niebuhr, into English (Stark 1999: 28). She must have declined his offer, since Mrs Gaskell came to his aid and referred him on to Susanna Winkworth, who accepted the work and later offered Bunsen other administrative assistance (Stark 1999: 28). Her sister Catherine also translated for him and even accompanied him to Bonn on his return to Germany in 1854. Bunsen, who was encouraging of Anna Swanwick's activities as a translator (Stark 1999: 28), clearly approved of women's contributions to Anglo-German cultural transfer through translation, and would have doubtless supported Elizabeth Sabine in her activities as one of the mediators of Humboldt's ideas to a British audience.

Bunsen's precise role in the translation of *Cosmos* is harder to define: Sabine announced in a notice to the second volume that the translation 'in its progress through the press, has had the advantage of being compared with the original by the CHEVALIER BUNSEN' (*CS* II, n.p.), although it is unclear quite how invasive his corrections were, and we also do not know whether his involvement was only with this volume or with later ones as well. Bunsen's family reported that, initially at least, he spent the evenings carefully comparing every single sheet of the English translation with the German original (Humboldt 2006: 105). Certainly, he swiftly transmitted material between Humboldt in Germany and the Sabines in London, perhaps facilitated by his position in the diplomatic service. He offered valuable expertise in those areas in which Sabine might have deemed himself less proficient, notably the historical section of volume II, and was credited by Humboldt with getting the project back on its feet when it seemed to have utterly stalled (Humboldt 2006: 80, 88, 90, 94).

While the Sabines together undertook the translation, editing and annotation work for the Longman and Murray edition of *Cosmos*, they were also pivotal figures in the dissemination of their translation amongst Britain's scientific elite. 'We all heartily thank you both for your very kind & acceptable present of your translation of Cosmos,' wrote the British geologist and inventor Robert Were Fox in September 1846, adding 'I intended to send for the work, but now that it has been presented to us by the translator & editor, & by you; it is trebly valuable to us.'[34] Sabine had already instructed Murray in late August to send out courtesy copies, hot off the press, to the Scottish botanist Robert Brown and the English geologist David Ansted, to thank them for their assistance with queries arising from translating the first volume of *Kosmos*.[35] To Ross, who had been so helpful throughout the summer, Sabine wrote in September 1846, now that the first volume was out and the second under way:

> Elisa & I wish to give you a copy of Cosmos in token of our ancient & very true regard. [. . .] The translation has been a great pleasure to us though we have often been hard pressed to find the time for it: but it has certainly not been prejudicial to her health, which is the main consideration. I am not a little curious to hear what is said of her style – it is very agreeable to me, but I suppose that is no criterion of what unprejudiced people will think –

Humboldt has written her a warm eulogium both on the fidelity & on the style.[36]

Edward Sabine's letter reveals just how important style was as an evaluative category. He could be certain that the scientific detail was accurately conveyed. But translating style was less clear-cut, and this aspect of the translation obviously gave him greater cause for concern, even if he had Humboldt's reassurance that the Sabines' English edition was an elegant, accurate and faithful rendering of the original.

'The singular difficulties of his task': Prichard's Translation of *Kosmos*

'I was reading Cosmos in the railway carriages, have you seen it?' enquired Hooker of Darwin in September 1845.[37] 'I cannot understand many pages of it at all,' he groused, adding, 'I can send you my copy if you have it not, for such a translation is never worth buying I should think.' A fortnight or so later, Darwin confirmed Hooker's view: 'I have just begun the introduction & groan over the style, which in such parts is full half the battle.'[38] The first edition to appear on the British market, Prichard's translation had been awaited with eagerness and was read just as assiduously. It appeared with Hippolyte Baillière, an established scientific publisher (like his Parisian cousin Jean-Baptiste Baillière, who was a specialist in medical books). He had also issued the *London Journal of Botany* (1842–8), George Fleming Richardson's *Geology for Beginners* (1842) and, significantly, Cowles Prichard's *The Natural History of Man* (1845).

Although Darwin and Hooker were scathing of Augustin Prichard's translation of *Kosmos*, they appear to have been in a critical minority. The reviews that appeared in *The Athenæum* heralding the appearance of the two volumes of the Baillière edition in 1845 and 1848 praised the 'lucid manner in which every scientific fact is narrated by our illustrious author' and gushed that a 'beautiful tone prevails'.[39] Prichard went unmentioned as the translator because his name was absent from the title page of the first volume: it would appear only in the second volume of 1848. Even Humboldt was not aware of his identity (Humboldt 2009: 284). Yet if Prichard was not present in the front matter, then he was highly 'visible' in a Translator's Preface that was illuminating about his aims, aspirations and anxieties. He had sought to 'preserve the lofty tone and imaginative style of the Author' and had recognised that the Introduction was 'composed in the manner of an oration or popular discourse', but found this near-impossible to convey in English (*KP* I, n.p.). The second section, on the limitations and methods of exposition of the universe, he found 'extremely abstruse, and cost the Translator no small pains to render it, he trusts, intelligibly, into English' (*KP* I, n.p.). Style was not the only problem with which Prichard struggled. Legged down by the 'use of several compound words, formed after the German originals', which had compelled him to devise a few neologisms, and by 'the translation of technical or conventional scientific terms' ranging across disciplines in which he was not deeply versed, Prichard was rapidly out of his depth (*KP* I, n.p.).

Fearful that Prichard's mangling of his text would tarnish the reputation of *Cosmos* in the Anglophone world, Humboldt wrote an incandescent letter to Edward Sabine (who was currently weighing up the pros and cons of preparing a competing translation):

> Je me suis efforcé de prouver que la Weltbeschreibung n'est pas le Weltgeschichte, que l'histoire des révolutions qu'a subies notre planète, vu la voie lactée, ne doit pas être confondue avec la description de la terre et des espaces célestes, ce qui n'empêche pas M. Baillière de pervertir jusqu'au titre de mon livre: au lieu de physical description of the Universe il en fait un physical history. Je n'ose me plaindre de ce qui doit m'humilier devant le public anglais.[40]
> [I have striven to demonstrate that Weltbeschreibung is not Weltgeschichte, that the history of the revolutions which our planet has undergone, as exemplified by the Milky Way, should not be confused with the description of the earth and of the celestial spaces, which does not hinder Mr. Baillière from adulterating them, even right down to the title of my book: instead of physical description of the Universe he has made of it a physical history. I hardly dare complain about what will surely humiliate me before an English public.]

Humboldt presents himself here as the powerless author, at the mercy of his translator, who has insufficiently grasped the difference between a 'description of the world' (Weltbeschreibung) and a 'history of the world' (Weltgeschichte) – errors that will, in turn, have repercussions for his scientific reputation in Britain. This was not entirely the case, though. These issues had clearly been addressed by the time *ΚΟΣΜΟΣ: A General Survey of the Physical Phenomena of the Universe* went to press, which suggests that Baillière did seek Humboldt's critical input before this edition was exposed to the public eye. Humboldt had also complained bitterly in an earlier letter to Bunsen that his own Preface had been omitted, although this too had been restored by the time the translation came out (Humboldt 2006: 43).

Humboldt's criticism of Prichard's version revolved not only around mistranslations of key terms. The style of his prose, Humboldt complained, was such that all the grace of the living language had been lost in the English. It now sounded more like Sanskrit (Humboldt 2006: 43). Humboldt noted in subsequent letters to Edward Sabine that it was the lack of sensitivity in Baillière's edition towards those sections related to the imagination, the depiction of sites, and the exalted feelings excited by the power and fecundity of nature that pained him most.[41] Prichard's translation does abound with unusual collocations and awkward phrasings. In the opening section of ΚΟΣΜΟΣ, for example, he described Humboldt's intentions as being 'to develop [sic] the connection of the forces which actuate the universe', and he later portrayed the philosophy of nature as 'a thinking impersonation of things observed' (KP I, 3, 5). In the closing paragraphs Humboldt was made to comment, through Prichard, that it was 'therefore, felt as a want, in the frame of mind which leads us to look on nature [. . .] that we pursue the physical phenomena which present themselves to us on earth' (KP I, 373–4). These strings of prepositional phrases and disconnected subclauses obscured Humboldt's meaning and generated a disjointed prose style that was a far cry from the elegant, if complex, rhetoric of the German original.

'Quite a different book to read': Cosmos Emerging in the Sabines' Translation

'Have you seen the new Cosmos? It is excellently well done & quite a different book to read – I will send or bring it you if you care,' exclaimed Hooker to Darwin just over a year after his rather different tidings about the first translation of *Kosmos*.[42] The eagerness with which the British public snapped up the first of these later translations is well documented: Longman's first print run of 1,000 copies sold out within a day and went through five editions in the next three years (Humboldt 2009: 359). The confidence shown by British readers in the Sabines' translation belied its rather shaky start. Despite the obvious weaknesses in Prichard's translation (and Humboldt's own misgivings), Longman and Murray were initially cautious about taking the project on. 'It was with feelings somewhat akin to dismay, that I received two days ago the accompanying copy of the first part of a Translation of Cosmos – already published – and to be completed in parts,' wrote Murray to Sabine in July 1845, shortly after the Baillière edition had appeared.[43] While neither he nor Sabine was in any doubt about the poor quality of Prichard's translation, Murray was concerned it would 'withdraw all chances of success from any more elaborate & accurate Translation whose appearance shall be delayed for two or 3 months'.[44] But what irritated Murray most was that Humboldt had reneged on an informal promise to give his publishing house first refusal:

> I am the more annoyed in the present case because I have been for more than 12 years in correspondence with M. V. Humboldt respecting this work – which was announced at least as long ago – to be published by my late Father.[45]

Sabine replied a couple of days later to Murray, signalling his agreement:

> I have done little more as yet than dip into the translation, which I must confess is inferior in a greater degree than I had imagined probable: and I fear M. de Humboldt will be greatly mortified. [. . .] Mrs. Sabine had made some progress, but of course she has stopped.[46]

By December 1845, though, Bunsen had breathed new life into the project and things were moving apace at Woolwich. The Sabines were making progress on the first volume of their *Cosmos* (with an occasional glance at Charles Galusky's French rendering for Gide and Company, when they were in two minds about how to translate a term) and sheets of the translation were sent to Humboldt in Berlin for his perusal. 'He is quite sufficiently versed in our language to judge both of the sense and of the style,' Sabine assured Murray.[47] Humboldt responded just after Christmas that he was not sure he could also manage to read through the English proofs, since he kept on being lumbered with court duties.[48] As Sabine later realised, Humboldt's 'age, eminence & multifarious occupations' would require them to wait too long, losing valuable time.[49] By asking Humboldt to read a draft of their translation, the Sabines were seeking his authorisation of their work. They were essentially also enabling him, as Williams had done some thirty years before, to be involved in the creation of the English translation. In fact, Sabine originally suggested this as an advertising feature to Murray: '"This translation is made at the request of the Baron von Humboldt, & the sheets will be seen by him before their publication" or, something to that effect.'[50]

Longman and Murray were begging the Sabines by mid-December 1845 to 'proceed with the translation with all the speed that your convenience will allow of', and waving under their noses the promise of copyright ownership of the translation of the first German volume and £100 in notes nine months from the day of publication.[51] But the Sabines were not to be browbeaten. Time was an important factor, but so too was quality. As Edward Sabine remarked:

> Baillière's translation is generally a tolerably correct rendering into English of the german [sic] sentences; its failure so generally recognized as an adequate translation is owing to the peculiarity of the work itself. To make even a

moderately good translation of Cosmos must be in every hand a work of very considerable <u>labour</u>.⁵²

He reiterated this point as pressure from the publishers increased. 'It is a work of no ordinary difficulty, & requires time, as is seen by the delay of the french [*sic*] edition begun so long before ours,' he stressed in April of the next year.⁵³

As the Sabines progressed to the second volume in 1847, Edward continued to assure his publishers that deadlines could be met.⁵⁴ In mid-November they had received a complete copy of the second volume and were now under pressure to finish its English counterpart. 'There remain to be translated and printed 287 pages. Health permitting I think Mrs Sabine w^d finish this by the middle of February,' wrote Sabine optimistically, working on the premise that in the next three months (the Christmas festivities notwithstanding), his wife would complete just under 100 pages per month, or just over three pages of translated text a day.⁵⁵ This hardly seems demanding by today's standards, but it required her constant application, day in, day out, to a project on which she had already been labouring for the previous two years. Moreover, the solid teamwork that had hitherto been the Sabines' successful formula could no longer see them through:

> I cannot afford you a stronger proof of our anxiety to complete the work at the earliest possible period than the fact of Mrs Sabine translating the 53 pages since the last sheets were received, during a severe illness from which I have been suffering which has so affected my eyes that I have been unable to use them since the letters to you & Mr Longman on the receipt of the German sheets.⁵⁶

The sheer scale, speed and complexity of the undertaking were already taking their toll, as much on Humboldt himself as on those producing his English translation, and it would be this that, gradually, would cause them to be overtaken by their rivals working for Bohn.

Cosmos Emerging in the Bohn Edition

How well did Otté cope with translating the first volumes of the edition for Bohn's 'Scientific Library' series, how did she make her mark on the translation and how was it received by a British public? Otté's rendering of *Kosmos* has been seen by some modern critics as inferior. Margarita Bowen considers it 'rather badly translated, often in pompous and inaccurate prose, with its theory of science filtered through a rather crude form of materialist sense-empiricism' (Bowen 1981: 260). This ill reflects contemporary opinion. Bohn purported to have sold 5,000 copies of his first edition alone by March 1849, and the importance of the 1850–9 edition on the American market cannot be overemphasised (Humboldt 2009: 359). In correspondence with Prescott in 1849, Otté gives us a glimpse of the concentration and total commitment it required:

> I have been so assiduously engaged in translating, that to own the truth, I seldom do write a note or letter, that can by any pretext, good or bad, be avoided –
> My last work has been the translation of the venerable Humboldt's 'Cosmos' [. . .] – I fear from its scientific character that it may not prove intrinsically interesting to you, [. . .] but as Humboldt so peculiarly possesses the power of reflecting the grace and brightness of his mind on all the subjects that he treats of, I am confident that if you have not read the work, its perusal will afford you some hours calm enjoyment.[57]

Unlike Elizabeth Sabine, Otté was a far more visible figure in her translation, both through the 'Translator's Preface' and the additional footnotes, explicitly marked as by the translator, which referred the reader to related scientific texts. Almost three-quarters of the four-page Preface constituted a biographical account of Humboldt. The remainder was devoted to the rather imprudent assertion that the work would form only three volumes (of which Humboldt had apparently informed Bohn in a letter). This Preface therefore successfully conveyed the impression that the three Bohn volumes in preparation constituted a whole, imposing upon the narrative a sense of completeness. Otté, more openly than the Sabines, would acknowledge the help of other scientists in completing the translation, including, of course, George Day, but also the Professor of Natural History at Edinburgh, Edward Forbes, and the English meteorologist and balloonist James Glaisher.

Otté's footnotes were appended to the first volume only – an irregularity that suggests these were deliberately deployed by Bohn in the initial skirmish (point 2 of the 'Distinctive Features of My Edition' in Bohn's 'Rejoinder') and thereafter quietly forgotten.[58] These fifty-two footnotes were not provocative, in the way that Black's had been, nor were they quite so self-promotional. Rather, Otté used them to alert readers to related secondary material, including 'popular' geological works such as Gideon Mantell's *Wonders of Geology* (1838) and Lyell's *Principles of Geology*, and French and German journals such as Jean-Baptiste Dumas's *Annales de chimie* [*Annals of Chemistry*] and Johann Christian Poggendorff's *Annalen der Physik* [*Annals of Physics*]. Implicitly, these demonstrated that Otté was well versed in the various types of scientific literature on the European market. Explicitly, they pointed Humboldt's

British followers towards scientific works aimed at a general readership as well as more specialist journals, indicating that Otté presumed her target audience to comprise a heterogeneous group of people with diverging intellectual aims and interests.

One note highlights an aspect of Otté's engagement with science that looks beyond her role purely as the mediator of Humboldt's ideas. In a section on cosmic bodies, Humboldt had described the shape of comets, the size and definition of their nucleus and the length of their tails of vapour (CO I, 85–6). He had commented that, in a few rare cases, the nucleus was so bright that it 'appears like a star of the first or second magnitude, and has even been seen in bright sunshine; as, for instance, in the large comets of 1402, 1532, 1577, 1744, and 1843' (CO I, 85). As Otté eagerly noted, she herself had witnessed the latest of the large comets Humboldt had listed. On 28 February 1843 she had been in New Bedford, Massachusetts, and 'distinctly saw the comet, between 1 and 2 in the afternoon', noting further that 'the sky at the time was intensely blue, and the sun shining with a dazzling brightness unknown in European climates' (CO I, 86). This account makes her a key witness of an important scientific event and emphasises her direct engagement with scientific knowledge-making since she can confirm the brightness of the comet's nucleus. She therefore enters into a different kind of collaboration with Humboldt from that of previous translators, by guaranteeing the veracity of his account through her own testimony, and, by injecting her own experiences into the paratext of his narrative, can heighten the imaginative appeal of his writing.

Opposites of Attraction: *Cosmos* by the Sabines and by Otté

Apart from these obvious paratextual differences, what characterised the early volumes of these translations published by Longman and Murray and by Bohn, and how did they differ from Baillière's edition? If we compare the versions produced by Prichard, by the Sabines and by Otté, it is clear that the first English translation was improved upon as later versions honed sentence structure and style. Prichard's cautious, literal approach had led to inelegant phrasing and mistranslations that generated a foreignising text. Nevertheless, it could still have usefully served as a 'negative norm' (Frank and Schultze 1988: 107) for Humboldt's later translators and alerted them to the pitfalls of translating *Kosmos* too literally. Sabine and Otté made Humboldt's prose vivid, spontaneous and dynamic through the adoption of translational strategies that largely coincided, even if their translations themselves could not. Metaphor enhanced the visual power of Humboldt's narrative, direct speech lent Humboldt's prose greater immediacy than paraphrase, and a first-person narrative voice (rather than passive constructions) heightened the apparent directness and intimacy of the prose. These techniques were all employed at various points in the editions by Longman and Murray and by Bohn, but they occurred contrapuntally, so that where one feature was present in Sabine's translation, the converse was selected by Otté. They therefore confirm Delabastita's assertion that one translation can have a marked influence *in absentia* upon its successors.

Kosmos was not directly at the forefront of a new scientific revolution, but it did reiterate Humboldt's ideas about how the universe could be seen in terms of a vast network of interacting forces. Metaphor played a crucial role in forging analogies, explaining new and unfamiliar models in terms of the familiar and more generally giving scientific writing greater imaginative appeal. Thomas Kuhn has argued that metaphorical usage patterns are helpful in conveying new scientific ideas because metaphors work by 'creating or calling forth the similarities' between phenomena we know and those that are presented to us as new (Kuhn 1993: 533). They are invaluable in structuring the way that we visualise and categorise scientific ideas. While Sabine and Otté were both aware of this, their translations did not always reflect the importance of metaphor and imagery to the same degree. As Humboldt outlined his intentions to study natural phenomena in their relation to humankind's needs and its general intellectual progress, he stated (in Sabine's translation) that 'its highest result is found in the knowledge of those mutual relations which link together the various powers of nature' (*CS* I, 4). Sabine's solution was a faithful rendering of Humboldt's intention to offer 'Einsicht in den Zusammenhang der Erscheinungen' [insight into the interconnection of phenomena]. Otté's solution was rather different: she had taken the idea of the connections between the various elements one stage further, making of this phrase the 'knowledge of the chain of connection, by which all forces are linked together' (*CO* I, 1). Her version therefore pressed the point home – using 'chain', 'connection' and 'linked' – that in Humboldt's concept of the universe, the forces of nature were coupled to each other within one great system.

As Humboldt reflected on how the knowledge scientists had acquired of the relationship between natural phenomena had developed at the same pace as the vast increase in individual scientific facts, he noted that gaps between areas of knowledge were now being filled: 'die Klüfte zwischen den Wesen werden ausgefüllt' (*K* 23). Sabine made ample use of this imagery to translate it as: 'The chasms which divide facts from each other are rapidly filling up' (*CS* I, 34). In so doing, she faithfully rendered the image of the 'Klüfte' – 'gulfs' dividing two entities and hindering them from being able to communicate with each other – as 'chasms', which were now swiftly being filled, such that the distinctions between different areas of science were now longer so clear-cut and interdisciplinary exchange and research could be encouraged. On this occasion, Otté dispensed altogether with the image of the 'gulf' or 'chasm', writing 'there has also arisen a more intimate knowledge of the connection existing among all phenomena' (*CO* I, 31). Perhaps she did not consider that the boundaries between disciplines were so sharply defined; perhaps she had difficulty drawing on similar imagery without using the same term as Sabine, which would make Otté's translation seem merely to be borrowing from the one that preceded it.

As Otté selected turns of phrase and imagery that would distance her rendering from the edition by Longman and Murray, she was often forced not only to rethink questions of imagery, but also to shift the register of her writing. Humboldt acknowledged the advantage his travels had brought him over other travellers: namely, 'that of having seen not merely coasts', as Sabine put it (*CS*, I, n.p.), but also the interior of the lands he described. Otté appears to have been compelled to rephrase this by drawing on less common vocabulary: namely, of 'having seen not only littoral districts' (*CO* I, x). Likewise, where Sabine had described early Greek poetry as 'semi-priestly' (*CS* II, 9), Otté opted for 'half-sacerdotal' (*CO* II, 375), and in the edition by Longman and Murray Oedipus approached the grove of the 'Furies' (*CS* II, 11), where in the Bohn edition the less common term, the 'Eumenides', was used (*CO* II, 377). Evidence of the converse – Otté adopting a lower register to contrast with Sabine's more formal, specialist vocabulary – can also be found. Where Sabine talked of the 'sidereal heavens' (*CS* I, 138) in a passage on the relative position of stars and nebulae, Otté opted for 'starry heavens' (*CO* I, 139), and where her rival had contrasted 'the telluric and the celestial' (*CS* II, 43), she used 'the earth itself, and the regions of space' (*CO* II, 38).

Similar oscillations between the two translations can be seen in the use of direct or indirect speech. In a passage in which he argued that an increased understanding of the natural world would not lessen our experience of nature, Humboldt had quoted the philosopher Burke, writing: 'Wir können daher dem geistreichen Burke nicht beipflichten, wenn er behauptet, daß „aus der Unwissenheit von den Dingen der Natur allein die Bewunderung und das Gefühl des Erhabenen entstehe"' (*K* 18). Sabine had offered: 'I cannot therefore agree with Burke when he says, that our ignorance of natural things is the principal source of our admiration, and of the feeling of the sublime' (*CS* I, 20). Otté came up with a slightly different version: 'I cannot, therefore, agree with Burke when he says, "it is our ignorance of natural things that causes all our admiration, and chiefly excites our passions"' (*CO* I, 19). Although her citation of Burke in quotation marks does not actually reproduce what he wrote, it did give her translation an apparent immediacy, since the philosopher appeared to be directly addressing the reader. On numerous occasions, Otté would style her translation in this way, allowing a series of figures – from the British aesthetician Burke, to the Greek philosopher Hesiod and the German mathematician Friedrich Wilhelm Bessel – to 'speak' in her text. Conversely, the French mathematician Pierre-Simon Laplace and the British philosopher Roger Bacon, who were quoted in direct speech in the Sabines' translation, were 'silenced' as they slipped into paraphrase in the Bohn edition.

A stylistic feature of Sabine's translation that it was easier for Otté to circumvent was what we have previously described as Sabine's 'double' translations: her use of two words to convey one in the German. In Humboldt's opening sentence, he reflected that *Kosmos* was appearing 'am späten Abend eines vielbewegten Lebens', which Sabine cast as in the 'evening of a varied and active life' (*K* 3; *CS* I, n.p.). She therefore used 'varied and active' to convey the compound adjective 'vielbewegt', where Otté had simply taken one term, 'active' (*CO* I, ix). Similarly, Sabine would speak of a 'vegetation fanned by soft and warm breezes' to describe a painting by Jan van Eyck that Humboldt had described as 'von lauen Lüften umweht', where Otté would use just a single adjective for 'lau' in her 'gentle breezes' (*CS* II, 79; *CO* II, 445; *K* 228). Sabine's doubling up of terms in her translation was hardly a sign of linguistic insecurity on the part of this highly experienced translator: rather, it was an attempt to convey the richness of the original, a text abounding with meanings and associations that she sought to convey as comprehensively as possible in English.

Another point of divergence between the two translations lay in their handling of nationhood and identity. Elizabeth Sabine's translation repeatedly reminded its British readership that the source text and its author were German. Humboldt's brief pointers to people and places in *Kosmos* were translated literally in the edition by Longman and Murray as 'by my own countryman, Carl Ritter', 'in our own country (Germany)' and 'In our German fatherland' (*CS* I, 15, 54; II, 66). Otté, by contrast, was careful to produce a domesticating translation, so that the reference to Ritter's nationality was completely suppressed, the explicit naming of Humboldt's home country disappeared, and references to the 'fatherland' were replaced by 'Germany' (*CO* I, 16, 49; II, 434).

It was not only in these examples that Sabine reminded her British readers that this was a translation and that Humboldt was not a supranational figure but a specifically German cultural icon. She occasionally left German terms in her text, which highlighted the sheer untranslatability of some of the notions that Humboldt introduced. In an introductory passage on the enjoyment that could be gained from contemplating nature, Humboldt had observed:

> Ein anderer Naturgenuß, ebenfalls nur das Gefühl ansprechend, ist der, welchen wir, nicht dem bloßen Eintritt *in das Freie* (wie wir tief bedeutsam in unserer Sprache sagen), sondern dem individuellen Charakter einer Gegend, gleichsam der physiognomischen Gestaltung der Oberfläche unseres Planeten verdanken. (*K* 10)

Sabine translated this passage into English, leaving in the original language the expression that Humboldt had himself considered so quintessentially German that he had italicised it:

> The mere contact with nature, the issuing forth into the open air, – that which by an expression of deep meaning my native language terms *in das Freie*, – exercises a soothing and calming influence on the sorrows and on the passions of men. (*CS* I, 6)

Otté's translation, by contrast, removed any sense of foreignness and any reflection upon what could be conveyed in one language but only inadequately captured in another:

> Mere communion with nature, mere contact with the free air, exercise a soothing yet strengthening influence on the wearied spirit, calm the storm of passion, and soften the heart when shaken by sorrow to its inmost depths. (*CO* I, 3)

The fundamental significance of Humboldt's comment turned upon the possibilities inherent in the German adjectival noun 'das Freie' to speak in more wide-ranging terms about 'freedom' than the more common (and decidedly politicised) noun 'die Freiheit'. For Humboldt, the domain of nature had always been coterminous with personal liberty, and looking back on his travels to South America from his perspective in the late 1840s, it must also have become synonymous with freedom from the constraints of court life, political turmoil and the pressures of writing and publication. Alice Jenkins argues that Otté's translation significantly downplayed Humboldt's reflections on language, a recurring interest in *Kosmos*, and the political associations they embodied (Jenkins 2007b: 101). The sense of movement conveyed in the accusative 'in das Freie' [into the open air] was certainly emphasised in Sabine's version through her English translation 'issuing forth into the open air'. Nature therefore mobilised the (scientific) spirit in ways less clear in the Bohn edition that simply read 'with the free air', which did not carry the range of associations that Humboldt had made with this term in German.

Although Elizabeth Sabine often conveyed sensitively how difficult it was to achieve equivalence between the German original and her English translation, on one occasion she almost admitted defeat. In a rare translator's footnote, she commented:

> In every language, views entertained in the infancy of nations have led to a confusion of the ideas of *earth* and *world*: the common expressions of 'voyages round the world', 'map of the world', 'new world', are instances of this confusion. The more accurate and more noble expressions* of 'system of the world', 'creation of the world', and others of a similar nature, relate either to the whole of the bodies with which celestial space is filled, or to the origin of the entire universe.
> *Our language does not possess all the expressions referred to by M. de Humboldt. We have no direct English equivalents for the expressive German terms 'Weltgebäude', 'Weltraum', and 'Weltkörper'. (CS I, 55–6)

Her footnote is astonishingly open in its scrutiny of the untranslatability of Humboldt's narrative. Rather than quietly passing over these difficulties – Otté's rendering italicised English versions, with no glimpse of the complexities inherent in trying to convey the meaning of compound nouns – Sabine opted to reveal the difficulties to her readers (CO I, 50). Like Prichard, then, she made no secret of her mediatory role in the text, but she would earn credit rather than contempt from her readers for pointing up the difficulties she had encountered.

Cosmos and Science

Where the editions by Longman and Murray and by Bohn were most likely to overlap was in the use of scientific terminology. As a specialist in one of the fields that Humboldt addressed, and with connections to experts in many others, Edward Sabine unintentionally aided Otté in negotiating some of the trickier aspects of *Kosmos*. While he went to some lengths to ensure that the correct terms were used to describe scientific apparatus, methods and phenomena, he also expended considerable energy on dealing with apparently straightforward terms. His correspondence with Herschel is instructive in understanding the amount of deliberation on such points that went into the Sabines' translation. A discussion of how to translate the German 'Kräfte' – 'forces' or 'powers' – is a case in point. Sabine opted in the first and second editions for 'powers'. Herschel's alternative suggestion, 'forces', made in October 1847, came too late for the second edition, which would be published around the end of that year. Sabine responded:

> You did not before notice the translation of the word Kräfte by Forces: if you had, it would assuredly have rec^d [...] very serious consideration, as it shall now, – (but it cannot be in time for the 2^d Edn which is printed and sold). I apprehend that Forces is the strict and usual translation of Kräfte in mechanics & generally in scientific writings. It is the usual accepted term in English; we speak of the mechanical, chemical, polar & vital forces; which, generically, constitute M. de H's Kräfte. The french [sic] translation used the word 'Forces' also in the same places.[59]

Galusky's French translation of *Kosmos* now came into its own for the Sabines as a way of double-checking that they had made the same decisions as him regarding terminology. But the discussion with Herschel regarding the possible meanings of 'powers' and 'forces' did not revolve solely around which term was correct in a scientific context. One of Sabine's concerns related to style, the other to the vexed question of the image it might convey to his more fervently religious readers:

> It is true that the frequent repetition (which is one of M. de H's characteristics) of the word Forces may have a tendency to produce in some minds a dry mechanical expression, such as you describe; / it does not do so in my mind, since I know the forces to be no less the creation of the Divinity than is matter itself) – but it appears to me there might be danger also in the substitution of the word Powers; and that in certain minds, it would give rise to the impression of an indwelling, self-existing life in created matter.[60]

The Sabines were therefore constantly preoccupied by the worry that Humboldt's *Cosmos* would be read not only as a scientific but also as a spiritual work, and that any implication, however small, that the universe was governed by scientific 'forces' rather than divine 'powers' could invite an interpretation with disastrous consequences for Humboldt's British reputation. Following what he described as 'the suspicions of my own mind', Sabine opted for the 'scientific word Forces', acknowledging that 'the impressions which may be created in others are important, & I will take time for reflection & consultation'.[61]

Herschel's reply by return post argued for the opposite on grounds that were not chiefly religious, but scientific:

> As you defend the use of 'forces' and consider that there may be danger in substituting 'powers' I can only hope that the further consideration you propose to give it will cause you to take a different view. It is right however for one to add that my own objection to it is not so much its materialistic appearance in a religious view (for of course force as well as power is equally a delegation

from & manifestation of the Creator) but that <u>Science has outgrown it</u> and that it is impossible to conceive mere mechanical <u>forces</u> to be the only agents in producing the phenomena of the material universe (to say nothing of those of organic life) without superadding <u>causes</u> antecedent to force & capable of producing forces, destroying, suspending or altering their intensity. To such causes, I conceive the word <u>powers</u> to apply naturally & properly.[62]

The depth of the debate between Sabine and Herschel, and the time and ink they expended in discussing potential English translations of this one German word, might seem surprising. But it indicates that terms had acquired quite precisely defined fields of meaning in scientific writing at mid-century. Scientific readers of *Cosmos* would have expected the Sabines to be alert to this and to ensure that these terms were correct in translation. Humboldt's translators also needed to be sensitive to the wider (non-scientific) context in which a word could be used and think about the associations that other readers in their presumed target audience would make.

Ticklish Subjects: *Cosmos* and Religion

The social theorist, Whig writer and mesmerist Harriet Martineau was singularly unimpressed by Humboldt's magisterial work. She devoted a full twelve pages to his life in her *Biographical Sketches* (1869) and observed, 'there can be no difference of opinion about his failure in his highest effort, as exhibited in his "Kosmos"' (Martineau 1869: 149). His rambling argumentation, clouds of words 'which look philosophical, but will not bear a plain rendering', and inability to present coherently his great 'Scheme of the Universe' were essential stylistic defects (Martineau 1869: 150). But it was really the work's failure to tackle religious matters head-on that, in Martineau's opinion, caused it to fall disastrously short.

> On the essential topics, for instance, of Creation, of Spontaneous Generation, and of the basis and scope of Mental Philosophy, with some other such ticklish subjects [she observed], he either keeps back his views, or permits them to be discoverable only by a process of inference of which none but the highly qualified are capable. (Martineau 1869: 150)

Having converted from Unitarianism to atheism as an adult, Martineau was delighted by works like Darwin's *Origin of Species* (1859), a book she described to her fellow Malthusian and atheist George Holyoake as 'overthrowing (if true) revealed Religion on the one hand, & Natural [theology] (as far as Final Causes & Design are concerned) on the other'

(Logan 2007: 208–9). What she disliked about *Cosmos* was Humboldt's opacity on the issue of creation. As she astutely noted, the Sabines' translation could do little to hide what she perceived as the inadequacies of the original:

> The word 'creation' used repeatedly for the universe, is misleading, or at least perplexing to readers who have duly attended to some preceding passages; and if they turn to the original they find that Humboldt spoke of the frame of things, the universe, the collective phenomena of nature, or the like. (Martineau 1869: 150–1)

Martineau, who had herself been translating Auguste Comte's work on positive philosophy, the *Cours de philosophie positive* (1839) [*Course in Positive Philosophy* (trans. 1853)], while the Sabines laboured over the third volume of Humboldt's *Kosmos*, was well aware of the translator's power to modify, manipulate and, if necessary, rework. Of all Humboldt's British readers and commentators, Martineau was definitely one of the most thorough. She did not read the Sabines' translation 'in isolation' but took the time to go back to the German original and compare the texts word for word. As a result, she discovered that the Sabines had repeatedly used the religiously connoted word 'creation' where no such term could be found in *Kosmos* – adjustments that, to her way of thinking, clouded Humboldt's line of argument. To the Sabines, they were necessary changes made to deflect criticism from parts of the British establishment that Humboldt's work lacked appropriate references to God.

Not everyone saw a discrepancy between science and religion in Humboldt's writings. Journals with a demonstratively religious orientation, like the *Wesleyan-Methodist Magazine*, had already cited lengthy passages from Humboldt's *Personal Narrative*, in which he had found seashells on the Andes at an elevation of 14,120 feet above the level of the sea, as evidence of God's work. In an article titled 'Geological Confirmations of the Universal Deluge', the reviewer had used such evidence to conclude 'that the researches of Geology confirm the fact of a Universal Deluge, and thus afford a *sensible* proof of the credibility of the Sacred Historian, and consequently, of the truth of the doctrines of Divine Revelation'.[63] The same journal would record a couple of years later, in its 'Select List of Books Recently Published, Chiefly Religious', Humboldt's *Personal Narrative* (in Williams's translation) as a 'very scientific and valuable volume'.[64] While radical atheists had unsuccessfully tried in the 1840s to promulgate a materialistic vision of the sciences, a Christian interpretation of science still dominated throughout the nineteenth century. It is in this light that the many enthusiastic British reviews of *Cosmos* can be viewed, not least the one published in the *Eclectic Review*, which observed in 1852: 'The larger our acquaintance with facts, [. . .] the profounder and devouter will be the sentiment with which we adore the Invisible Being, of whose presence they are the signs, and of whose wisdom and goodness they are glorious manifestations.'[65]

But this did nothing to stop the vociferous protests at Humboldt's *Kosmos*. Forbes, whose piece for the *Quarterly Review* would remain burned into Humboldt's memory, felt compelled to ask: Where was God in Humboldt's mapping of the universe? How could this work be positioned in relation to an ever-widening corpus of work on natural theology? While Forbes detested those works of natural science that dragged in religious material at every turn, he also abhorred pedantry of the opposite kind. 'We conceive it to be impossible for any well-constituted mind to contemplate the sum and totality of creation', he admonished, 'without referring more or less to the doctrine of final causes, and to the *design* of a superintending Providence.'[66] This 'pedantry of intellect' had stifled those thoughts that would naturally arise as one contemplated the remoteness of space and the connection between human beings and the universe. Humboldt appeared to have silenced God.

Thomas Carlyle thought likewise. Reading *Kosmos* in the original German, he remarked to the Presbyterian mathematician Thomas Chalmers in February 1847,

> I also have to say that Humboldt's *Kosmos* gave me the same sad feeling that it has given you; the feeling namely that this view of Nature was an altogether *un*worshipping, and therefore as I think, unworthy, lamed, and indeed inhuman one! (Ryals and Fielding 1993: 165)

Carlyle's associates were still decrying the religious shortcomings of *Cosmos* some two decades later. The leading English art critic and evangelical Christian John Ruskin, writing in March 1876 (just before his own 'unconversion'), noted wryly:

> I've been looking at Humboldt. I see he 'defines' the vital force as 'that which prevents the original affinities from acting'. Not at all which acts itself! What a lovely and cheerful view of life! (Mortal and other). God – as the Preventor of Original Affinities from acting – and omnipotent Drag upon Originality? *isn't* it nice? (Cate 1982: 229)

For Humboldt's more conservative Victorian readers expecting affirmation of the Divine in the grand scheme of things, *Cosmos* would fall rather short of the mark.

The timing of the appearance of English translations of *Kosmos* on the British book market is crucial in understanding why some gave it such a prickly reception. The edition by Longman and Murray started to appear in 1846 (or mid-1845, if Bohn's Rejoinder is to be believed), just a year or so after the anonymously published *Vestiges of the Natural History of Creation*. More controversial than any scientific or philosophical book of its day, this work by the Scotsman Robert Chambers combined astronomy, geology, anthropology, psychology, physiology and theology into a general theory of creation (Secord 2000: 1). Starting from the notion of a primeval nebulous fire-mist in which suns and planets had originated, it described the spontaneous generation of life and its subsequent diversification, working through from the development of apes to end with human society. An evolutionary work that paved the way for Darwin's *Origin* (even if there was much science in the *Vestiges* that Darwin noted was deeply flawed), it was a visionary, daring piece that greatly appealed to those in progressive Whig circles (Secord 2000: 117). It gained immediate popularity on its initial appearance and was even being read to the young Queen Victoria in early 1845 by Prince Albert, who was particularly interested in supporting scientific endeavour (Secord 2000: 168). But the *Vestiges* lost its prestige just as quickly as it had gained it. Denounced by the more zealous as atheistic, it also came to be viewed as materialistic and threatening to religion (Knight 2004: 101). Critical articles in the *Edinburgh Review* and the *North British Review* tried to limit its success in 1845, and by summer 1846 it was no longer the subject of fashionable conversation (Secord 2000: 38).

Such was the context within which Humboldt's *Kosmos* and its English editions were reviewed in the mid-1840s. It is hardly surprising that the *Westminster Review* ran a mammoth, fifty-page, double critique of Humboldt's *Kosmos* together with the *Vestiges*, still anonymous and now in its fourth edition, in 1845. Crosse's comparison of the two works concluded that they acted 'as explanatory or corrective of each other', each in some way responding to the reality that the accumulation of scientific facts had 'outrun any reasonable arrangement of them'.[67] Both were also more 'popular' scientific works: they deliberately encouraged the invasion of the hallowed halls of science by the 'clamorous cries of the thronging half-educated masses'.[68] However, Crosse was at pains to distinguish the *Vestiges* – a book that 'tickled the dull repose of collegiate orthodoxy' but would 'perish instantly' – from *Kosmos*, a work that he felt did make some valuable contributions to science.[69] But it was unclear to him what its spiritual message might be:

> There is an occasional union of science and sentiment which will appear strangely unnecessary to many a hard inductive reasoner; and a sketch of the

universe in which the *word* 'God' appears nowhere, but the *spirit of God* is supposed everywhere, will perhaps be regarded as dangerously Atheistical by the stickler for *the word*, and as 'rankly material' by many a repentant candidate for lost religious reputation.[70]

Any suggestions that Humboldt's work could be considered 'dangerously Atheistical' would have alarmed both Humboldt's British publishers, worried about a loss in sales, and his translators. The challenge that the Sabines and Otté continued to face was how they could subtly employ translational creativity to steer *Cosmos* clear of the obstacles that had caused *Vestiges* to run aground, yet present British readers with an accurate rendering of the German original.

Humboldt registered these assaults on *Cosmos* with some surprise. Yet, as early as 1805, when he had been discussing the very first English translations with Pictet while living in Paris, Humboldt had given him *carte blanche* to change anything that seemed to him too local, too much in favour of France and insufficiently Christian (Humboldt 1869: 45). Humboldt therefore sensed that his work might be deemed implicitly irreligious by some. He knew all too well the conservative nature of his British readers – he frequently used the word 'puritanical' and 'British' in the same breath – but he was ill prepared for the reaction that *Kosmos* would provoke in his own country. The Catholic fortnightly periodical, the *Historisch-politische Blätter für das katholische Deutschland* [*Historical-Political Papers for Catholic Germany*], seemed resigned to the fact that he had at no point stressed that God had created and sustained the world, and added darkly that it did not want to draw too many conclusions about his own inner convictions.[71] God was not much in evidence for the reviewers of the *Die Grenzboten* [*The Border Messengers*], which cast Humboldt as a radical, rampantly progressive scientific atheist.[72] As Humboldt noted wryly to Arago, South German clerics now thought him involved in all sorts of anti-Christian machinations and demagogical plots (Humboldt 1908: 265). To combat such disapproval, Humboldt added a sentence to the French translation in which he expressly stated that everything beyond the laws of science belonged to another, higher, realm, and referred to Kant's famous openly Christian essay 'Die Allgemeine Naturgeschichte und Theorie des Himmels' ['Universal Natural History and Theory of the Heavens'] (1755).

The Sabines were not oblivious to the attacks made on Humboldt following the appearance of the first volume of *Kosmos* in German in 1845. Although they did not tackle the question of religion in the first volume, they were certain of its repercussions for the translation of the second. Edward Sabine corresponded at length with Herschel over how to resolve this problem. Sabine 'felt less and less disposed to concur in the severity' with which Forbes had attacked *Cosmos* in the *Quarterly Review* but he did recognise that something needed to be done.[73] Elizabeth Sabine had already tried to raise this question with Humboldt:

> On the completion of the 1st. vol, and the receipt of the 2d, Mrs Sabine [. . .] wrote to the effect, that whilst engaged in the translation after the 1st vol, her mind had been frequently led by his descriptions / as when in presence of the grand & beautiful in nature / to the reflections so appropriately expressed in the 24.th verse of the same psalm (the 104th) from which he has so largely quoted in his remarks on the poetry of the Hebrews, but which did not happen to be one of the verses quoted; 'O Lord, how manifold are thy works; in wisdom has thou made them all; the earth is full of thy riches.'[74]

Humboldt's response to Elizabeth had been important both for its affirmation of his own spiritual persuasions and for the freedoms he would accord his translators. As Edward Sabine noted:

> I have not M. de H's reply with me, but it conveyed an evident and decided gratification [. . .] and (wholly unsolicited) invited her to introduce in the translation the verse to which she had referred. The sheet in which the other verses are quoted had sometime been printed; moreover the addition in that place would have had no particular suitability; she now thinks of introducing it instead in the preface to the 2d vol, (of course with an express reference to M. de H's sanction) and where it might be brought in more suitably as expressing generally the acknowledgements suggested by the perusal of M. de H's work.[75]

Humboldt therefore sanctioned a translation by the Sabines that would offer a more overt religious interpretation of the workings of the universe. Edward Sabine did not see this as a belated attempt to curry favour with British readers. Rather, as he noted ruefully to Herschel, its aim was to correct the picture that Forbes had felt duty-bound to paint of *Cosmos*, even if 'the impression which he received from passages such as the one referred to, was not what M. de H. intended to convey, or that which he himself approves'.[76] The problem was exacerbated by the rather traditional attitudes of a British reading public. 'I quite agree with you', he wrote to Herschel in 1847, 'that the absence of any notice of the evidences of thought, intention & design in the works of creation is even painful; to english [*sic*] readers in general, it must form a most striking & serious defect', adding, 'I doubt whether a feeling of that nature exists in Germany in reference to Cosmos.'[77] The Sabines' translation was, then, to act as a corrective, rehabilitating Humboldt as a scientist who would never strive 'to deprive the Creator of his praise'.[78]

Martineau's attack on the Sabines' liberal use of the words 'creation' and 'Creator' was not unfounded. In the introductory observations in the first volume of *Cosmos*, Elizabeth had specifically added a reference to God – with Humboldt's sanction[79] – so that in her translation she remarked: 'It is this necessity, this occult but permanent connection, this periodical recurrence in the progressive development of forms, of phænomena and of events, which constitute nature obedient to the first-imparted impulse of the Creator' (*CS* I, 33). Otté's translation carried no such reference to 'the Creator' at this point because there was, quite simply, none in the German original (*CO* I, 30).

This is not to say that those working on the Bohn edition eschewed the idea of injecting religiosity into their version of Humboldt's *Cosmos*. Statistically, the editions by Bohn and by Longman and Murray of the first two volumes used the words 'creation' and 'Creator' to approximately the same degree – and generally in the same places – to inject a greater religious feel into Humboldt's work. This was particularly true of the first volume where both Otté and the Sabines described the impression that the heavens and the rich plant life made on the scholarly eye as 'more worthy of the majesty of creation' than on minds untrained to recognise causal relations in the natural world (*CS* I, 20; *CO* I, 19). The German original had simply suggested that such sights offered 'einen großartigeren Anblick' ['a more splendid sight'] (*K* 18). And where in *Kosmos* it was 'Die Fülle der lebendigen Gestalten' ['the wealth of animate forms'], which Humboldt discussed in terms of zones of habitation, his English translators made of this either 'the multitude of organised bodies which embellish creation' or 'the innumerable multitude of organised bodies, which form so large a portion of the beauties of creation' (*CS* I, 48; *CO* I, 43; *K* 29).

The Sabines subtly employed addition to make *Cosmos* appetising fare for those with more orthodox religious tastes but they also used omission to free it of more awkward passages. Humboldt's humorous aside on English observatories not being allowed to make observations on a Sunday was deftly excised from the first volume, as was a section that alluded to the 'cradle of humanity'. The sentence that appeared to be most offensive was actually taken from Wilhelm von Humboldt's unpublished work ['Über die Verschiedenheit der Sprachen und Völker' ['On the Varieties of Languages and Nations']. In Otté's translation it ran thus: 'The separate mythical relations found to exist independently of one another in different parts of the earth, appear to refute the first hypothesis, and concur in ascribing the generation of the whole human race to the union of one pair' (*CO* I, 364). This, Bohn roundly observed in his Rejoinder, alluded to 'Adam and Eve and Paradise' and he accused the Sabines of suppressing information included in the original.

Bohn, who would soon be busy extending his 'Scientific Library' series to include the Bridgewater Treatises – works of popular science commissioned to explore 'the Power, Wisdom, and Goodness of God, as manifested in the Creation' – had a good sense of what the religious impact of Humboldt's *Cosmos* needed to be. In his edition one other subtle, but significant, change was made to those sections in which Humboldt had paraphrased or quoted from the Bible. Humboldt had gestured towards the relevance of Psalm 104 in presenting a picture of the entire cosmos (*K* 211), but had paraphrased rather than quoted directly from the Bible the nine verses to which he explicitly referred. In the Sabines' translation, this section starts with 'The Lord covereth himself with light as with a garment, He hath stretched out the heavens like a canopy' (*CS* II, 45). Otté's version takes the relevant passage straight from the King James Version: 'Who coverest thyself with light as with a garment: who stretchest out the heavens like a curtain' (*CO* II, 413). Since Humboldt's own rendering was itself a loose paraphrase, Elizabeth Sabine saw no need to quote directly. But Otté forged an explicit link between *Kosmos* and the Bible that would better suit conservative British readers expecting to hear the word of God in Humboldt's work.

'A Pure Flame Expiring': The Final Volumes of *Kosmos*

In summer 1849 Humboldt outlined in a letter to the German philologist August Böckh how the third (and, at that stage, final) volume of *Kosmos* would captivate readers by describing all the attempts made over the previous 2,500 years to improve human understanding of the natural world (Humboldt 2011: 191). The main body would merely enumerate these in a simple descriptive style; the introduction, by contrast, called for a warmth of manner to smooth the transition from the almost wholly literary second volume. 'You will smile', he noted drily to Böckh, 'at an eighty-year-old man placing such importance on a drama that is played out before the most inattentive of audiences' (Humboldt 2011: 192). Although Humboldt appeared in control of the third volume of *Kosmos*, it swiftly gained the upper hand and he was able to complete its second part on time only by relying heavily on the help of the secretary and librarian Eduard Buschmann (Humboldt 2009: 478). *Kosmos* was expanding as fast as Humboldt could write it. As he watched it spill over into a fourth volume, he wrote dispiritedly to Bunsen that he could remain master of his materials – 'Herr der Materialien' – only by sorting his notes into a series of cardboard boxes and registers (Humboldt 2006: 146). In April 1852, Humboldt informed Cotta that he hoped to finish the fourth volume himself, but should it remain incomplete on his death, then von Buch, Dove, the zoologist Wilhelm Peters and the botanist Johann Friedrich Klotzsch would be able to tie up unfinished ends (Humboldt 2009: 482). By Christmas 1856, the sheer mass of data Humboldt was trying to tame, coupled with his own advancing age and a recurring skin irritation, had become too much. As he reflected to his old friend Bunsen, it was discomforting to witness the loss of his own 'phosphorus of thought' (Humboldt 2006: 199).

By 1850, Longman was becoming increasingly ruthless in his efforts to forestall Bohn and imposing ever-tighter deadlines on his translators. The Sabines had intended to take a summer break, but were now compelled either to shelve their travel plans or to place the translation in another's hands. As Sabine wrote courteously to Murray early in September 1850, they had opted for the latter and wanted no more part in translating or overseeing the third volume of *Kosmos*: 'With respect to the revision of a translation to be made by any other person, I beg, without the least disrespect to whoever the Gentleman finally undertaking it may be, to decline it.'[80] For Humboldt, this abrupt and ill-timed decision by his British publishers was an immense blow. The Sabines were a couple whose combined scientific and linguistic expertise made their translations elegant and scientifically accurate. Together, they gave *Kosmos* a homogeneity that ensured terminological consistency and a harmony of style and tone that characterised Humboldt's 'voice' in English translation. Most importantly, he counted them among his British friends. Incensed at their treatment by Longman, Humboldt instructed Cotta to write to London forthwith, to protest against any 'amateur astronomer' being used in the place of Colonel Sabine (Humboldt 2009: 447). Oil was poured on troubled waters, since the Longman ledgers for April 1850 to February 1851 indicate that money had been paid to a 'J. Johns' for doing the corrections and a 'M. Poullain' for translating part of the third volume, even if the volume itself bears the characteristic phrase 'Translated under the Superintendence of Colonel Edward Sabine'.[81]

In 1853, things took another turn for the worse. Elizabeth Sabine was struggling with illness as she worked on the opening sections of the fourth volume and with her increasingly disgruntled publisher (Humboldt 2009: 516). Longman must have asserted that his financial outlay for the third volume had been excessive and he was no longer prepared to pay for the proof sheets to be sent over from Germany, so Sabine was forced to consider paying for them himself (Humboldt 2009: 519). The Sabines were certainly remunerated generously for their translation – they received £6 17s 6d per sheet and therefore a total of around £96 for the second part of the third volume, which was more than for the first and second volumes of *Cosmos*, paid at £5 17s 0d a sheet.[82] Considering the highly specialised nature of their undertaking, the author's obvious satisfaction with their translation and the guarantee that copies would sell, Longman's complaints seem hard to justify. Incensed at the unjust treatment of the Sabines, Humboldt took the matter into his own hands and suggested that Cotta deliver the sheets directly to them, bypassing Longman, who would have no choice but to wait patiently for the translators to deliver (Humboldt 2009: 521).

Longman's nervousness was perhaps justified. Dassow Walls remarks that the later volumes, detailing scientific results, had fewer readers (Dassow Walls 2009: 249). Longman's ledgers show that while the first two volumes appeared in print runs of 1,000 each time – and there is even a record of 2,000 copies of the seventh edition of the cheap two-volume version in April 1849, followed by a further 1,000 in January 1850 and another 1,000 in October of the same year – print runs for the later volumes selling in the mid-1850s were lower.[83] With the exception of the fourth volume (of which 1,000 copies were printed in May 1858), all other print runs remained well below this mark: the first part of the third volume came on to the market in batches of only 500 in 1851. There is no evidence that larger numbers of copies followed. The *Athenæum* and the London *Literary Gazette* did review the Sabines' *Cosmos* in its third volume, even if, as the critic in the *Athenæum* observed, Humboldt's zeal was now 'like a pure flame expiring'.[84]

But the publication of the first part of the fourth volume in the edition by Longman and Murray suddenly galvanised British reviewers into action. Instead of eliciting polite exhaustion, it energised them into a show of patriotism as they lauded Edward Sabine for his 'Editor's Notes', which made a seminal contribution to the work. The strong focus on terrestrial magnetism – Sabine's own specialism – meant that Humboldt had urged him to 'make many insertions and additions in dealing with this portion of his work', which resulted, according to the reviewer in the *Examiner*, in the Sabines' English translation having 'a great advantage over every other'.[85] Sabine turned his Notes into a *laudatio* to the British government, the British Trigonometrical Survey and the Board of Longitude. Not one to hide his light under a bushel, he also reminded readers of his own part in recommending to the Board that the series of gravitational measurements completed in 1819 under the physicist Captain Henry Kater should be pursued across a wider geographical area – and thus fulfilled an important role 'in matters appertaining to the advancement of science and to the national duty in its promotion'

(*CS* IV:1, 472). Some credit did go to the German Magnetic Association, overseen by Gauss and Wilhelm Eduard Weber, for its work on the irregular disturbances in magnetic reading and magnetic variation. However, as Sabine did not fail to recall:

> The scheme of the British colonial observatories, adopted by the British Government on the joint recommendation of the Royal Society and of the British Association for the advancement of science [*sic*], was much more comprehensive than that of the German Association, inasmuch as it included the investigation of the laws of all the magnetic variations, secular, periodical, and occasional, sensible at the surface of the globe; and embraced as distinct subjects of inquiry the phenomena of each of the three magnetic elements, the declination, the inclination, and the intensity of the magnetic force. (*CS* IV:1, 484)

Sabine's rhetoric of colonial superiority is hard to ignore, as indeed is his emphasis on the thoroughness with which the British worked to gather the separate data that underpinned Humboldt's important discoveries regarding the larger forces at work across the globe.

Although the Sabines were largely protected from the caprices of their publishers during Humboldt's lifetime, after his death the project immediately began to falter. The fifth volume of *Kosmos* was left incomplete but Buschmann's efforts ensured that a good 150 pages of text and notes were published, together with a brief passage from Sabine on variations in magnetic inclination, and an index to the contents of *Kosmos* in 1862. Longman, sensing that this would not be a profitable exercise, quickly put the dampers on any further translation work. 'We are afraid of going on with Cosmos,' he noted to Sabine just before Christmas 1860, justifying this decision by stating, 'I am very sorry the "commercial view" will not allow us to complete the book, but Bohn's competition has spoiled the market.'[86]

Cosmos and Closure

The collaborative efforts required to complete the translation in the Bohn edition suggest that Otté was under the same kind of strain as the Sabines in the race to finish. While the third Bohn volume is attributed to her as sole translator, the chemist Benjamin Horatio Paul (1827–1917) assisted her with volume four and the zoologist William Sweetland Dallas (1824–90) with volume five. Paul had gained his PhD with the German chemist Justus von Liebig in Giessen in 1848 and also studied under the Scottish chemist Thomas Graham, a pioneer in dialysis and the diffusion of gases. He was also an 'accomplished and versatile linguist', with an 'intimate knowledge of Continental life and thought', who counted among his friends Dante Gabriel Rossetti and others belonging to the Pre-Raphaelite brotherhood. [87] Like Prichard, he had a good knowledge of scientific German and was well versed in modern scientific developments. While Paul's specialisms covered more of the areas of science treated in *Kosmos* than Prichard's had ever done, it is strange that Paul aided Otté with the fourth volume. With its heavy emphasis on astronomy, it seems far less suited to his expertise than the fifth volume of the Bohn edition, which tackled subjects closer to Paul's interests, notably the chemical composition of rock, gaseous springs and naphtha emissions.

Dallas – now best known as Darwin's painstaking, pedantic indexer – was largely self-taught in natural history. In 1849 he was elected Fellow of the Linnean Society and later became curator of the Yorkshire Philosophical Society's museum. A specialist in German, he translated foreign articles for journals such as the *Chemical Gazette* (1852–9) and the *Annals and Magazine of Natural History* (1852–90). Henry Huxley would commend him to the Council of the Geological Society as 'one of the hardest workers that I know'.[88] Industriousness was doubtless what Bohn was seeking in a translator to help complete the editions of *Cosmos*, and Dallas's geological interests would have been useful when working on the fifth volume, likewise his eye for detail.

Since neither Bohn nor Longman and Murray tackled the final volume of the German edition (which broke off in the middle of a discussion of porphyry in St Petersburg), the narrative was essentially left in mid-air. It lacked the polished ending that Humboldt would have endorsed. The closing sentence of *Cosmos* in the Bohn edition reads only a little differently in the Sabines' translation:

> We are probably to include among these phænomena [sic] the old masses of trap of the lower Silurian formation of the south-west of England, by whose exact chronometric determination my illustrious friend Sir Roderick Murchison has so greatly enlarged the scope and heightened the character of our knowledge of the geological construction of the globe. (CS IV, 1, 448)

Humboldt's long, involved, sentence that acknowledged the findings of the British geologist Sir Roderick Impey Murchison was no great rhetorical flourish, even if the last word 'globe' was apposite in a work that had aimed to chart the earth and all beyond it. But in some ways it did neatly round off this unfinished work. Particularly in the edition for Longman and Murray it could be interpreted by British readers as Humboldt's closing recognition of the great achievements of British scientists and scientific institutions, which Sabine would immediately reinforce in the unashamedly patriotic 'Editor's Notes' that followed.

Conclusions

A series of tensions were played out in the translation of *Kosmos* that were essential in shaping the legacy of this edition in the minds of Humboldt's British readers. Elizabeth Sabine's production of a foreignising translation repeatedly drew attention to the fact that the source text and its author were German, despite the stylistically elegant English of her work. Yet Edward's editorial notes were deeply assimilatory, locating Humboldt's achievements within a distinctly British scientific tradition and therefore conceptually reclaiming parts of *Kosmos* as the result of British interests in the advancement of science. The volumes published by Bohn made no such claims. They cast Humboldt's *Kosmos* as a supranational work committed to demonstrating how the scientific community worked across linguistic, cultural and political divides to further knowledge of the universe.

It was the paratext more than the text itself that sundered any similarities between the rival translations. Given that Humboldt had fully endorsed Edward Sabine's intervention in the text, it is difficult to cast the edition by Longman and Murray as a misappropriation of the original. But by highlighting his own authorship of key parts of the work and his centrality as Humboldt's British mediator, Sabine played an important part in giving his *Cosmos* a different character from the version issued by Bohn. Its scientific significance, reinforced through heavily fact-laden notes, corrections and additions, implicitly oriented it towards a more specialist audience. Bohn, strategic to the last, issued an edition that was light in touch in its annotations, and therefore appealed stylistically to a wider British readership. Its immediate reappearance with the publishing house of Bell and Daldy, which bought up Bohn's titles on his retirement from the trade in 1864, indicates the continuing saleability of the translation completed by Otté and her associates. Reprints of this edition appeared regularly through the 1880s and 1890s, and were brought to a halt only in 1913, on the eve of the First World War. They stand as testament to its durability in conveying Humboldt's ideas to a British readership more than half a century after his death.

Humboldt's beginnings on the British publishing market hardly compare with the dynamics of the relationship he enjoyed with his translators and publishers at the end of his life. While Black had not seen fit to develop any intense working relationship with his author as he worked on the *Essai politique*, Humboldt's subsequent translators realised that he would take a keen interest in how they put his work into English, and keep a close eye not only on scientific accuracy but also on their prose style. In Elizabeth Sabine, Humboldt had a truly 'dedicated' translator: dedicated to producing work of continuously high quality and committed to ensuring that the style and voice across the *Aspects of Nature* and *Cosmos* would be consistent. Few other scientists of the Victorian era were as fortunate as Humboldt in having their works so well translated by figures like the Sabines. The majority had to find translators by activating their own social or professional networks, waiting to be approached by an enthusiastic (if sometimes misguided) individual, or drawing on freelance translators recruited by their publishing houses. The outcomes were often hit and miss, and scientists were right to be anxious about their work in a new translator's hands. Reliable, thorough and industrious, the Sabines must have given Humboldt valuable peace of mind as he struggled to complete *Kosmos*, his final work that defied containment.

The surviving archival material has enabled us to present a clear picture of the challenges that the Sabines faced in translation Humboldt's *Cosmos*. The same has not been possible for Otté and her fellow translators, yet their achievements should not be overlooked. Precisely because the translation by their competitors, the Sabines, was so elegant and accurate, it compelled them to rise to new challenges as they likewise struggled to capture the rich imagery of Humboldt's writing and at the same time convey its scientific complexity, across hundreds of pages of text. New life was breathed into the Bohn edition by the arrival of men like Paul – representatives of a new generation of male scientists who were well versed in foreign languages, particularly German, had studied on the Continent and were aware of their place in the wider world of European science. Over the course of the nineteenth century, translating Humboldt's writing into English seemed to require ever broader and more efficient collaborative working practices to contend with the sheer volume of material embodied in the works he produced and the breadth of knowledge they encompassed. The English translations of Humboldt's *Kosmos* represented the culmination of four decades of skills and expertise acquired by British translators, editors and advisors. These individuals themselves represented a diverse range of intellectual cultures, social backgrounds and disciplines, and they all, in their way, were ambassadors for Humboldt's writing in the Anglophone world.

Notes

1. *Literary Gazette and Journal of the Belles Lettres, Arts, Sciences, &c.*, 1677 (1849), p. 161.

2. I use here the dates of publication given on the translations themselves. Bohn asserted in 'Henry G. Bohn's Rejoinder' that the Baillière translations had actually appeared in July 1845 and December 1847, the Longman and Murray translations in September 1846 and December 1847, and his own first two volumes on 1 February 1849. See page 1 of the seven-page 'Henry G. Bohn's Rejoinder' of 28 February 1849, bound amongst the additional material at the back of the *Views of Nature*, trans. by Elise C. Otté and Henry Bohn (London: Bohn, 1850).
3. *Literary Gazette and Journal of the Belles Lettres, Arts, Sciences, &c.*, 1677 (1849), pp. 162–3.
4. First editions of the initial volume of the Sabines' translation were quickly sold out and are now scarce. I work here with the fourth edition of volume I (1849) – because it includes Longman's response to Bohn – and then the first or second edition of each subsequent volume, so II (1848), III:1 (1852), III:2 (1852) and IV:1 (1858).
5. *Metropolitan Magazine*, 54 (1849), p. 474.
6. *Examiner*, 2143 (1849), p. 117.
7. *Examiner*, 2287 (1851), p. 754.
8. *The Athenæum*, 1275 (1852), p. 374.
9. *Examiner*, 2029 (1846), p. 803; *The Athenæum*, 1619 (1858), p. 589.
10. *Quarterly Review*, 77 (1845), pp. 156–7.
11. *Quarterly Review*, 94 (1853), p. 50.
12. *Art Journal*, March 1849, p. 100.
13. *The Athenæum*, 1114 (1849), p. 223.
14. JMA, Ms.41052, letter of 18 April 1846 from Humboldt to Edward Sabine: 'Il y a dans votre version une telle élévation du style; une telle délicatesse dans les nuances les plus fines du langage.'
15. American Philosophical Society Library, B:D25.L, letter of 27 April 1847 from Alexander von Humboldt to Charles Lyell: 'J'avois depuis 30–40 ans l'imprudence de croire à la possibilité d'un tableau de la nature, réunissant l'ensemble des phénomènes dans la liaison qui leur est reconnue à une époque donnée.'
16. *Quarterly Review*, 77 (1845), p. 158.
17. *Westminster Review*, 44.1 (1845), p. 193.
18. See the contract of 20 October/1 November 1849 drawn up with Cotta regarding the third volume of *Kosmos* and the second volume of the *Ansichten der Natur*, Document 17 (Humboldt 2009: 646).
19. Humboldt originally asked the ardent anglophile Samuel Heinrich Spiker, translator of Shakespeare's *Macbeth* (trans. 1826) and *Henry VIII* (trans. 1837), to put it into German. However, Varnhagen von Ense considered Spiker's translation 'bad from every point of view' and it was never included in *Kosmos* (Humboldt 1860. 70–1).
20. *Edinburgh Review*, 87 (1848), p. 170.
21. Royal Society Centre for History of Science, London, hereafter 'RSCHS', Edward Sabine Correspondence, MS258, 500, letter of 12 December 1848 from Adolf Erman to Edward Sabine.
22. *Literary Gazette and Journal of the Belles Lettres, Arts, Sciences, &c.*, 1677 (1849), p. 164.

23. JMA, Ms.41052, letter of 4 February 1849 from Edward Sabine to John Murray.
24. *Literary Gazette and Journal of the Belles Lettres, Arts, Sciences, &c.*, 1677 (1849), p. 164.
25. *The British Medical Journal*, 1 (1898), pp. 250–1.
26. JMA, Ms.40590, letter of 5 May 1845 from Alexander von Humboldt to John Murray: 'L'ouvrage est très difficile à traduire. Il faut surtout des connoissances astronomiques, magnétiques, géologiques, botaniques.'
27. RSCHS, Edward Sabine Correspondence, MS258, 417, letter of 1 December 1852 from Heinrich Wilhelm Dove to Edward Sabine: 'Ich hätte mir nie träumen lassen, dass meine Arbeiten eine Uebersetzerin finden würden, am wenigsten die des Kosmos.'
28. RSCHS, Edward Sabine Correspondence, MS258, 407, letter of 14 June 1847 from Heinrich Wilhelm Dove to Edward Sabine: 'Die herzlichsten Grüße an Mad. Sabine. [...] Wenn ich meiner Frau ein Beispiel gebe, wie die eines Gelehrten seyn soll, entlehne ich immer mein Beispiel in Woolwich. . . .'
29. JMA, Ms.40590, letter of 15 December 1846 from Humboldt to John Murray.
30. JMA, Ms.41052, letter of 13 February 1846 from Edward Sabine to John Murray.
31. RSCHS, Edward Sabine Correspondence, MS260, 1171, letter of 26 May 1846 from Edward Sabine to James Clark Ross.
32. RSCHS, Edward Sabine Correspondence, MS260, 1173, letter of 2 September 1846 from Edward Sabine to James Clark Ross.
33. RSCHS, Edward Sabine Correspondence, MS260, 1172, letter of 11 July 1846 from Edward Sabine to James C. Ross. The letter gives different measurements, namely 'from depths of 1320 to 1726 English feet', suggesting that Ross corrected this before the text went to press. Passage now at CO I, 341.
34. RSCHS, Edward Sabine Correspondence, MS258, 575, letter of 12 September 1846 from Robert Were Fox to Edward Sabine.
35. JMA, Ms.41052, letter of 29 August 1846 from Edward Sabine to John Murray.
36. RSCHS, Edward Sabine Correspondence, MS260, 1173, letter of 2 September 1846 from Edward Sabine to James C. Ross.
37. Darwin Correspondence Project, available at <https://www.darwinproject.ac.uk/letter/DCP-LETT-910.xml>, letter of 1 September 1845 from Joseph Dalton Hooker to Charles Darwin (last accessed 18 October 2017).
38. Darwin Correspondence Project, available at <https://www.darwinproject.ac.uk/letter/DCP-LETT-917.xml>, letter of 18 September 1845 from Charles Darwin to Joseph Dalton Hooker (last accessed 18 October 2017).
39. *The Athenæum*, 929 (1845), p. 805.
40. Letter of 14 August 1845 from Humboldt to Edward Sabine, quoted in JMA, Ms.41052, Letter of 27 October 1846 from Edward Sabine to John Murray (spelling and punctuation corrected).
41. Letter of 14 August 1845 from Humboldt to Edward Sabine, quoted in JMA, Ms.41052, letter of 27 October 1846 from Edward Sabine to John Murray: 'je gémis surtout de ce manque de goût dans la partie littéraire en

tout ce qui tient à l'imagination, à la peinture des sites, à l'élévation des sentiments qui se réflètent en nous au milieu d'une nature puissante et féconde'.

42. Darwin Correspondence Project, available at <https://www.darwinproject.ac.uk/letter/DCP-LETT-998.xml>, letter of 28 September 1846 from Joseph Dalton Hooker to Charles Darwin (last accessed 31 July 2017).
43. JMA, Ms.4190, 218, letter of 2 July 1845 from John Murray to Edward Sabine.
44. JMA, Ms.4190, 218, letter of 2 July 1845 from John Murray to Edward Sabine.
45. JMA, Ms.4190, 218, letter of 2 July 1845 from John Murray to Edward Sabine.
46. JMA, Ms.41052, letter of 5 July 1845 from Edward Sabine to John Murray.
47. JMA, Ms.41052, letter of 2 December 1845 from Edward Sabine to John Murray.
48. JMA, Ms.40590, letter of 31 December 1845 from Alexander von Humboldt to John Murray.
49. JMA, Ms.41052, letter of 23 April 1846 from Edward Sabine to John Murray.
50. JMA, Ms.41052, letter of 13 February 1846 from Edward Sabine to John Murray.
51. JMA, Ms.41910, 229, letter of 17 December 1845 from John Murray to Edward Sabine.
52. JMA, Ms.41052, letter of 2 December 1845 from Edward Sabine to John Murray.
53. JMA, Ms.41052, letter of 23 April 1846 from Edward Sabine to John Murray.
54. JMA, Ms.41052, letter of 6 January 1847 from Edward Sabine to John Murray.
55. JMA, Ms.41052, letter of 13 November 1847 from Edward Sabine to John Murray.
56. JMA, Ms.41052, letter of 13 November 1847 from Edward Sabine to John Murray.
57. Massachusetts Historical Society, William Hickling Prescott Papers, letter of 15 April 1849 from E. C. Otté to William Hickling Prescott.
58. See 'Bohn's Rejoinder', at the back of *VN*, p. 2.
59. RSCHS, John Herschel Papers, HS 15, 208, letter of 8 October 1847 from Edward Sabine to John Herschel.
60. RSCHS, John Herschel Papers, HS 15, 208, letter of 8 October 1847 from Edward Sabine to John Herschel.
61. RSCHS, John Herschel Papers, HS 15, 208, letter of 8 October 1847 from Edward Sabine to John Herschel.
62. RSCHS, John Herschel Papers, HS 15, 209, letter of 9 October 1847 from Edward Sabine to John Herschel.
63. *Wesleyan-Methodist Magazine*, 3 (1824), p. 540.
64. *Wesleyan-Methodist Magazine*, 5 (1826), p. 260.
65. *Eclectic Review*, 3 (1852), p. 219.
66. *Quarterly Review*, 77 (1845), pp. 163–4.
67. *Westminster Review*, 44.1 (1845), p. 152.

68. *Westminster Review*, 44.1 (1845), p. 152.
69. *Westminster Review*, 44.1 (1845), p. 153.
70. *Westminster Review*, 44.1 (1845), p. 154.
71. *Historisch-politische Blätter für das katholische Teutschland*, 16 (1845), p. 455.
72. *Die Grenzboten: Zeitschrift für Politik und Literatur*, 4 (1845), p. 591.
73. RSCHS, John Herschel Papers, HS 15, 207, letter of 13 September 1847 from Edward Sabine to John Herschel.
74. RSCHS, John Herschel Papers, HS 15, 207, letter of 13 September 1847 from Edward Sabine to John Herschel.
75. RSCHS, John Herschel Papers, HS 15, 207, letter of 13 September 1847 from Edward Sabine to John Herschel.
76. RSCHS, John Herschel Papers, HS 15, 207, letter of 13 September 1847 from Edward Sabine to John Herschel.
77. RSCHS, John Herschel Papers, HS 15, 208, letter of 8 October 1847 from Edward Sabine to John Herschel.
78. RSCHS, John Herschel Papers, HS 15, 207, letter of 13 September 1847 from Edward Sabine to John Herschel.
79. RSCHS, John Herschel Papers, HS 15, 207, letter of 13 September 1847 from Edward Sabine to John Herschel: 'The modification was not only made with M. de H's sanction but the special reference to the Creator was introduced at his own desire, with the remark that such was strictly his original meaning.'
80. JMA, MS41052, letter of 9 September 1850 from Edward Sabine to John Murray.
81. URSC, Longman ledgers, 4D, p. 180.
82. RSCHS, Edward Sabine Correspondence, MS259, 804, letter of 22 March 1854 and MS259, 800 of 25 March 1852 from Longman to Edward Sabine and pages from the Longman ledgers at MS259, 801. These payments are certainly quite high when contextualised across the industry as a whole. The Richard Bentley Papers held at the British Library show that Maria Innes was paid £14 14s for the translation of fourteen sheets from volume six of the Duchess d'Abrantès's *Memoirs* in October 1833 (ADD. 46649 f. 34) and W. Gerard £9 16s for translating nine sheets from volume two of Manuel de Godoy's *Memoirs* in 1835 (ADD. 46649 f. 91). Thanks to Susan Pickford for drawing this information to my attention.
83. URSC, Longman ledgers, H11f140v, H11f172r, H11f210r.
84. *The Athenæum*, 1224 (1851), p. 401.
85. *Examiner*, 2627 (1858), p. 356.
86. RSCHS, Edward Sabine Correspondence, MS259, 814, letter of 10 December 1860 from Longman to Edward Sabine.
87. *Journal of the Chemical Society*, 113 (1918), p. 335.
88. Quoted in the obituary in the *Geological Magazine*, Series III, 7 (1890), p. 335.

Conclusions

Today, Humboldt's works have sloughed off their Victorian skins to emerge on to the Anglophone market in sleek, fresh translations. Lavishly illustrated and carrying a wealth of annotations, these acclaimed critical editions emphasise the centrality of Humboldt's writing to the cosmopolitan thought, global interconnections and environmental anxieties that shape our world.[1] The new translations are visually enticing works that conform sufficiently in their dustjacket design to have the look of a series. Bohn would have approved. The misgivings he might hold about their price could be assuaged by reflecting that their target audience is now a scholarly market, and the reviewers' comments on the dustjackets derive almost exclusively from an academic elite. Yet room has always been found in each of these weighty editions for a Translator's Note. Fundamental to the translation itself, it draws critical attention to the essential mediating role in these truly modern renderings of Humboldt's work. It also confronts us directly with the complicated business of transforming one series of ideas and images into a different language and culture, and, as twenty-first-century readers, into a different time.

These modern translators bemoan Humboldt's 'massive, knotty, chewy subclauses', his repeated over-layering of information and the elevated, formal character of his language (Humboldt 2014: 5). Yet they convince us that putting his works into clear and fluent English has demanded significant linguistic agility. They have risen admirably to this challenge. Tacitly, though, they testify still more persuasively to the extraordinary achievements of their nineteenth-century counterparts. Quite apart from the fact that today's translators can hone their word-processed documents with ease, use spell-checkers to avoid typing errors and make quick text searches to ensure continuity, they can also consult electronic dictionaries and thesauri at the touch of a button, call up search engines to hunt down rare terminology and leaf virtually through the latest scientific periodicals. The working environment for

Humboldt's first translators was radically different. We can only stand in awe of those who wrote every single word long-hand with pen and ink before a translation went for typesetting, doubtless working late into the evening in dimly lit rooms as they met increasingly punishing deadlines. And we must also applaud their ability to decipher Humboldt's missives, penned in an increasingly crabbed, spiky hand, like the plottings of a frenetic seismograph.

The mediation of knowledge is by no means an abstract, impersonal undertaking. We have seen how the humble work of translators, editors, illustrators, publishers and reviewers was of paramount importance in placing a text in the public sphere and influencing how it was received and by which segment of the market. Translators not only remoulded the language of the scientists whose work they were putting into English, but they also reshaped their styles of enquiry, the rhetorical devices by which they established credit and authority, and the grounds on which their testimony would be accepted by the wider scientific community. In collaboration with others involved in the business of publishing, translators also influenced the organisation and layout of scientific print media – the footnotes, illustrations, indices – which guided the reading experience. Above all, as the discussions in the preceding chapters have emphasised, translators were involved in shaping what Marina Frasca-Spada and Nick Jardine call the 'handiness and heftiness of books' (Frasca-Spada and Jardine 2000: 9): that is, what it actually felt like for the British nineteenth-century reader to take a volume of Humboldt's works off the shelf and appreciate it as a print object in its material solidity.

This study has, unashamedly, placed Humboldt's nineteenth-century translators centre-stage. It has done so to understand how, in an era before professionalisation or formal qualifications, individuals with good foreign language skills – particularly women – could make significant contributions to the development of science. The case studies here suggest that we have been exploring a highly specific moment in the history of scientific translation in Britain. The period in which Humboldt's works came on to the British market coincided with moves towards greater inclusiveness in scientific activities, in line with wider ambitions to encourage mass literacy in science and appeal to emerging reading audiences (Secord 2014: 4). This opened a window of opportunity for women such as Williams, Ross, Otté and Sabine to put their linguistic talents to good use as they transformed foreign texts into editions that were accessible by British readers. This window appears to have been closing by the mid-1860s, the end of the period covered here, as science became professionalised, much more highly specialised and, ultimately, a male domain – something that only gradually began to change towards the end of the nineteenth century, with the admission of women to university degree programmes.

Digging down to reveal something of the day-to-day working patterns and practices of Humboldt's translators, this study has deliberately emphasised that translating his work figured among their more mundane activities. Theo Hermans has stressed that in order 'to appreciate the role translation has played in a particular historical configuration, we need to have a sense of how translation was practised and perceived at the time' (Hermans 2012: 244). While, from a modern perspective, we might consider that to translate Humboldt's writing was to be working at the cutting edge of science with one of the most brilliant minds of the age, the reality, at times, felt rather different to his nineteenth-century translators. They may well have felt honoured to put the works of this revered scientist into English, and they certainly found it an intellectually all-consuming exercise, but we have also witnessed just how invasive an undertaking it was in their lives. Williams lurched from financial stability to dire hardship, and the tenacious Sabines from enthusiasm to exhaustion, as they invested considerable amounts of their time (and money) in putting Humboldt's multivolume narratives into English.

This study has also shown that 'translating Humboldt' was about much more than simply transferring the text of his French and German works into English. All of his translators added paratextual material – sometimes brief, sometimes quite extensive – to their translations, which potentially overshadowed the accomplishments of the author whose work they were essentially promoting. This material is revealing of the public persona they wished to project, but also of the assumptions they made about who would be buying and reading their books. Outi Paloposki is right to suggest that more work still needs to be done if we are to understand how translation incorporates processes of addition and explanation intended to ensure that readers actually understand the issues being discussed (Paloposki 2017: 64). These processes are, ostensibly, about creating meaning and ensuring that a text's substance, reasoning, specific line of argument, ideas and hypotheses are accessible by the intended readership. They also cannot be divorced from key phenomena in the nineteenth-century publishing trade, notably the emergence of quite clearly defined reading communities, and of developments in marketing strategy to target scientific books at specific audiences characterised by a particular social status, purchasing power and set of intellectual aspirations.

The information needs of translators themselves also merit further investigation. Pekka Kujamäki has tried to reflect on how exactly individual translators' 'workbenches' of reference works might have looked and which other publications they might have readily had to hand to come by terms they did not know (Kujamäki 1997: 584). Humboldt, who on several occasions attempted to translate parts of his own work before consigning the task to a more experienced translator, knew at first hand the practical difficulties of transferring a text from one language into another. A good thesaurus was an essential tool of the trade for any wordsmith, he acknowledged to Varnhagen von Ense: 'If only we had, in German, a book of synonyms as good and as simply arranged as the [French] one I sent you,' it would save 'a vast deal of time in the event of one's having to look for an equivalent. You see at once the word which may be substituted' (Humboldt 1860: 20). However, a thesaurus would not have stood on the writing desks of Black or Williams, for example. Peter Roget's *Thesaurus of English Words and Phrases* did not come on to the market until April 1852, too late in the day to help Otté and the Sabines with anything but the final volumes of *Cosmos* or Ross with the second volume of the *Personal Narrative*. With its conceptual, rather than alphabetical, organisation of word classification, it might not, in any case, have suited their working style.

Nugent's French–English dictionary had been on the market since the 1760s and other French–English dictionaries were well established by the time Black came to tackle the *Essai politique*. With the appearance of later reference works such as Professors Fleming and Tibbins's *Royal Dictionary English and French and French and English* (Paris, 1841) and Alexandre Spiers's *General English and French Dictionary* (London, 1846), quite apart from more manageably sized 'miniature' dictionaries on the market, Ross would have had an even greater amount of choice.

Similarly, by the 1840s, when Otté and the Sabines set to work on the *Ansichten der Natur* and *Kosmos*, multivolume German–English dictionaries, pocket dictionaries and 'practical' dictionaries were flooding the market. Hilpert and Kärcher's two-volume *Dictionary of the English and German, and the German and English Language* (Karlsruhe: Braun, 1846) was on sale for eight shillings. Alternatively, they could have consulted the two thick octavo volumes of *Flügel's Complete Dictionary of the German and English Languages*, retailing at thirty-six shillings in 1845. Although these dictionaries now included collocations – giving the usage of a word in context – they were not specialised, technical dictionaries. The Sabines' deliberations about how 'Kraft' could best be translated were not solved by consulting reference volumes such as these. The emergence on the market of works like Tauchnitz's trilingual *Technological Dictionary in French, German and English* in the 1880s did eventually begin to fill this gap in the market, but even then, such dictionaries might not always have been useful aids.

This study has emphasised that translation is an interpretative act, and style a cognitive entity closely related to the choices and decisions made by those agents involved in the process of translation, who are themselves governed by various norms. We have seen that Humboldt's British translators drew on unique and individual sets of strategies to put his writing into English. Yet each translator responded to three main needs: to be creative, contemporary and critical. Black, working on the *Essai politique*, arguably poured most of his creative energies into his fierce translator's footnotes. However, he was also the mediator of whom the most linguistic creativity was required, as he tackled scientific terminology in the French original for which there was still no obvious current English equivalent. This was far less problematic for Williams, since her translation of the *Relation historique* was completed together with Humboldt, who was himself well versed in the scientific vocabulary of the language into which she was translating. Williams's creative input shaped less the scientific than the aesthetic qualities of the *Personal Narrative*. The second translation of this work, by Ross, would seem, with all its parings and omissions, to be anything other than 'creative'. It required considerable skill, however, to reduce Humboldt's account so radically for a mid-century audience, while fashioning a narrative that remained coherent, engaging and meaningful. The creativity required of Elizabeth Sabine and Otté as they tackled the *Ansichten der Natur* and *Kosmos* in their rival translations is equally subtle. While both were committed to conveying the original in as faithful a manner as possible, they were disadvantaged by the presence of Prichard's rendering of the first two volumes of *Kosmos*, from which they needed to distance themselves stylistically as well as commercially, if their translations were to be successful. Where the Sabines were confronted merely with the problems posed by Humboldt's own prose in the *Ansichten der Natur*, Otté again had to attempt the double feat of accurately conveying the source text, yet producing a demonstrably different text from that of her rivals.

Humboldt's British translators each faced a rather different set of challenges. It was doubtless easier for Black and Williams to work from the French, with its similar sentence structure and vocabulary to English, than for Prichard, Otté and Sabine to work from the German, with its complex chains of interlinked subclauses. His translators were not successful to an equal degree in putting his narrative into fluent nineteenth-century English. But such 'foreignised' renderings would at least have reminded perceptive readers that they were not hearing Humboldt's own voice, unmediated and unadulterated, and Humboldt's

British critics were quick to pass sentence on the quality of the translations they read. Judgements about whether the translations produced by Black, Williams, Ross, Prichard, Otté and the Sabines were 'good' or 'bad' rarely take into account the prevailing norms, or indeed the limited range of translation solutions that might have been open to translators, particularly when working in competition with each other. They also often turn upon rather subjective expectations of what constitutes the 'best' translation – expectations themselves governed by the historical moment in which such assessments are being made. For the purposes of this investigation, Prichard's supposedly 'poor' rendering has actually been most instructive in highlighting the difficulties that Humboldt's translators faced as they grappled, under immense pressure of time, with the complexities of his prose.

Style was central to the international communication and reception of scientific knowledge for three reasons: it shaped how scientific hypotheses, results and conclusions were presented to and received by the target culture audience; it influenced how the narrating author was cast as a scientific traveller; and it enabled those agents involved in the process of intercultural transfer (translators, editors, publishers) to impress their own aesthetic, political and confessional agenda upon the source material. This examination of how Humboldt's texts fared in English translation has ranged across a half-century of material. It has demonstrated that even in the apparently 'style-less' genre of scientific narrative, style had strong historical contingency because it reflected the external forces acting upon Humboldt and his Anglophone translators, editors and publishers. It was also connected to the internal 'history' of Humboldt's writing in British translation. In the retranslations and subsequent editions of his works, style was deployed by later mediators to distance their editions from previous versions and cast their own as new, improved and superior.

This study has focused necessarily on one canonical figure and his canonical works. But in shedding new light on the teams of translators, editors and other technicians of print working for Longman, Murray and Bohn, we have observed at close hand how canonisation is not a 'given', but rather a historical process. Precisely because Humboldt's works were able to withstand sustained activities of reviewing, editing, (re)translating and abridgement over an extended period of time, and, in these different forms appealed to diverse markets, they acquired a supranational reputation. Secord has noted that scientific books that became publishing 'sensations' are a good 'cultural tracer' of where such works were read, by whom and with what intellectual intent (Secord 2000: 3). The tireless efforts of his scientific associates, translators, vocal critics and warring publishers primarily ensured that Humboldt's books attracted public attention, but they also played a significant role in targeting his work at different segments of the British market, in conditioning how his British audience read his work and in influencing how these readings perpetuated discussion and meaning-making around these texts.

We have uncovered some of the networks in which Humboldt's British translators operated and gained intriguing insights into where scientific knowledge was generated, mediated and circulated. If material production and knowledge production are intimately connected, then it is also important to understand where the translations took place and how these sites can be linked to the 'geographies of knowledge' (Rupke 2005; Hesse 2006; Withers and Ogborn 2010) that scholars have been constructing for the Enlightenment and Romantic periods. Charles Withers has argued for the long eighteenth century that the 'dethronement of the universal and transcendental nature of science' brought increased emphasis 'on its local making', since science was as much conditioned by its social and historical context as by its geographical circumstances (Withers 2007: 9). The four source texts by Humboldt and their translations forged links between individuals who were socially and culturally dissimilar, and scattered across a wide geographical area. Humboldt's texts were 'displaced' not only culturally and linguistically, but also spatially, by their British translators.

London, of course, held a central position: it was here that the two main rival publishers Longman and Bohn waged their war over the translations of the *Relation historique*, the *Ansichten der Natur* and *Kosmos*. It was also where Black completed his translation of the *Essai politique*, where Ross, with her family background in Fleet Street journalism, would finalise the Bohn edition of the *Travels*, where the Sabines would labour at putting the *Ansichten der Natur* and *Kosmos* into English and where Bunsen would correct some parts of *Cosmos* in the translation for Longman and Murray. But the topography of translation revealed by this analysis of Humboldt's writing in English was by no means exclusively focused on Britain's capital, even if the Sabines' house at Woolwich in southeast London was a focal point for the meeting of scientific minds and the exchange of scientific ideas.

The importance of Scotland in the translation of Humboldt's writing into English should not be overlooked. Several of Humboldt's translators, editors, advisors, advocates and fiercest critics were active participants in the scientific culture of Edinburgh, Glasgow and St Andrews. To name but a few, Black, MacGillivray, Edward Forbes, James D. Forbes and Baillie Warden either were Scottish, had been educated in Scotland

and moved down to London, or were teaching at Scottish universities and integrated into the scientific community there. Black in particular injected a strong sense of Scottish patriotism into his translation of the *Essai politique* that could have enhanced its appeal to a British, rather than just to an English, audience. The house in St Andrews where Otté completed part of her translation of *Kosmos* could not have been more distant from the hubbub of Berlin court life that invigorated and infuriated Humboldt by turns. Yet, as a bustling university city, St Andrews also brought Otté into contact with greater and humbler figures of Scottish science, which would provide her with the intellectual stimulation and encouragement from which she would have benefited as she was translating Humboldt's work. The London-based Scottish publisher John Murray, whose name stood as an index of quality, was another powerful force in ensuring that Humboldt's works successfully targeted a segment of the reading public interested in 'popular' non-fiction.

Paris likewise occupied a key role in the international dissemination of Humboldt's writing, since it was there that Humboldt spent a highly productive part of his professional life and Stone's printing business issued his earliest works. It was also where Williams hosted her salon, in which various currents of intellectual, political and economic thought could flow freely, and where Humboldt presented and discussed parts of the *Vues des Cordillères* and the *Relation historique* to a select audience during the time that Williams was putting them into English. The intense exchange between Humboldt and Williams as he corrected drafts of her work suggests that it was also in Paris that the closest of working relationships between Humboldt and one of his translators was forged.

This book raises a number of questions that invite further investigation by scholars of women's studies, translation studies and the history of science. The stylistic input of Humboldt's female translators was essential in defining the particular place that his scientific prose held in the British popular consciousness well into the next century. Elizabeth Sabine, working at the forefront of the geological and physical sciences, translated prestigious texts for her husband's French and German colleagues. Yet of the four female translators featured in this study, she is the least vocal and least 'visible'. It is difficult to endorse Healy's view that Edward Sabine's status as a famous scientist and explorer aids us in retrieving 'a certain sense of Elizabeth's role and her reputation' (Healy 2004: 269). In fact, it is easier to reconstruct the lives of female translators who were single and not overshadowed by their spouses' achievements – in this study Williams, Ross and Otté – since they had to liaise with authors and publishers directly, leaving the archival paper trail by which we now gain some feel for their lives. Much more needs to be done, however, if we are to gain a fuller picture of how women were engaged in scientific translation in the nineteenth century. Was such inclusiveness a particularly British phenomenon that can be linked to certain institutional outlooks on education in the Victorian era? If not, then where else were women turning their hands so energetically to translating scientific prose? And were they likewise tackling works by the scientific figureheads of their age, or did their translation activities centre upon more marginal figures and less influential ideas?

In this study, we have focused specifically on the first English translations of Humboldt's works as general conduits for the transmission of his ideas into the wider Anglophone world. A number of other avenues would repay further exploration, notably the circulation and consumption of these British editions beyond the British Isles. How quickly did they reach British India, for example (a part of the world that Humboldt would have given his eye-teeth to visit), which bookshops, libraries, clubs and other institutions stocked his works, and in which edition? How accessible were these copies by which audiences of readers? Other forms of translation being undertaken in the English-speaking world also deserve greater attention: for example, the transatlantic crossing made by Humboldt's works in translation between Britain and America. It is worth remembering that Prichard's rendering of *Kosmos* appeared with Harper and Brothers in New York in 1845, followed by the Otté edition of *Cosmos* with Harpers in 1847 (Belgum 2005: 125). Did the reception history in America echo that in England or was there more tolerance of Prichard's rendering? Why did the Longman and Murray edition not gain a firmer foothold, given Humboldt's own transatlantic networks? And which marketing strategies were applied to the Otté edition to ensure it would gain Humboldt lasting recognition across the United States?

While this study is devoted to the translation of Humboldt's writing into English, his works have, of course, also been translated into a host of other languages in the two centuries since their initial publication. This analysis of the translations into English gives insights into the complex set of connections that have facilitated the international transmission and reception of his works in competing editions across linguistic and cultural boundaries. Writing – and the related creative practice of translation – is a never-ending process of adjustment and revision. As the community of Humboldt scholars itself becomes truly global in its own interconnections and exchanges, the time is ripe for further explorations of both the long-distance and the local encounters that enabled Humboldt's works to reach new audiences, invite new readings and promote new ideas. Only by revealing the asymmetries of knowledge exchange, the individual peculiarities of communication flow and the very specific cultures of translation practice, can we arrive at a better understanding of how Humboldt's translators acted as significant intermediaries in bringing disparate worlds together.

Note

1. To date, the following translations have appeared with the University of Chicago Press: *The Political Essay on the Island of Cuba* (2011; trans. J. Bradford Anderson and Vera M. Kutzinski, Anja Becker), *Views of the Cordilleras* (2012; trans. J. Ryan Poynter) and *Views of Nature* (2014; trans. Mark W. Person). A new translation of the *Political Essay on the Kingdom of New Spain* (trans. J. Ryan Poynter, Ken Berry and Vera M. Kutzinski) is in press.

Bibliography

Abir-Am, Pnina and Dorinda Outram (eds) (1987), *Uneasy Careers and Intimate Lives: Women in Science, 1789–1979*, New Brunswick, NJ: Rutgers University Press.
Agorni, Mirella (2005), 'A Marginal(ized) Perspective on Translation History: Women and Translation in the Eighteenth Century', *Meta*, 50.3: 817–30.
Aikin, Lucy (1864), *Memoirs, Miscellanies and Letters of the Late Lucy Aikin: Including those Addressed to the Rev. Dr. Chaning, from 1826 to 1842*, ed. P. H. Le Breton, London: Longman, Green, Longman, Roberts and Green.
Allen, David Elliston (2001), *Naturalists and Society: The Culture of Natural History in Britain, 1700–1900*, Aldershot: Ashgate.
Allen, Esther (2013), 'Footnotes *sans Frontières*: Translation and Textual Scholarship', in Brian Nelson and Brigid Maher (eds), *Perspectives on Literature and Translation: Creation, Circulation, Reception*, London: Routledge, pp. 210–20.
Alter, Peter (1984), 'Alexander von Humboldt und England', *Berichte der Wissenschaftsgeschichte*, 7: 118–20.
Altick, Richard D. (1957), *The English Common Reader: A Social History of the Mass Reading Public 1800–1900*, Chicago: University of Chicago Press.
— (1958), 'From Aldine to Everyman: Cheap Reprint Series of the English Classics 1830–1906', *Studies in Bibliography*, 11: 3–25.
— (1978), *The Shows of London*, Harvard: Belknap Press of Harvard University.
Anon. (1813), *A Citizen of Maryland, An Abridgement of Humboldt's Statistical Essay on New Spain Being a Geographical, Philosophical and Political Account of the Kingdom of Mexico, or the Vice-Royalty of Mexico, and the Internal Provinces Subject to the Commandment or Governor General Residing at Chihuahua*, Baltimore: Wane and O'Reilly.
Anon. (1823), *Public Characters of All Nations, Consisting of Biographical Accounts of Nearly Three Thousand Eminent Contemporaries, Alphabetically Arranged*, 3 vols, London: Lewis.
Atkinson, Dwight (1999), *Scientific Discourse in Sociohistorical Context: The Philosophical Transactions of the Royal Society of London, 1675–1975*, Mahwah, NJ: Lawrence Erlbaum Associates.
Babbage, Charles (1830), *Reflections on the Decline of Science in England, and on Some of its Causes*, London: Fellowes.
Bachleitner, Norbert (2013), 'From Scholarly to Commercial Writing: German

Women Translators in the Age of the "Translation Factories"', *Oxford German Studies*, 42.2: 173–88.

Baker, Mona (2000), 'Towards a Methodology for Investigating the Style of a Literary Translator', *Target*, 12.2: 241–66.

— (2006), *Translation and Conflict: A Narrative Account*, Abingdon: Routledge.

Barker, Anna (2011), 'Helen Maria Williams' *Paul and Virginia* and the Experience of Mediated Alterity', in Luise von Flotow (ed.), *Translating Women: Gender and Translation in the 21st Century*, Ottawa: University of Ottawa Press, pp. 57–70.

Bassnett, Susan and Peter France (2006), in Peter France and Kenneth Haynes (eds), 'Translation, Politics, and the Law', *The Oxford History of Literary Translation into English. Vol. 4: 1790–1900*, Oxford: Oxford University Press, pp. 48–58.

Beck, Hanno (1966), *Alexander von Humboldt und Mexiko: Beiträge zu einem geographischen Erlebnis*, Bad Godesberg: Inter Nationes.

Beer, Gillian (1983), *Darwin's Plots: Evolutionary Narrative in Darwin, George Eliot and Nineteenth-Century Fiction*, London: Routledge and Kegan Paul.

— (1996), *Open Fields: Science in Cultural Encounter*, Oxford: Clarendon Press.

Belgum, Kirsten (2005), 'Reading Alexander von Humboldt: Cosmopolitan Naturalist with an American Spirit', in Lynne Tatlock and Matt Erlin (eds), *German Culture in Nineteenth-Century America: Reception, Adaptation, Transformation*, Rochester, NY: Camden House, pp. 107–27.

Benjamin, Walter (1992), 'The Task of the Translator', trans. Harry Zohn, in Rainer Schulte and John Biguenet (eds), *Theories of Translation*, Chicago: University of Chicago Press, pp. 71–82.

Berman, Antoine (1995), *Pour une critique des traductions: John Donne*, Paris: Gallimard.

Bernardin de Saint-Pierre, Jacques-Henri (1795), *Paul and Virginia*, trans. Helen Maria Williams, London: Robinson.

Biermann, Kurt-Reinhardt (1982), 'Alexander von Humboldts berühmter Bericht über seine amerikanische Forschungsreise 1799–1804 – ein Torso', *NTM Schriftenreihe für Geschichte der Naturwissenschaften, Technik und Medizin*, 19.2: 59–66.

Blumenbach, Johann Friedrich (1803), *Manuel d'histoire naturelle*, trans. Soulange Artaud, 2 vols, Metz: Collignon.

Boase-Beier, Jean (2006), *Stylistic Approaches to Translation*, Manchester: St Jerome.

Bohls, Elizabeth (1995), *Women Travel Writers and the Language of Aesthetics 1716–1818*, Cambridge: Cambridge University Press.

Boswell, James (1828), *The Life of Samuel Johnson LL.D.*, 4 vols, London: J. Richardson & Co.

Bourguet, Marie-Noëlle (2002), 'Landscape with Numbers: Natural History, Travel and Instruments in the Late Eighteenth and Early Nineteenth Centuries', in Marie-Noëlle Bourguet, Christian Licoppe and H. Otto Sibum (eds), *Instruments, Travel and Science: Itineraries of Precision from the Seventeenth to the Twentieth Century*, London/New York: Routledge, pp. 96–125.

Bowen, Margarita (1981), *Empiricism and Geographical Thought: From Francis Bacon to Alexander von Humboldt*, Cambridge: Cambridge University Press.

Bravo, Michael T. (1999), 'Precision and Curiosity in Scientific Travel:

James Rennell and the Orientalist Geography of the New Imperial Age (1760–1830)', in Jás Elsner and Joan Pau Rubiés (eds), *Voyages and Visions: Towards a Cultural History of Travel*, London: Reaktion, pp. 162–83.

Bret, Patrice (2012), 'Sciences et Techniques', assisted by Norbert Verdier, in Yves Chevrel, Lieven D'Hulst and Christine Lombez (eds), *Histoires des traductions en langue française: XIXe siècle*, Lagrasse: Éditions Verdier, pp. 927–1007.

Brock, Claire (2012), 'Introduction', in Claire Brock (ed.), *New Audiences for Science: Women, Children, Labourers, Victorian Science and Literature: Vol. 5*, general eds Gowan Dawson and Bernard Lightman, London: Pickering and Chatto, pp. ix–xxii.

Brock, William H. (1993), 'Humboldt and the British: A Note on the Character of British Science', *Annals of Science*, 50: 365–72.

Brownlie, Siobhan (2006), 'Narrative Theory and Retranslation Theory', *Across Languages and Cultures*, 7.2: 145–70.

Bruhns, Karl (1873), *The Life of Alexander von Humboldt*, 2 vols, London: Longmans.

Buckland, Adelene (2013), *Novel Science: Fiction and the Invention of Nineteenth-Century Geology*, Chicago: University of Chicago Press.

Burt, Thomas (1924), *Thomas Burt, M. P., D. C. L., Pitman & Privy Councillor: An Autobiography, with Supplementary Chapters by Aaron Watson, Author of 'A Great Labour Leader', etc. and a Foreword by Wilfrid Burt*, London: Fisher Unwin.

Buzelin, Hélène (2007), 'Translations "in the Making"', in Michaela Wolf and Alexandra Fukari (eds), *Constructing a Sociology of Translation*, Amsterdam/Philadelphia: Benjamins, pp. 135–69.

Carlton, William J. (1955), 'Dickens and the Ross Family', *The Dickensian*, 51: 58–66.

Cate, George Allan (ed.) (1982), *Correspondence of Thomas Carlyle and John Ruskin*, Stanford, CA: Stanford University Press.

Cervantes, Miguel de (1849), *El Buscapié by Miguel de Cervantes; with the Illustrative Notes of Don Adolfo de Castro*, trans. Thomasina Ross, London: Bentley.

Chamberlain, Lori (1988), 'Gender and the Metaphorics of Translation', *Signs*, 13.3: 454–72.

Clayden, Peter William (1889), *Rogers and his Contemporaries*, 2 vols, London: Smith, Elder and Co.

Clemm, Sabine (2009), *Dickens, Journalism, and Nationhood: Mapping the World in Household Words*, New York: Routledge.

Coleridge, Samuel Taylor (1830), *On the Constitution of the Church and State, According to the Idea of Each: With Aids towards a Right Judgement on the Late Catholic Bill*, London: Hurst, Chance and Co.

Colles, William Morris (1889), *Literature and the Pension List: An Investigation*, London: Incorporated Society of Authors.

Cordasco, Francesco (1951), *The Bohn Libraries: A History and a Checklist*, New York: Burt Franklin.

Cosgrove, Peter W. (1991), 'Undermining the Text: Edward Gibbon, Alexander Pope, and the Anti-Authenticating Footnote', in Stephen A. Barney (ed.), *Annotation and Its Texts*, Oxford: Oxford University Press, pp. 130–51.

Cronin, Michael (2000), *Across the Lines: Travel, Language, Translation*, Cork: Cork University Press.

Daston, Lorraine (1991), 'Baconian Facts, Academic Civility, and the Prehistory of Objectivity', *Annals of Scholarship*, 8: 337–63.
— (1995), 'The Moral Economy of Science', *Osiris*, 2nd series, 10: 2–24.
— (2011), 'The Empire of Observation, 1600–1800', in Lorraine Daston and Elizabeth Lunbeck (eds), *Histories of Scientific Observation*, Chicago: University of Chicago Press, pp. 81–113.
Dassow Walls, Laura (2009), *The Passage to Cosmos: Alexander von Humboldt and the Shaping of America*, Chicago: University of Chicago Press.
Daum, Andreas (2002), *Wissenschaftspopularisierung im 19. Jahrhundert: Bürgerliche Kultur, naturwissenschaftliche Bildung und die deutsche Öffentlichkeit, 1848–1914*, Munich: Oldenbourg.
Dawson, Gowan (2007), *Darwin, Literature and Victorian Respectability*, Cambridge: Cambridge University Press.
Deane-Cox, Sharon (2014), *Retranslation: Translation, Literature and Reinterpretation*, London: Bloomsbury-Continuum.
Delabastita, Dirk (2008), 'Status, Origin, Features: Translation and Beyond', in Anthony Pym, Miriam Shlesinger and Daniel Simeoni (eds), *Beyond Descriptive Translation Studies: Investigations in Homage to Gideon Toury*, Amsterdam: John Benjamins, pp. 233–46.
Delamétherie, Jean-Claude (Messidor An XII [=1804]), 'Notice d'un voyage aux tropiques, exécuté par MM. Humboldt et Bonpland en 1799, 1800, 1801, 1802, 1803 et 1804', *Journal de physique, de chimie, d'histoire naturelle et des arts*, 59: 122–39.
De Quincey, Thomas (2000–3), *The Works of Thomas De Quincey*, general. ed. Grevel Lindop, 21 vols, London: Pickering and Chatto.
Derrida, Jacques (1991), 'This Is Not an Oral Footnote', in Stephen A. Barney (ed.), *Annotation and Its Texts*, Oxford: Oxford University Press, pp. 192–205.
Dettelbach, Michael (2001), 'Alexander von Humboldt between Enlightenment and Romanticism', *Northeastern Naturalist*, Special Issue: *Alexander von Humboldt's Natural History Legacy and its Relevance for Today*, 8: 9–20.
— (2005), 'The Stimulations of Travel: Humboldt's Physiological Construction of the Tropics', in Felix Driver and Luciana Martins (eds), *Tropical Visions in an Age of Empire*, Chicago: University of Chicago Press, pp. 43–58.
Dickens, Charles (1965–2002), *The Letters of Charles Dickens*, ed. Madeline House, Graham Storey and Kathleen Tillotson, 12 vols, Oxford: Clarendon Press.
Eliot, Simon (1995), 'Some Trends in British Book Production, 1800–1919', in John O. Jordan and Robert L. Patten (eds), *Literature in the Marketplace: Nineteenth-Century British Publishing and Reading Practices*, Cambridge: Cambridge University Press, pp. 19–43.
Ellegård, Alvar [1958] (1990), *Darwin and the General Reader: The Reception of Darwin's Theory of Evolution in the British Periodical Press, 1859–1872*, with a new foreword by David L. Hull, Chicago, University of Chicago Press.
Elshakry, Marwa and Carla Nappi (2016), 'Translations', in Bernard Lightman (ed.), *A Companion to the History of Science*, Oxford, Wiley Blackwell, pp. 372–86.
Encyclopaedia Britannica, or, a Dictionary of Arts, Sciences, and Miscellaneous Literature, (1797) 3rd edn, 18 vols, Edinburgh: Bell.
Escott, Thomas H. S. (1911), *Masters of English Journalism: A Study of Personal Forces*, London: Fisher Unwin.

Ette, Ottmar (1991), 'Der Blick auf die Neue Welt', *Reise in die Äquinoktial-Gegenden des Neuen Kontinents*, ed. and epilogue by Ottmar Ette, 2 vols, Leipzig: Insel, vol. 2, pp. 1563–97.
— (1996), 'Von Surrogaten und Extrakten: Eine Geschichte der Übersetzungen und Bearbeitungen des amerikanischen Reisewerks Alexander von Humboldts im deutschen Sprachraum', in Karl Kohut, Dietrich Briesemeister and Gustav Siebenmann (eds), *Deutsche in Lateinamerika – Lateinamerika in Deutschland*, Frankfurt am Main: Vervuert, pp. 98–126.
— (1999) 'Alexander von Humboldt heute', in Frank Holl and Petra Kruse (eds), *Alexander von Humboldt: Netzwerke des Wissens*, Ostfildern-Ruit: Hatje Cantz, pp. 19–31.
— (2001a), 'Eine "Gemütsverfassung moralischer Unruhe" – *Humboldtian Writing*: Alexander von Humboldt und das Schreiben in der Moderne', in Ottmar Ette, Ute Hermanns, Bernd M. Scherer and Christian Suckow (eds), *Alexander von Humboldt – Aufbruch in die Moderne*, Berlin: Akademie, pp. 33–55.
— (2001b), *Literatur in Bewegung: Raum und Dynamik grenzüberschreitenden Schreibens in Europa und Amerika*, Weilerswist: Velbrück Wissenschaft.
— (2002), *Weltbewußtsein: Alexander von Humboldt und das unvollendete Projekt einer anderen Moderne*, Weilerswist: Velbrück Wissenschaft.
— (2009), *Alexander von Humboldt und die Globalisierung*, Frankfurt: Insel.
Fahnestock, Jeanne (1999), *Rhetorical Figures in Science*, Oxford: Oxford University Press.
Fara, Patricia (2004), *Pandora's Breeches: Women, Science and Power in the Enlightenment*, London: Pimlico.
Favret, Mary A. (1993), 'Spectatrice as Spectacle: Helen Maria Williams in the Revolution', *Studies in Romanticism*, 32.2: 273–95.
Fiedler, Horst and Ulrike Leitner (2000), *Alexander von Humboldts Schriften: Bibliographie der selbständig erschienenen Werke*, Berlin: Akademie.
Flint, Kate (1993), *The Woman Reader 1837–1914*, Oxford: Clarendon Press.
Flotow, Luise von (2011), 'Preface', in Luise von Flotow (ed.), *Translating Women*, Ottawa: University of Ottawa Press, pp. 1–10.
Forster, Georg (1777), *A Voyage Round the World, in His Britannic Majesty's Sloop, Resolution, Commanded by Capt. James Cook, During the Years 1772, 3, 4, and 5*, 2 vols, London: B. White, J. Robson and P. Elmsly.
Fosbrooke, Thomas Dudley (1822), *The Wye Tour, or Gilpin on the Wye, with Picturesque, Historical, and Archaeological Additions*, New Enlarged Edition, Ross: Farror, 1822.
Frank, Armin Paul and Brigitte Schultze (1988), 'Normen in historisch-deskriptiven Übersetzungsstudien', in Harald Kittel (ed.), *Die literarische Übersetzung: Stand und Perspektiven ihrer Erforschung*, Berlin: Erich Schmidt, pp. 96–121.
Frasca-Spada, Marina and Nick Jardine (2000), 'Introduction: Books and the Sciences', in Marina Frasca-Spada and Nick Jardine (eds), *Books and the Sciences in History*, Cambridge: Cambridge University Press, pp. 3–10.
Freddi, Maria, Barbara Korte and Josef Schmied (2013), 'Developments and Trends in the Rhetoric of Science', *European Journal of English Studies*, 17.3: 221–34.
Fulford, Tim, Debbie Lee and Peter J. Kitson (2004), *Literature, Science and Exploration in the Romantic Era: Bodies of Knowledge*, Cambridge: Cambridge University Press.

Fyfe, Aileen (2000), 'Young Readers and the Sciences', in Marina Frasca-Spada and Nick Jardine (eds), *Books and the Sciences in History*, Cambridge: Cambridge University Press, pp. 276–90.

—— (2008), 'Science and Religion in Popular Publishing in 19th-Century Britain', in Peter Meusburger, Michael Welker and Edgar Wunder (eds), *Clashes of Knowledge: Orthodoxies and Heterodoxies in Science and Religion*, Heidelberg: Springer, pp. 121–32.

—— and Bernard Lightman (2007), 'Science in the Marketplace: An Introduction', in Aileen Fyfe and Bernard Lightman (eds), *Science in the Marketplace: Nineteenth-Century Sites and Experiences*, Chicago: University of Chicago Press, pp. 1–19.

Gates, Barbara T. (1998), *Kindred Nature: Victorian and Edwardian Women Embrace the Living World*, Chicago: University of Chicago Press.

Gerbi, Antonello (1973), *The Dispute of the New World: The History of a Polemic, 1750–1900*, trans. Jeremy Moyle, Pittsburgh: University of Pittsburgh Press.

Godard, Barbara (1990), 'Theorizing Feminist Theory/Translation', in Susan Bassnett and André Lefevere (eds), *Translation: History and Culture*, London: Pinter, pp. 87–96.

Godwin, William (1993), *Political and Philosophical Writings of William Godwin. Vol. 5: Educational and Literary Writings*, ed. Pamela Clemit, London: Pickering.

Goethe, Johann Wolfgang von (1993), *Die letzten Jahre: Briefe, Tagebücher und Gespräche von 1823 bis zu Goethes Tod, Teil II: Vom Dornburger Aufenthalt 1828 bis zum Tode, Sämtliche Werke*, vol. 11, ed. Horst Fleig, Frankfurt am Main: Deutscher Klassiker.

Gosse, Edmund (1931), *Life and Letters of Sir Edmund Gosse*, ed. Evan Charteris, 6 vols, London: Heinemann.

Görbert, Johannes (2014), *Die Vertextung der Welt: Forschungsreisen als Literatur bei Georg Forster, Alexander von Humboldt und Adelbert von Chamisso*, Berlin: de Gruyter.

Graczyk, Annette (2004), *Das literarische Tableau zwischen Kunst und Wissenschaft*, Munich: Wilhelm Fink.

Grafton, Anthony (1997), *The Footnote: A Curious History*, London: Faber and Faber.

Grant, James (1871), *The Newspaper Press: Its Origin – Progress – and Present Position*, 2 vols, London: Tinsley.

Greppi, Claudio (2005), '"On the Spot": Traveling Artists and the Iconographic Inventory of the World, 1769–1859', in Felix Driver and Luciana Martins (eds), *Tropical Visions in an Age of Empire*, Chicago: University of Chicago Press, pp. 23–42.

Grésillon, Almuth (1994), *Eléments de critique génétique: Lire les manuscrits modernes*, Paris: Presses Universitaires de France.

Gross, Alan G. (2006), *Starring the Text: The Place of Rhetoric in Science Studies*, Carbondale: Southern Illinois University Press.

Grün, Wolf-Dieter (1983), 'Die englischen Übersetzungen von Alexander von Humboldts *Kosmos*', *Cosmographia Spiritualis: Festschrift für Hanno Beck*, Bonn: Hafiz, pp. 83–98.

Guillory, John (1993), *Cultural Capital: The Problem of Literary Canon Formation*, Chicago: University of Chicago Press.

Hamel, Jürgen and Klaus-Harro Tiemann (eds) (1993), *Über das Universum:*

Die Kosmosvorträge 1827/28 in der Berliner Singakademie, assisted by Martin Pape, Frankfurt am Main/Leipzig: Insel.
Haney, John Louis (1970), *Early Reviews of English Poets*, New York: Lennox Hill.
Harvey, Joy (1997), *'Almost a Man of Genius': Clémence Royer, Feminism and Nineteenth-Century France*, New Brunswick, NJ: Rutgers University Press.
Haynes, Kenneth (2006), 'Translation and British Literary Culture', in Peter France and Kenneth Haynes (eds), *The Oxford History of Literary Translation into English. Vol. 4: 1790–1900*, Oxford: Oxford University Press, pp. 3–19.
Helmreich, Christian (2009), 'Science et éloquence dans le *Voyage aux régions équinoxiales du Nouveau Continent* (1807–1838) d'Alexandre de Humboldt', *Cahiers de l'association internationale des études françaises*, 61: 293–308.
Hermans, Theo (1996), 'The Translator's Voice in Translated Narrative', *Target*, 8.1: 23–48.
— (1999), 'Translation and Normativity', in Christina Schäffner (ed.), *Translation and Norms*, Clevedon: Multilingual Matters, pp. 50–71.
— (2007), *The Conference of the Tongues*, Manchester: St Jerome.
— (2012), 'Translation as an Approach to History: Response', *Translation Studies*, 5.2: 242–5.
Herschel, John Frederick William (1831), *A Preliminary Discourse on the Study of Natural Philosophy*, new edn, London: Longman.
Hesse, Carl (2006), 'Towards a New Topography of Enlightenment', *European Review of History/Revue européenne d'histoire*, 13.3: 499–508.
Hey'l, Bettina (2007), *Das Ganze der Natur und die Differenzierung des Wissens: Alexander von Humboldt als Schriftsteller*, Berlin/New York: De Gruyter.
Higgitt, Rebekah and Charles W. J. Withers (2008), 'Science and Sociability: Women as Audience at the British Association for the Advancement of Science, 1831–1901', *Isis*, 99.1: 1–27.
Hilger, Stephanie (2009), *Women Write Back: Strategies of Response and the Dynamics of European Literary Culture, 1790–1805*, Amsterdam/New York: Rodopi.
Holmes, John and Sharon Ruston (eds) (2017), *The Routledge Companion to Nineteenth-Century British Literature and Science*, London: Routledge.
Hooker, Joseph Dalton (1918), *Life and Letters of Sir Joseph Dalton Hooker Based on Materials Collected and Arranged by Lady Hooker*, ed. Leonard Huxley, 2 vols, London: J. Murray.
Horner, Francis (1843), *Memoirs and Correspondence of Francis Horner, M.P.*, ed. Leonard Horner, 2 vols, London: John Murray.
Humboldt, Alexander von (1811a), *Essai politique sur le royaume de la Nouvelle-Espagne*, 2 vols, Paris: Schoell.
— (1811b), *Political Essay on the Kingdom of New Spain*, trans. John Black, 4 vols, London: Longman, Hurst, Rees, Orme and Brown.
— (1814), *Researches Concerning the Institutions and Ancient Monuments of the Ancient Inhabitants of America, with Descriptions and Views of Some of the Most Striking Scenes in the Cordilleras!*, trans. Helen Maria Williams, 2 vols, London: Longman, Hurst, Rees, Orme and Brown, J. Murray and H. Colburn.
— (1814–29), *Personal Narrative of Travels to the Equinoctial Regions of the New Continent, During the Years 1799–1804*, trans. Helen Maria Williams,

7 vols, London: Longman, Hurst, Rees, Orme, and Brown, J. Murray and H. Colburn.

— (1845–8), *ΚΟΣΜΟΣ: A General Survey of the Physical Phenomena of the Universe*, trans. by Augustin Prichard, M.D., M.R.C.S., 2 vols, London: Hippolyte Baillière.

— (1846–58), *Cosmos: Sketch of a Physical Description of the Universe*, trans. under the Superintendence of Lieut.-Col. [from 1858 'Major-General'] Edward Sabine, London: Longman, Brown, Green and Longmans; and John Murray.

— (1849–58), *Cosmos: A Sketch of a Physical Description of the Universe*, trans. E. C. Otté (and B. H. Paul [vol. 4] and W. S. Dallas [vol. 5]), 5 vols, London: Bohn.

— (1849), *Aspects of Nature, in Different Lands and Different Climates; with Scientific Elucidations*, trans. Mrs Sabine, 2 vols, London: Longman, Brown, Green and Longmans; and John Murray.

— (1850), *Views of Nature: Or Contemplations on the Sublime Phenomena of Creation*, trans. Elise C. Otté and Henry G. Bohn, London: Bohn.

— (1851), *Tableaux de la nature*, trans. C. Galusky, 2 vols, Paris: Baudry.

— (1852–3), *Personal Narrative of Travels to the Equinoctial Regions of America, During the Years 1799–1804*, trans. Thomasina Ross, 3 vols, London: Bohn.

— (1859–60), *Reise in die Aequinoctial-Gegenden des neuen Continents*, trans. Hermann Hauff, 4 vols, printed in 2, Stuttgart: Cotta.

— (1860), *Letters of Alexander von Humboldt, Written Between the Years 1827 and 1858, to Varnhagen von Ense. Together with Extracts from Varnhagen's Diaries, and Letters from Varnhagen and Others to Humboldt. Authorised Translation from the German, with Explanatory Notes and a Full Index of Names*, trans. Ludmilla Assing, London: Trübner.

— (1869), *Lettres d'Alexandre de Humboldt à Marc-Auguste Pictet 1798–1824, Publiées dans le Journal* Le Globe, ed. Albert Rilliet, Geneva: Carey Frères.

— (1908), *Correspondance d'Alexandre de Humboldt avec François Arago*, ed. Ernest-Théodore Hamy, Paris: Guilmoto.

— (1957), *Political Essay on the Kingdom of New Spain*, trans. and annot. Hensley C. Woodbridge, Lexington: University of Kentucky Library.

— (1970), *Relation historique du voyage aux régions équinoxiales du nouveau continent*, 3 vols [Paris: Dufour, 1814 ; Maze, 1819 ; Smith and Gide, 1825], repr. Stuttgart: Brockhaus, ed. Hanno Beck.

— (1988), *Political Essay on the Kingdom of New Spain: The John Black Translation [Abridged]*, ed. and intro. Mary Maples Dunn, Norman: University of Oklahoma Press.

— (1995), *Personal Narrative of a Journey to the Equinoctial Regions of the New Continent*, trans. Jason Wilson, London: Penguin Classics.

— (2004a), *Kosmos: Entwurf einer physischen Weltbeschreibung*, ed. Ottmar Ette and Oliver Lubrich, Frankfurt am Main: Eichborn.

— (2004b), *Alexander von Humboldt und die Vereinigten Staaten von Amerika: Briefwechsel*, ed. Ingo Schwarz, Berlin: Akademie.

— (2004c), *Ansichten der Natur, mit wissenschaftlichen Erläuterungen*, ed. Anette Selg, Frankfurt am Main: Eichborn.

— (2006), *Briefe von Alexander von Humboldt an Christian Carl Josias Bunsen*, ed. Ingo Schwarz, Berlin: Rohrwall.

— (2007) *Alexander von Humboldt – Samuel Heinrich Spiker: Briefwechsel*, ed. Ingo Schwarz, assisted by Eberhard Knobloch, Berlin: Akademie.
— (2009), *Alexander von Humboldt und Cotta: Briefwechsel*, ed. Ulrike Leitner, assisted by Eberhard Knobloch, Berlin: Akademie.
— (2010), *Briefwechsel Alexander von Humboldt und Carl Ritter*, ed. Ulrich Päßler, assisted by Eberhard Knobloch, Berlin: Akademie.
— (2011), *Alexander von Humboldt – August Böckh: Briefwechsel*, ed. Romy Werther, assisted by Eberhard Knobloch, Berlin: Akademie.
— (2014), *Views of Nature*, trans. Mark W. Person, Chicago: University of Chicago Press.
Hume, David (1748), *Essays, Moral and Political*, 3rd edn, London: Millar.
James, Frank A. J. L. (ed.) (1991–2011), *The Correspondence of Michael Faraday*, 6 vols, Stevenage: Institute of Electrical Engineers.
Janes, R. M. (1978), 'On the Reception of Mary Wollstonecraft's *A Vindication of the Rights of Woman*', *Journal of the History of Ideas*, 39: 293–302.
--, James A. Secord and Emma C. Spary (eds) (1996), *Cultures of Natural History*, Cambridge: Cambridge University Press.
Jenkins, Alice (2007a), *Space and the 'March of Mind': Literature and the Physical Sciences in Britain, 1815–1850*, Oxford: Oxford University Press.
— (2007b), 'Alexander von Humboldt's *Kosmos* and the Beginnings of Ecocriticism', *Interdisciplinary Studies in Literature and Environment*, 14.2: 89–105.
— (ed.) (2008), *Michael Faraday's Mental Exercises: An Artisan Essay-Circle in Regency London*, Liverpool: Liverpool University Press.
Jerdan, William (1852–3), *The Autobiography of William Jerdan, with his Literary, Political, and Social Reminiscences and Correspondence during the Last Fifty Years*, 4 vols, London: Arthur Hall.
Johnston, Judith (2013), *Victorian Women and the Economies of Travel, Translation and Culture, 1830–1870*, Abingdon: Routledge.
Jones, Stanley (1989), *Hazlitt: A Life from Winterslow to Frith Street*, Oxford: Clarendon Press.
Jones, Vivien (1992), 'Women Writing Revolution: Narratives of History and Sexuality in Wollstonecraft and Williams', in Stephen Copley and John Whale (eds), *Beyond Romanticism: New Approaches to Texts and Contexts 1780–1832*, London/New York: Routledge, pp. 178–99.
Kehlmann, Daniel (2005), *Die Vermessung der Welt*, Reinbek bei Hamburg: Rowohlt.
Keighren, Innes M., Charles W. J. Withers and Bill Bell (2015), *Travels into Print: Exploration, Writing and Publishing with John Murray, 1773–1859*, Chicago: University of Chicago Press.
Kelly, Gary (1993), *Women, Writing, and Revolution 1790–1827*, Oxford: Clarendon Press.
Kennedy, Deborah (1991), '"Storms of Sorrow": The Poetry of Helen Maria Williams', *Man and Nature: Proceedings of the Canadian Society for Eighteenth-Century Studies*, 10: 77–91.
— (2002), *Helen Maria Williams and the Age of Revolution*, Lewisburg, PA: Bucknell University Press.
Klancher, Jon (1987), *The Making of English Reading Audiences*, Madison: University of Wisconsin Press.
Knapp, Oswald G. (ed.) (2005), *The Intimate Letters of Piozzi and Pennington*, Stroud: Nonsuch.

Knight, David M. (2004), *Science and Spirituality: The Volatile Connection*, London: Routledge.
Korte, Barbara (2000), *English Travel Writing from Pilgrimages to Postcolonial Explorations*, trans. Catherine Matthias, Basingstoke: Macmillan.
Kuhn, Thomas S. [1979] (1993), 'Metaphor in Science', in Andrew Ortony (ed.), *Metaphor and Thought*, Cambridge: Cambridge University Press, pp. 533–42.
Kujamäki, Pekka (1997), 'Was ist ein Übersetzungsfehler? – Gefragt anhand mehrerer deutscher Übersetzungen eines finnischen Romans', in Eberhard Fleischmann, Wladimir Kutz and Peter A. Schmitt (eds), *Translationsdidaktik. Grundfragen der Übersetzungswissenschaft*, Tübingen: Gunter Narr, pp. 580–6.
Kutzinski, Vera M. (2009), 'Translations of Cuba: Fernando Ortiz, Alexander von Humboldt, and the Curious Case of John Sidney Thrasher', *Atlantic Studies*, 6.3: 303–26.
— (2010), 'Alexander von Humboldt's Transatlantic Personae', *Atlantic Studies*, 7.2: 100–12.
— (2012), 'Introduction: Alexander von Humboldt's Transatlantic Personae', in Vera Kutzinski (ed.), *Alexander von Humboldt's Transatlantic Personae*, London/New York: Routledge, pp. 1–14.
Landes, Joan B. (1988), *Women and the Public Sphere in the Age of Revolution*, Ithaca, NY: Cornell University Press.
Latour, Bruno (1987), *Science in Action: How to Follow Scientists and Engineers Through Society*, Cambridge, MA: Harvard University Press.
Leask, Nigel (1999), '"The Ghost in Chapultepec": Fanny Calderón de la Barca, William Prescott and Nineteenth-Century Mexican Travel Accounts', in Jás Elsner and Joan Pau Rubiés (eds), *Voyages and Visions: Towards a Cultural History of Travel*, London: Reaktion, pp. 184–209.
— (2001), 'Salons, Alps and Cordilleras: Helen Maria Williams, Alexander von Humboldt, and the discourse of Romantic travel', in Elizabeth Eger, Charlotte Grant, Clíona Ó Gallchoir and Penny Warburton (eds), *Women, Writing and the Public Sphere, 1700–1830*, Cambridge: Cambridge University Press, pp. 217–35
— (2002), *Curiosity and the Aesthetics of Travel Writing, 1770–1840: 'From an Antique Land'*, Oxford: Oxford University Press.
— (2003), 'Darwin's "Second Sun": Alexander von Humboldt and the Genesis of *The Voyage of the Beagle*', in Helen Small and Trudi Tate (eds), *Literature, Science, Psychoanalysis, 1830–1970: Essays in Honour of Gillian Beer*, Oxford: Oxford University Press, pp. 13–36.
Lefevere, André (1992), *Translation, Rewriting, and the Manipulation of Literary Fame*, London/New York: Routledge.
Leighton, Gerald R. (1901), *The Life-History of British Serpents and Their Local Distribution in the British Isles*, Edinburgh: Blackwood.
Leitner, Ulrike (1997), 'Die englischen Übersetzungen Humboldtscher Werke', *Acta Historica Leopoldina*, 27: 63–74.
Lessing, Gotthold Ephraim (1836), *Laocoon; or the Limits of Poetry and Painting*, trans. William Ross, London: J. Ridgway and Sons.
Levine, George (ed.) (1987), *One Culture: Essays in Science and Literature*, Madison: University of Wisconsin Press.
— (2011), *Darwin the Writer*, Oxford: Oxford University Press.
Lightman, Bernard (1997), '"The Voices of Nature": Popularizing Victorian

Science', in Bernard Lightman (ed.), *Victorian Science in Context*, Chicago: University of Chicago Press, pp. 187–211.
— (2015), 'Scientific Naturalists and Their Language Games', *History of Science*, 53.2: 395–416.
Logan, Deborah Anna (2007), *The Collected Letters of Harriet Martineau. Vol. 4: Letters, 1856–1862*, London: Pickering & Chatto.
Lohrli, Anne (1973), *Household Words: A Weekly Journal 1850–1859, Conducted by Charles Dickens*, Toronto: University of Toronto Press.
Lyell, Charles (1997), *Principles of Geology*, ed. and intro. James Secord, London: Penguin.
Lyell, Katherine (ed.) (1881), *Life, Letters and Journals of Sir Charles Lyell, Bart.*, 2 vols, London: John Murray.
McCrory, Donald (2010), *Nature's Interpreter: The Life and Times of Alexander von Humboldt*, Cambridge: Lutterworth Press.
McGann, Jerome J. (1992), *A Critique of Modern Textual Criticism*, Charlottesville: University Press of Virginia.
MacGillivray, William (1832), *The Travels and Researches of Alexander von Humboldt: Being a Condensed Narrative of his Journeys in the Equinoctial Regions of America, and in Asiatic Russia; Together with Analyses of his More Important Investigations*, Edinburgh/London: Oliver and Boyd/ Simpkin, Marshall and Co.
Maistre, Xavier de (1817), *The Leper of the City of Aoste*, trans. Helen Maria Williams, London: Cowie.
Martin, Alison E. (2006), 'Annotation and Authority: Georg Forster's Footnotes to the *Nachrichten von den Pelew-Inseln* (1789)', *Translation and Literature*, 15: 177–201.
— (2007), 'Die Rolle von Georg Forsters Übersetzungen in den intellektuellen Netzwerken seiner Zeit: Thomas Forrests *Voyage to New Guinea* (1779)', *Georg-Forster-Studien*, 12: 59–75.
Martineau, Harriet (1869), *Biographical Sketches*, New York: Leypoldt and Holt.
Martins, Luciana and Felix Driver (2005), '"The Struggle for Luxuriance": William Burchell Collects Tropical Nature', in Felix Driver and Luciana Martins (eds), *Tropical Visions in an Age of Empire*, Chicago: University of Chicago Press, pp. 59–74.
Mellor, Anne K. (2002), 'Mary Wollstonecraft's *A Vindication of the Rights of Woman* and the Women Writers of Her Day', in Claudia L. Johnson (ed.), *The Cambridge Companion to Mary Wollstonecraft*, Cambridge: Cambridge University Press, pp. 141–59.
Mill, John Stuart (1996), *The Collected Works of John Stuart Mill*, 33 vols, vol. 15, London: Routledge.
Milton, John and Paul Bandia (eds) (2009), *Agents of Translation*, Amsterdam: Benjamins.
Montgomery, Scott L. (2000), *Science in Translation: Movements of Knowledge through Cultures and Time*, Chicago: University of Chicago Press.
— (2010), 'Scientific Translation', in Yves Gambier and Luc van Doorslaer (eds), *Handbook of Translation Studies*, Amsterdam/Philadelphia: Benjamins, vol. 1, pp. 299–305.
Morgan, Lady Sydney (1830), *France in 1829–30*, 2 vols, London: Saunders and Otley.
Mortimer, Thomas (1801), *Lectures on the Elements of Commerce, Politics and*

Finances; Intended as a Companion to Blackstone's Commentaries on the Laws of England, London: Longman and Rees.

Mueller, Judith C. (2003), 'A Tale of a Tub and Early Prose', in *The Cambridge Companion to Jonathan Swift*, ed. Christopher Fox, Cambridge: Cambridge University Press.

Munday, Jeremy (2008), *Style and Ideology in Translation: Latin American Writing in English*, London/New York: Routledge.

Myers, Greg (1989), 'The Pragmatics of Politeness in Scientific Articles', *Applied Linguistics*, 10.1: 1–35.

Nangle, Benjamin Christie (1995), *The Monthly Review Second Series 1790–1815: Index of Contributors and Articles*, Oxford: Clarendon Press.

Neeley, Kathryn A. (2001), *Mary Somerville: Science, Illumination, and the Female Mind*, Cambridge: Cambridge University Press.

Nelson, Brian and Brigid Maher (2013), 'Introduction', in Brian Nelson and Brigid Maher (eds), *Perspectives on Literature and Translation: Creation, Circulation, Reception*, London: Routledge, pp. 1–10.

Nicolson, Malcolm (1990), 'Alexander von Humboldt and the Geography of Vegetation', in Andrew Cunningham and Nicholas Jardine (eds), *Romanticism and the Sciences*, Cambridge: Cambridge University Press, pp. 169–85.

Olohan, Maeve (2013), 'Gate-keeping and Localizing in Scientific Translation Publishing: The Case of Richard Taylor and *Scientific Memoirs*', *British Society for the History of Science*, 47.3: 433–50.

— (2014), 'History of Science and History of Translation: Disciplinary Commensurability?', *The Translator*, 20.1: 9–25.

Oppitz, Ulrich-Dieter (1969), 'Der Name der Brüder Humboldt in aller Welt', in Heinrich Pfeiffer (ed.), *Alexander von Humboldt: Werk und Werkgeltung*, Munich: Piper, pp. 277–430.

Orr, Mary (2015), 'The Stuff of Translation and Independent Female Scientific Authorship: The Case of *Taxidermy* . . . anon. (1820)', *Journal of Literature and Science*, 8.1: 27–47.

Ortiz, Fernando (2009), 'Humboldt's Translator in the Context of Cuban History', trans. Vera M. Kutzinski, *Atlantic Studies*, 6.3: 327–43.

O'Sullivan, Carol (2009), 'Translation Within the Margin: The "Libraries" of Henry Bohn', in John Milton and Paul Bandia (eds), *Agents of Translation*, Amsterdam/Philadelphia: Benjamins, pp. 107–29.

Otis, Laura (2002), 'Introduction', in Laura Otis (ed.), *Literature and Science in the Nineteenth Century*, Oxford. Oxford University Press, pp. xvii–xxviii.

Palmeri, Frank (1990), 'The Satiric Footnotes of Swift and Gibbon', *The Eighteenth Century*, 31: 245–62.

Paloposki, Outi (2017), 'Translators and the Task of Popular Education: Furthering Knowledge in Finland in the 19th Century', in Kristiina Taivalkoski-Shilov, Liisa Tiittula and Maarit Koponen (eds), *Communities in Translation and Interpreting*, Quebec: Vita Traductiva, pp. 64–85.

Pengelly, Hester (ed.) (1897), *A Memoir of William Pengelly, or Torquay, F.R.S., Geologist, with a Selection from his Correspondence*, London: Murray.

Phillips, Patricia (1990), *The Scientific Lady: A Social History of Women's Scientific Interests 1520–1918*, New York: St Martin's Press.

Pickford, Susan (2012), 'Writing with "Manly Vigour": Translatorial Agency in Two Early Nineteenth-Century English Translations of François Pouqueville's *Voyage en Morée, à Constantinople et en Albanie* (1805)', in Alison E. Martin and Susan Pickford (eds), *Travel Narratives in Translation, 1750–1830: Nationalism, Ideology, Gender*, London/New York: Routledge, pp. 197–217.

Pitman, Thea (2007), 'Mexican Travel Writing: The Legacy of Foreign Travel Writers in Mexico, or Why Mexicans Say They Don't Write Travel Books', *Comparative Critical Studies*, 4: 209–23.

Polwhele, Richard (1798), *The Unsex'd Females*, London: Cadell and Davies.

Pratt, Mary Louise (1992), *Imperial Eyes: Travel Writing and Transculturation*, London/New York: Routledge.

Prelli, Lawrence J. (1989), *A Rhetoric of Science: Inventing Scientific Discourse*, Columbia: University of South Carolina Press.

Prüfer Leske, Irene (2001), 'Übersetzungen, Manipulation und Neuübersetzung des *Essai politique sur l'île de Cuba* Alexander von Humboldts', in Ottmar Ette and Walther L. Bernecker (eds), *Ansichten Amerikas: Neuere Studien zu Alexander von Humboldt*, Frankfurt: Vervuert, pp. 219–30.

Pym, Horace N. (ed.) (1882), *Memories of Old Friends: Being Extracts from the Journals and Letters of Caroline Fox of Penjerrick, Cornwall, from 1835 to 1871*, 2 vols, Leipzig: Tauchnitz.

Quatrefages de Bréau, Jean Louis Armand de (1857), *The Rambles of a Naturalist on the Coasts of France, Spain, and Sicily*, trans. E. C. Otté, 2 vols, London: Longman, Brown, Green, Longmans & Roberts.

Raj, Kapil (2013), 'Beyond Postcolonialism . . . and Postpositivism: Circulation and the Global History of Science', *Isis*, 104: 337–47.

Rauch, Alan (2001), *Useful Knowledge: The Victorians, Morality, and the March of Intellect*, Durham, NC: Duke University Press.

Richardson, George Fleming (1855), *An Introduction to Geology, and its Associate Sciences Mineralogy, Fossil Botany, and Palæontology*, ed. Thomas Wright, London: Bohn.

Ross, Janet (1893), *Three Generations of English Women: Memoirs and Correspondence of Susannah Taylor, Sarah Austin, and Lady Duff Gordon*, London: Fisher Unwin.

Rousseau, George S. (2003), 'Science, Culture, and the Imagination: Enlightenment Configurations', in Roy Porter (ed.), *The Cambridge History of Science. Vol. 4: Eighteenth-Century Science*, Cambridge: Cambridge University Press, pp. 762–99.

Rupke, Nicolaas (2000), 'Translation Studies in the History of Science: The Example of *Vestiges*', *British Journal for the History of Science*, 33: 209–22.

— (2005), 'Alexander von Humboldt and Revolution: A Geography of Reception of the Varnhagen von Ense Correspondence', in David N. Livingstone and Charles W. J. Withers (eds), *Geography and Revolution*, Chicago: University of Chicago Press, pp. 336–50.

— (2007), *Alexander von Humboldt: A Metabiography*, Chicago: University of Chicago Press.

Ryals, Clyde De L. and Kenneth J. Fielding (1993), *The Collected Letters of Thomas Carlyle and Jane Welsh Carlyle. Vol. XXI: August 1846–June 1847*, Durham, NC: Duke University Press.

Sachs, Aaron (2006), *The Humboldt Current: Nineteenth-Century Exploration and the Roots of American Environmentalism*, New York: Viking.

Safier, Neil (2008), *Measuring the New World: Enlightenment Science and South America*, Chicago: University of Chicago Press.

Sandford, Margaret E. (1888), *Thomas Poole and his Friends*, 2 vols, London: Macmillan.

Schaffer, Simon, Lissa Roberts, Kapil Raj and James Delbourgo (eds) (2009), *The Brokered World: Go-Betweens and Global Intelligence, 1770–1820*, Sagamore Beach, MA: Watson Publishing International.

Schäffner, Christina (1999), 'The Concept of Norms in Translation Studies', in Christina Schäffner (ed.), *Translation and Norms*, Clevedon: Multilingual Matters, pp. 1–8.

Schiebinger, Londa (1987), 'The History and Philosophy of Women in Science: A Review Essay', *Signs*, 12.2: 305–32.

Schleiermacher, Friedrich (1838), *Sämmtliche Werke*, vol. 2, Berlin: Reimer.

Schuchard, Barbara (1998), 'Ausschweigen und Vermuten: Zu den deutschen Übersetzungen von Alexander von Humboldts *Relation historique*', in Martin Hummel and Christina Ossenkop (eds), *Lusitanica et Romanica*, Hamburg: Buske, pp. 211–25.

Schwarz, Ingo (2003), '"Ein beschränkter Verstandesmensch ohne Einbildungskraft": Anmerkungen zu Friedrich Schillers Urteil über Alexander von Humboldt', *Alexander von Humboldt im Netz*, IV.6, available at <http://www.hin-online.de/index.php/hin/article/view/38> (last accessed 12 May 2012).

Secord, James (2000), *Victorian Sensation: The Extraordinary Publication, Reception, and Secret Authorship of 'Vestiges of the History of Creation'*, Chicago: Chicago University Press.

— (2004), 'Knowledge in Transit', *Isis*, 95.4: 654–72.

— (2014), *Visions of Science: Books and Readers at the Dawn of the Victorian Age*, Oxford: Oxford University Press.

Seville, Catherine (1999), *Literary Copyright Reform in Early Victorian England*, Cambridge: Cambridge University Press.

Shapin, Steven (1985), *A Social History of Truth: Civility and Science in Seventeenth-Century England*, Chicago: University of Chicago Press.

Shattock, Joan (1989), *Politics and Reviewers: The* Edinburgh *and the* Quarterly *in the Victorian Age*, Leicester: Leicester University Press.

Shteir, Ann B. (1996), *Cultivating Women, Cultivating Science: Flora's Daughters and Botany, 1760–1860*, Baltimore: Johns Hopkins University Press.

— (2003), 'Finding Phebe: A Literary History of Women's Science Writing', in Katherine Binhammer and Jeanne Wood (eds), *Women and Literary History: 'For There She Was'*, Newark, DE: University of Delaware Press, pp. 152–66.

Shuttleworth, Sally (1981), *The Language of Science and Psychology in George Eliot's* Daniel Deronda, New York: New York Academy of Sciences.

Simeoni, Daniel (1998), 'The Pivotal Status of the Translator's Habitus', *Target*, 10.1: 1–39.

Simon, Sherry (1996), *Gender in Translation: Cultural Identity and the Politics of Transmission*, London: Routledge.

Slater, Michael (1983), *Dickens and Women*, London: Dent.

Sleigh, Charlotte (2011), *Literature and Science*, Basingstoke: Palgrave Macmillan.

Sonoda, Akiko (2007), 'Historical Overview of Formation of International Copyright Agreements in the Process of Development of International Copyright Law from the 1830s to 1960s', *Institute of Intellectual Property (IIP) Bulletin*, 22, 1–9.

Southey, Robert (1829), 'The State and Prospects of the Country', *Quarterly Review*, 39: 475–520.

— (1965), *New Letters of Robert Southey*, ed. Kenneth Curry, 2 vols, New York: Columbia University Press.
Sprat, Thomas (1667), *The History of the Royal-Society of London, For the Improving of Natural Knowledge*, London: Martyn.
Stark, Susanne (1999), *Behind Inverted Commas: Translation and Anglo-German Cultural Relations in the Nineteenth Century*, Clevedon: Multilingual Matters.
Stern, Madeleine B. (1980), 'The English Press in Paris and its Successors, 1793–1852', *Bibliographical Society of America*, Papers, 74: 307–59.
Strang, Harald (1992), *Einleitung – Fußnote – Kommentar: Fingierte Formen wissenschaftlicher Darstellung als Gestaltungselemente moderner Erzählkunst*, Bielefeld: Aisthesis.
Taylor, John (1824), *Selections from the Works of the Baron de Humboldt, Relating to the Climate, Inhabitants, Productions, and Mines of Mexico*, London: Longman, Hurst, Rees, Orme, Brown, and Green.
Théodoridès, Jean (1966), 'Humboldt and England', *British Journal for the History of Science*, 3.9: 39–55.
Thompson, Carl (2011), *Travel Writing*, Abingdon: Routledge.
Thwaite, Ann (1984), *Edmund Gosse: A Literary Landscape*, Oxford: Oxford University Press.
Ticknor, George (1876), *Life, Letters, and Journals of George Ticknor*, 2 vols, London: Sampson Low, Marston, Searle and Rivington.
Todd, Janet (1986), *Sensibility: An Introduction*, London: Methuen.
Toews, John Edward (2004), *Becoming Historical: Cultural Reformation and Public Memory in Early Nineteenth-Century Berlin*, Cambridge: Cambridge University Press.
Topham, Jonathan R. (2000), 'Scientific Publishing and the Reading of Science in Nineteenth-Century Britain: A Historiographical Survey and Guide to Sources', *Studies in History and Philosophy of Science Part A*, 31.4: 559–612
— (2004), 'Technicians of Print and the Making of Natural Knowledge', *Studies in History and Philosophy of Science*, 35.2: 391–400.
— (2009a), 'Rethinking the History of Science Popularization/Popular Science', in Faidra Papanelopoulou, Agustá Nieto-Galan and Enrique Perdiguero (eds), *Popularising Science and Technology in the European Periphery, 1800–2000*, Aldershot: Ashgate, pp. 1–20.
— (2009b), 'Scientific and Medical Books, 1780–1830', in Michael F. Suarez and Michael Turner (eds), *The Cambridge History of the Book in Britain*, 7 vols, Cambridge: Cambridge University Press, vol. 5, pp. 827–33.
Valle, Ellen (1999), *A Collective Intelligence: The Life Sciences in the Royal Society as a Scientific Discourse Community, 1665–1965*, Turku: University of Turku.
Venuti, Lawrence (2008), *The Translator's Invisibility: A History of Translation*, 2nd edn, London/New York: Routledge.
Walford, Edward (1856), *Hardwicke's Annual Biography for 1856*, London: Hardwicke.
Weedon, Alexis (2003), *Victorian Publishing: The Economics of Book Production for a Mass Market, 1836–1916*, Aldershot: Ashgate.
Weigl, Engelhard (1990), *Instrumente der Neuzeit: Die Entdeckung der modernen Wirklichkeit*, Stuttgart: Metzler.
Werner, Petra (2004), *Himmel und Erde: Alexander von Humboldt und sein Kosmos*, Berlin: Akademie.

Williams, Helen Maria (1803), *Political and Confidential Correspondence of Lewis the Sixteenth; with Observations on Each Letter by Helen Maria Williams*, 3 vols, London: G. & J. Robinson.
— [1790] (1995), *Julia, A Novel; Interspersed with some Poetical Pieces*, ed. Peter Garside, 2 vols, London: Routledge/Thoemmes.
Wilson, Leonard (1972), *Charles Lyell*, New Haven, CT: Yale University Press.
Withers, Charles W. K. (2007), *Placing the Enlightenment: Thinking Geographically about the Age of Reason*, Chicago: University of Chicago Press.
— and Miles Ogborn (eds) (2010), *Geographies of the Book*, Aldershot: Ashgate, 2010.
Wolf, Michaela and Alexandra Fukari (eds) (2007), *Constructing a Sociology of Translation*, Amsterdam/Philadelphia: Benjamins.
Woodmansee, Martha (1994), *The Author, Art, and the Market: Rereading the History of Aesthetics*, New York: Columbia University Press.
Woodress, James (1958), *A Yankee's Odyssey: The Life of Joel Barlow*, Philadelphia: Lippincott.
Woodward, Lionel D. (1930), *Une Anglaise amie de la Révolution Française: Hélène-Maria Williams et ses amis*, Paris: Honoré Champion.
Wordsworth, William (2007), *Lyrical Ballads*, ed. Michael Mason, 2nd edn, Harlow: Longman.
Wulf, Andrea (2015), *The Invention of Nature*, London: Murray.
Wyatt-Edgell, Edgell (1838), 'Moral Statistics of the Parishes of St. James, St. George, and St. Anne Soho, in the City of Westminster', *Journal of the Statistical Society of London*, 1: 478–92.
Yeo, Richard (2003), *Defining Science: William Whewell, Natural Knowledge and Public Debate in Early Victorian Britain*, Cambridge: Cambridge University Press.

Unpublished Theses

Healy, Michèle, *The Cachet of the 'Invisible' Translator: Englishwomen Translating Science (1650–1850)*, University of Ottawa, 2004.

Internet Sources

Darwin Correspondence Project, <https://www.darwinproject.ac.uk/>
Ette, Ottmar (2004), 'Die Ordnung der Weltkulturen. Alexander von Humboldts Ansichten der Kultur', *Alexander von Humboldt im Netz*, V.9 <https://www.uni-potsdam.de/romanistik/hin/hin9/ette.htm> (last accessed 7 October 2017).
Lindquist, Jason H. (2004), '"Under the Influence of an Exotic Nature ... National Remembrances are Insensibly Effaced": Threats to the European Subject in Humboldt's *Personal Narrative of Travels to the Equinoctial Regions of the New Continent*', *Humboldt im Netz* (HiN), V.9 <http://www.uni-potsdam.de/u/romanistik/humboldt/hin/hin9/lindquist.htm> (last accessed 20 July 2017).
Oxford English Dictionary <http://www.oed.com/> (last accessed 31 January 2018).

Index

Note: Page numbers in *italics* refer to figures. Page number followed by an n indicate end-of-chapter notes.

Abir-Am, Pnina, 11
Account of the Pelew Islands Situated in the Western Part of the Pacific Ocean (Keate), 55
Aikin, Lucy, 10, 69
Alexander von Humboldt: A Metabiography (Rupke), 7–8
Alexander von Humboldt's Transatlantic Personae (Kutzinski), 17
Allen, Esther, 56
Allgemeine Theorie des Erdmagnetismus (Gauss), 162
Altick, Richard, 19, 120
American indigenous peoples, 44–5, 62–3, 108, 143–5
American reception of Humboldt's work, 17, 21, 71, 241
Americas, 41–2, 44–5; *see also Essai politique sur le royaume de la Nouvelle-Espagne* (Humboldt)
annotations *see* footnotes; paratext
Ansichten der Natur (Humboldt, 1808), 7, 150
  Cotta as publisher of, 150, 154, 155, 158
  critics, 151–2
  scientific content, 158–9
  spirituality, 182, 183
  style, 33, 153–8, 174–9, 184
  translation, 20, 159–60, 237; *see also Aspects of Nature* (Humboldt, trans. Sabine); *Views of Nature* (Humboldt, trans. Otté, 1850)
Ansichten vom Niederrhein (Forster), 163

anti-imperialist stance, 45, 63–4
Arago, François, 19, 141, 158, 162, 164, 219
Art Journal, 189–90
Aspects of Nature (Humboldt, trans. Sabine), 7, 151, 160–1, 163–5
  critics, 153, 182
  Longman and Murray as publishers of, 151, 162, 163
  presentation and paratext, 173–4
  religion, 182–4
  style, 173, 175–82, 184
  see also Ansichten der Natur (Humboldt); *Views of Nature* (Humboldt, trans. Otté)
Athenæum, 117, 151, 152, 166, 168, 189, 202, 224
Atkinson, Dwight, 132
audience, 66–7, 68–9
Augustan Review, 77
Austin, Sarah, 12, 18, 159–60, 200

Babbage, Charles, 5, 47
Bachleitner, Norbert, 125–6
Baillie Warden, David, 93–4, 239
Baillière, Hippolyte, 7, 187, 189–90, 193, 197, 202, 203–5, 208
Baker, Mona, 36, 39
banana plant, 98–9, *100*
Barker, Anna, 84
Barrow, John, 50, 51, 76, 77, 80, 104–5, 110
Beck, Hanno, 44
Bell, Bill, 151
Bentham, Jeremy, 47
Bentley, Richard, 127, 232n

Bentley's Miscellany, 127
Berlin, 7, 29, 30, 37, 153, 190, 191, 195, 198–9, 205, 240
Berman, Antoine, 129
Bernadin de Saint-Pierre, Jacques-Henri, 50, 83–4
Biermann, Kurt-Reinhardt, 104
Black, John, 40, 237, 240
  professional background, 46–8
  recognition, 73n
  as translator, 47–8
  visibility in translation, 48–9
  see also Political Essay on the Kingdom of New Spain (Humboldt, trans. Black)
Blätter für literarische Unterhaltung, 147
Boase-Beier, Jean, 35–6
Böckh, August, 222
Bohls, Elizabeth, 78
Bohn (publisher)
  and the book trade, 16, 120–2
  Library Series, 119–22
  translators working for, 125
  see also Cosmos (Humboldt, trans. Otté); *Personal Narrative of Travels to the Equinoctial Regions of America* (Humboldt, trans. Ross); *Views of Nature* (Humboldt, trans. Otté)
Bohn, Henry, 119–20
  as translator, 120–1, 174–9
Bonpland, Aimé, 1, 31, 42, 43, 75, 80, 83, 87, 94, 96, 101, 110, 133, 158
book trade in Britain, 16, 19, 150; *see also* Baillère, Hippolyte; Bohn; Longman; Murray
book trade in Germany, 196; *see also* Cotta
Boswell, James, 76–7, 83
Bowen, Margarita, 206
Boyle, Robert, 132
Bravo, Michael T., 133–4
Bret, Patrice, 24
Brewster, David, 153, 167, 168, 182, 471
British Association for the Advancement of Science (BAAS), 11, 14, 198
Brock, Claire, 113
Brown, Robert, 18, 201
Brownlie, Siobhan, 129–30
Bunsen, Christian Carl Josias, 147, 192, 194, 200–1, 222

Burt, Thomas, 13
Buscapié (Cervantes), 127, 128
Buschmann, Eduard, 222, 225
Buzelin, Hélène, 92

Canary Isles, 105–8, 108–9, 136, 140
Carlyle, Thomas, 47, 217
Cervantes, Miguel de, 127
Chamberlain, Lori, 12
Chambers, Robert, 218
Coleridge, Samuel Taylor, 15
collaborative translation work, 78, 91–101, 174–9, 192–3, 199–202, 225–6
colonialism, 7, 45, 72, 99, 118, 127, 143–5, 225
comets, 208
commerce, 41, 46, 59–60, 69, 71
Cook, James, 55, 64, 65, 83, 93, 104, 106
copyright, 120, 170, 195–7, 205
Cordasco, Francesco, 120, 122
Cosmos (Humboldt, trans. Otté), 13, 14, 33, 206–8, 241
  Bohn as publisher of, 187–90
  collaborative work, 225–6
  comparison with Sabine's translation, 208–13
  critics, 197
  popular knowledge, 15
  price, 16, 189–90
  religion, 221, 222
  see also Cosmos (Humboldt, trans. Sabine); *Kosmos* (Humboldt)
Cosmos (Humboldt, trans. Sabine), 13, 14, 187, 188–9, 199–200, 201–2, 204–6, 213–14, 220, 223, 224–5, 227
  collaborative work, 199–202
  comparison with Otté's translation, 208–13
  critics, 193
  Longman and Murray as publishers of, 187–90, 197, 199, 204, 223–5
  omissions, 221
  religion, 216, 219–22
  scientific terminology, 213–15
  see also Cosmos (Humboldt, trans. Otté); *Kosmos* (Humboldt)
Cotta, Johan Friedrich, 43, 70, 86, 93, 145, 150, 154–5, 159, 160–1, 189, 190–2, 195
Critical Review, 40
Cronin, Michael, 118

Crosse, Andrew, 192, 218–19
Curiosity and the Aesthetic of Travel Writing, 1770–1840 (Leask), 3
Cuvier, Georges, 80, 140–1

Daily News, 14, 18, 117, 119, 152, 165
Dallas, William Sweetland, 225, 226
Darwin, Charles, 1, 5–6, 11, 14, 22–3, 31–2, 118, 132, 202, 204, 215, 218, 226
Dassow Walls, Laura, 3, 83, 110, 192, 193, 223–4
Daston, Lorraine, 54, 62, 136, 138
Day, George Edward, 166, 167, 207
De Pauw, Cornelius, 44–5, 62
De Quincey, Thomas, 26–7
Deane-Cox, Sharon, 130
Delabastita, Dirk, 160, 209
Delamétherie, Jean-Claude, 41–2
Derrida, Jacques, 56
Dettelbach, Michael, 30, 83, 96
Dickens, Charles, 14, 124, 125, 127–8
dictionaries, 236–7
direct speech, 32, 113, 209–11
Distribution of Heat over the Surface of the Globe (Dove, trans. Sabine), 162, 198
Dove, Heinrich Wilhelm, 162, 198, 223
Driver, Felix, 108
Dunciad Variorum (Pope), 53–4

East Germany, 8
Eclectic Review, 102, 217
Edinburgh, 16, 47, 48, 96, 207, 239
Edinburgh Review, 76, 77, 85, 102, 193, 218
El Buscapié (Cervantes), 127, 128
Ellegård, Alvar, 14
Elshakry, Marwa, 5
endnotes, 74n, 173
English Press, 87
Erman, Georg Adolf, 196
Essai politique sur le royaume de la Nouvelle-Espagne (Humboldt), 7, 9, 40, 43–6
  origins, 41
  second edition, 70
  technical vocabulary, 49
  see also Political Essay on the Kingdom of New Spain (Humboldt, trans. Black)
Ette, Ottmar, 3, 6, 9, 21n, 30, 33, 73n, 88, 146, 158
Examiner, 117, 119, 151, 189, 224–5

experimentation in science, 132, 136, 137, 138, 148, 180

Fahnestock, Jeanne, 64
Fara, Patricia, 10
Faraday, Michael, 29, 199
Favret, Mary A., 80
Fiedler, Horst, 8
Flotow, Luise von, 105
footnotes, 53–7
  Cosmos (Humboldt, trans. Otté), 207–8
  Cosmos (Humboldt, trans. Sabine), 213
  Kosmos (Humboldt), 195
  Political Essay on the Kingdom of New Spain (Humboldt, trans. Black), 42, 43, 52–3, 56–7, 58, 59–69, 72–3
  Views of Nature (Humboldt, trans. Otté), 173, 174
  see also paratext
Forbes, James D., 189, 192, 217
Forster, Georg, 4, 33–4, 55, 163
Fox, Caroline, 163
Frasca-Spada, Marina, 234
Freddi, Maria, 23
French language, 6, 27, 32
French translations of Humboldt's work, 205, 214
Fucus vitifolius, 97–8, 98
Fyfe, Aileen, 16, 112, 183

Galusky, Charles, 159, 185, 205
Gates, Barbara, 11
Gauss, Carl Friedrich, 162, 225
gender *see* women; women translators
General Theory of Terrestrial Magnetism (Gauss, trans. Sabine), 162
genetic criticism (*critique génétique*), 92
gentlemanly behaviour in science, 62, 132, 141
'Geographical-Political Tables of the Kingdom of New Spain' (Humboldt), 41
Gerbi, Antonello, 44, 62
German language, 6, 27, 237
German translations of Humboldt's works, 146–7
Gilpin, William, 175
Glasgow, 93, 124, 239
Gmelin, Ferdinand Gottlob, 146
Godwin, William, 26–7, 69

Goethe, Johann Wolfgang von, 30, 122, 151, 154
Görbert, Johannes, 3–4
Gosse, Edmund, 167–8
Gosse, Philip Henry, 167–8
Graczyk, Annette, 155, 156
Grafton, Anthony, 53, 54, 62
Grant, James, 47
Greek literature, 66–8
Grésillon, Almuth, 92
Gumilla, José, 62–3

habitus, 24
Harvey, Joy, 11
Hauff, Herman, 146–7
Haynes, Kenneth, 120
Healy, Michèle, 11, 162–3, 240
Helmreich, Christian, 31
Hermans, Theo, 37, 38, 56, 235
Herschel, John, 25–6, 27, 161, 193, 213–15, 220
Hey'l, Bettina, 154, 163
Higgett, Rebekah, 11
Holland, Henry, 189
Homer, 67
Hooker, Joseph Dalton, 1, 202, 204
Horner, Francis, 85–6
Horner, Joanna, 122
Horner, Leonora, 122
Household Words (Dickens), 128
Howitt, Mary, 121
Humboldt, Alexander von, 1
  changes in name, 103
  connections in British society, 17–20
  correspondence with John Hurford Stone, 86–7
  critical reception of his work, 14, 40, 77, 117, 152–3, 187, 189–90, 197, 202, 204
  English translations of major works, 2, 6–7, 8–10; see also *individual titles*
  financial difficulties, 87–8
  footnotes, 54–5
  foreign language proficiency, 5
  Helen Maria Williams's salon, 81–2
  involvement in *Aspects of Nature* (trans. Sabine), 159, 163–5
  involvement in *Cosmos* (trans. Sabine), 205, 220
  involvement in *Personal Narrative* (trans. Williams), 91–3, 94–101, 97, *100*
  on *Kosmos*, 22, 191–3
  on *Kosmos* translations, 190, 198, 203–4
  modern translations, 233
  public lectures, 29–30
  recognition of Elizabeth Sabine, 199, 223
  relationship between Helen Maria Williams and, 88–9, 90–1
  style, 3–4, 30–1, 32–5, 157
  used to support ideologies, 8
Humboldt, Wilhelm von, 8, 18, 157, 173, 193, 221
Humboldt Current, The (Sachs), 17
Humboldt's Statistical Essay on New Spain Abridged (Anon.), 70–1
Hume, David, 44
Hurford Stone, John, 77, 86–7, 93, 240

Illustrirte Zeitung, 147
imagery, 179, 181, 209–10
Imperial Eyes: Travel Writing and Transculturation (Pratt), 9
Imprimerie Anglaise, 87
indigenous Americans, 44–5, 62–3, 108, 143–5
instruments and instrumentation, 96, 101, 133–5, 136
International Copyright Act 1838, 196
Introduction to Geology, An (Richardson), 122
invisibility *see* visibility in translation

Jameson, Robert, 95
Jardine, Nicholas, 6, 234
Jenkins, Alice, 29, 212
Jerdan, William, 126
Johnston, Judith, 12
Jones, Vivien, 76
Julia, A Novel (Williams), 82–3

Keate, George, 55
Keighren, Innes M., 151
Kelly, Gary, 82, 84, 103
Kennedy, Deborah, 83
Knight, David, 152, 218
knowledge, popularisation of, 13–17, 29, 218, 222, 240
Korte, Barbara, 23
ΚΟΣΜΟΣ (Humboldt, trans. Prichard), 187, 190, 202–4, 208, 241
Kosmos (Humboldt), 7, 187, 191–5, 237
  copyright, 195–7
  Cotta as publisher of, 191

critics, 189
final volumes, 222–5
literary references, 34, 193
metaphors, 209
popular knowledge, 15
reference to *Natural History of Man* (Cowles Prichard), 198
religion, 192, 215–22
style, 22
used to support ideologies, 8
see also Cosmos (Humboldt, trans. Otté); *Cosmos* (Humboldt, trans. Sabine); *ΚΟΣΜΟΣ* (Humboldt, trans. Prichard)
Kuhn, Thomas, 209
Kujamäki, Pekka, 235–6
Kulturnation, 8, 200
Kutzinski, Vera, 9, 17, 21n

Landes, Joan B., 79–80
landscape aesthetics, 105–6, 175
Lanzarote, 106
Laplace, Pierre Simon, 11–12, 211
Latour, Bruno, 133
Le Lépreux de la cité d'Aoste (Maistre), 86
Leask, Nigel, 3, 32, 49, 55, 78, 79, 143
Lebenskraft, 157
lectures, 29–30, 153, 191
Lefevere, André, 4
Leisure Hour, 18
Leitner, Ulrike, 8, 78
Leper of the City of Aoste, The (Maistre, trans. Williams), 86
Letters from France (Williams), 76–7
Letters on the Events Which Have Passed in France Since the Restoration in 1815 (Williams), 86
libraries, 13–14
Library of Useful Knowledge, 14
Liebig, Justus von, 225
Lightman, Bernard, 5, 16, 24
literal translation, 57–9
Literary Gazette, 102, 118, 126, 151, 187, 188, 197, 224
Literary Panorama, 40–1
Literatur in Bewegung [Literature in Motion] (Ette), 3
literature, 3, 28–9, 34–5, 66–8, 193
travel literature, 3, 12–13, 50–1, 96, 151
see also Williams, Helen Maria: as writer
Literature and Science (Sleigh), 3

Lloyd, Hannibal, 126
London, 13, 16, 18, 19, 47, 66, 89–90, 95, 117, 120, 161, 195, 198, 200, 205, 239
Longman (publisher), 46, 69, 190; *see also Aspects of Nature* (Humboldt, trans. Sabine); *Cosmos* (Humboldt, trans. Sabine); *Personal Narrative of Travels to the Equinoctial Regions of the New Continent* (Humboldt, trans. Williams)
Louis XVI, 85–6
Lovelace, Ada, 11
Love's Labour's Lost (Shakespeare), 193
Lyell, Charles, 6, 17, 28–9, 35, 112, 114, 118, 122, 207, 229

McCrory, Donald, 184
McGann, Jerome, 92
MacGillivray, William, 111–13
Madgett, Nicholas, 85
Maher, Brigid, 4–5
Maistre, Xavier de, 86
Mantell, Gideon, 122, 207
Marcet, Jane Haldimand, 11
Martineau, Harriet, 215–16, 221
Martins, Luciana, 108
Marxism, 8
Measuring the World (Safier), 17
Mellor, Anne K., 69
Memoirs of the Author of Vindication of the Rights of Woman (Goodwin), 69
metaphors, 108, 179, 181, 209
Mexico, 41–2; *see also Essai politique sur le royaume de la Nouvelle-Espagne* (Humboldt)
Mill, John Stuart, 47
missionaries, 63
modern translations of Humboldt's works, 233
Montgomery, Scott, 4
Monthly Review, 40, 41, 76, 77–8
Moore, Thomas, 81
Morning Chronicle, 47
Mortimer, Thomas, 60
Munday, Jeremy, 9, 36, 37–8
Murchison, Roderick Impey, 226
Murray (publisher)150, 151, 159, 187–90, 197–9, 204–5, 240; *see also Aspects of Nature* (Humboldt, trans. Sabine); *Cosmos* (Humboldt, trans. Sabine)

Nachrichten von den Pelew-Inseln in der Westgegend des stillen Oceans (Keate, trans. Forster), 55
Nappi, Carla, 5
Narrative of an Expedition to the Polar Sea (Wangell, trans. Sabine), 161
National Socialism, 8
nationhood, 211–12
Natural History of Man (Cowles Prichard), 197–8
Naturphilosophie, 154
Neeley, Kathryn A., 11–12
Nelson, Brian, 4–5
New Universal Magazine, 40
Nicolson, Malcolm, 154
North British Review, 153, 168, 182, 218

Of National Character (Hume), 44
Olohan, Maeve, 10, 49
'On the Present State of the English Language' (De Quincey), 26–7
Orr, Mary, 12
O'Sullivan, Carol, 122
Otis, Laura, 28
Otté, Elise C., 7, 121, 237, 240
   photograph of, *169*
   professional background, 166–9
   scientific interests, 166–8
   as translator, 174–85
   see also Cosmos (Humboldt, trans. Otté); *Views of Nature* (Humboldt, trans. Otté)
Outram, Dorinda, 11

package psychology, 120
Paloposki, Outi, 235
paratext, 38, 173–4, 227, 233, 235; *see also* footnotes
Paris, 7, 77, 79–82, 87–88, 95–96, 161, 191, 219, 240
Paul, Benjamin Horatio, 225–6
Paul et Virginie (Bernadin de Saint-Pierre), 83–4
Pengelly, William, 167
personal narrative, 136–7, 141–3, 208
Personal Narrative of Travels to the Equinoctial Regions of America (Humboldt, trans. Ross), 19, 129–31, 147–8, 237
   Bohn as publisher of, 117, 118
   front matter, *123*
   omissions, 131–2, 133, 135, 136–7, 137–8, 139, 140–1

personal narrative, 142–3
racial Otherness, 143–5
reception, 117–19
see also Relation historique du voyage aux régions équinoxiales du nouveau continent (Humboldt)
Personal Narrative of Travels to the Equinoctial Regions of the New Continent (Humboldt, trans. Williams), 1–2, 7, 75, 77–9, 118, 216, 237
   added value of translation, 113–14
   comparison with Ross's translation, 130–1
   correspondence between Humboldt and Hurford Stone, 86–7
   critics, 102, 216
   draft version ('*avant-texte*'), 92–101
   final version, 102–11: main text, 105–11; translator's preface, 103–5
   Humboldt's feedback, 91–3, 94–101, *97, 100*
   instrumentation, 134, *135*
   Longman as publisher of, 75, 88–90, 111
   observation, 136–7
   personal narrative, 142
   racial Otherness, 143–4
   scientific community, 140, *141*
   scientific shortcomings, 138, 139–40
   witnessing, 137
   see also Longman; *Personal Narrative of Travels to the Equinoctial Regions of America* (Humboldt, trans. Ross); *Relation historique du voyage aux régions équinoxiales du nouveau continent* (Humboldt); *Travels and Researches of Alexander von Humboldt* (MacGillivray)
Phillips, Patricia, 11
Pickford, Susan, 12–13
Pictet, Marc-Auguste, 18, 43, 45, 49, 58, 61, 145, 219
Pinkerton, John, 95
Pitman, Thea, 41
Plantes équinoxiales (Humboldt and Bonpland), 97–8, *98*
politeness in scientific discourse, 132–3, 140–1, 149n
Political and Confidential Correspondence of Lewis the Sixteenth (Williams), 85–6

Political Essay on the Kingdom of New Spain (Humboldt, trans. Black), 7, 40
  colonialism, 45
  criticism of translation, 40–1
  footnotes, 42, 43, 52–3, 56–7, 58, 59–69, 67, 72–3
  later editions, 69–70
  literal translation, 57–9
  Longman as publisher of, 69
  'Preface by the translator,' 50–3
  reproductions, 70–2
  technical vocabulary, 49
  see also Longman; *Essai politique sur le royaume de la Nouvelle-Espagne* (Humboldt)
Polwhele, Richard, 76
Poole, Thomas, 80
Pope, Alexander, 53–4
popular science, 13–17, 23–4, 29–30, 207–8; *see also* audience; public lectures
Pratt, Marie Louise, 9
precision in science, 133–4
Preliminary Discourse on the Study of Natural Philosophy (Herschel), 25–6
Prelli, Lawrence J., 23
Prescott, William Hickling, 166, 169, 207
Prichard, Augustin, 7, 187, 190, 197, 198, 208, 240–1
Prichard, James Cowles, 197–8
Prince Albert, 218
Principles of Geology (Lyell), 28–9
public lectures, 29–30, 153, 191
public libraries, 13–14
Punch, 19, 120–1

quarterlies, 14
Quarterly Review, 76, 104, 192, 217
Quatrefages de Bréau, Jean Louis Armand de, 168
queer theory, 8

racial Otherness, 143–5
Raj, Kapil, 6
Rambles of a Naturalist on the Coasts of France, Spain and Sicily (Quatrefages de Bréau, trans. Sabine), 168
Rauch, Alan, 13
Raynal, Guillaume Thomas François (Abbé Raynal), 44

reference works, 236
Reflections on the Decline of Science in England (Babbage), 5
Reise in die Aequinoctial-Gegenden des neuen Continents (Humboldt, trans. Hauff), 146–7
Relation historique du voyage aux régions équinoxiales du nouveau continent (Humboldt), 1, 43–4, 75
  abridged editions, 145–7
  modesty, 139
  personal narrative, 141–2
  scientific community, 140
  style, 31
  translations, 7, 9, 78, 237; *see also Personal Narrative of Travels to the Equinoctial Regions of America* (Humboldt, trans. Ross); *Personal Narrative of Travels to the Equinoctial Regions of the New Continent* (Humboldt, trans. Williams); *Reise in die Aequinoctial-Gegenden des neuen Continents* (trans. Hauff)
religion, 63, 182–4, 192, 214–22
Researches Concerning the Institutions and Monuments of the Ancient Inhabitants of America (Humboldt, trans. Williams), 75, 92, 101
retranslation hypothesis, 129
rhetoric, 23–4; *see also* style
Rhodian Genius, 157–8
Richardson, George Fleming, 122
Ritter, Carl, 159, 211
Ross, Charles (brother of Thomasina Ross), 124–5
Ross, Georgina (sister of Thomasina Ross), 125
Ross, James Clark, 199–200, 201–2
Ross, Thomasina, 7, 19, 117, 237
  family background, 124–5
  as translator, 125–9
  year of birth, 149n
  see also Personal Narrative of Travels to the Equinoctial Regions of America (Humboldt, trans. Ross)
Ross, William (father of Thomasina Ross), 124
Ross, William junior (brother of Thomasina Ross), 124
Royal Society, 54, 76, 95, 161–2, 165, 198, 225
Royer, Clémence, 11

Rupke, Nicolaas A., 4, 7–8
Ruskin, John, 217

Sabine, Edward
   on *Ansichten der Natur*, 163, 164, 165
   on *Cosmos* (Humboldt, trans. Sabine), 187, 195, 197, 199–200, 201–2, 205–6, 213–14, 220, 223, 224–5, 227
   research expeditions, 161–2
Sabine, Elizabeth, 7
   Rambles of a Naturalist on the Coasts of France, Spain and Sicily (Quatrefages de Bréau, trans. Sabine), 168
   scientific content, 159
   as translator, 161–3, 198–9, 223–4, 228
   visibility in translation, 187, 240
   see also Aspects of Nature (Humboldt, trans. Sabine); *Cosmos* (Humboldt, trans. Sabine)
Sachs, Aaron, 17, 184
Safier, Neil, 17
St Andrews, 166, 168, 239
St Andrews Literary and Philosophical Society, 167
salon culture, 79–82
Santa Cruz, 136
Schäffner, Christina, 36
Schiebinger, Londa, 11
Schiller, Friedrich, 157–8
Schleiermacher, Friedrich, 37
Schmied, Josef, 23
science
   gentlemanly behaviour in, 62, 132, 141
   inclusiveness in, 234
   modesty in, 132–3, 139–40
   observation in, 135–8
   'polite' discourses in, 132–3, 140–1, 149n
   popularisation of, 13–17, 23–4, 29–30, 207–8; *see also* audience; public lectures
   and religion, 183, 197–8, 214–22
   role of women in, 10–11, 234–5
   and young readers, 112–3, 146
Science in Translation (Montgomery), 4
scientific community, 132–3, 140–1, 192–3, 199–200, 201–2
scientific development, 2–3
scientific discourse, 132–3, 140–1, 149n

scientific terminology, 57–9, 61, 93, 158, 213–15; *see also* technical vocabulary
scientific writing
   literary dimension of, 3–4, 34–5, 114
   literary references in, 28–9
   translation of, 4–5; *see also* Humboldt, Alexander von: English translations of major works
   see also scientific discourse; style
Scotland, 65–6, 239–40
Secord, James, 6, 28, 118, 238
Selections from the Works of the Baron de Humboldt (Taylor), 71–2
sensibility, 31, 82, 85, 108, 114, 131
Seville, Catherine, 196
Shakespeare, William, 34–5, 193
Shteir, Ann B., 11
Simeoni, Daniel, 24
Simon, Sherry, 84
Sleigh, Charlotte, 3, 23
snob appeal, 120
Somerville, Mary, 11–12, 18, 130
South America, 109–10
Southey, Robert, 1, 15
Souvenirs d'un naturaliste (Quatrefages de Bréau), 168
Spanish colonies, 109–10
Spary, Emma, 6
Spenersche Zeitung, 30
Spiker, Samuel Heinrich, 229
spirituality, 182–4, 214; *see also* religion
Sprat, Thomas, 25
Stark, Susanne, 12, 200
style, 22–7, 238
   Alexander von Humboldt, 3–4, 30–1, 32–5
   Ansichten der Natur (Humboldt), 33, 153–8, 174–9, 184
   Aspects of Nature (Humboldt, trans. Sabine), 173, 175–82, 184
   Charles Darwin, 31–2
   developments in, 25–7, 132–3
   Kosmos (Humboldt), 22
   literary references, 28–9, 34
   plain style, 27
   Relation historique du voyage aux régions équinoxiales du nouveau continent (Humboldt), 31
   translation of, 9, 35–8, 39
   Views of Nature (Humboldt, trans. Otté), 175–82, 184
   see also personal narrative;

scientific writing; stylistic choices, translators
Style and Ideology in Translation (Munday), 36
Stylistic Approaches to Translation (Boase-Beier), 35–6
stylistic choices, translators, 2, 85, 175–82, 184, 208–13
Swanwick, Anna, 122, 169, 201
Swift, Jonathan, 53
symbolic capital, 28

'Tablas Geográfico-Políticas del Reino de la Nueva-España' (Humboldt), 41
Tale of a Tub, A (Swift), 53
Taylor, John, 71–2
technical vocabulary, 49, 71, 95–6; *see also* scientific terminology
Tenerife, 107–8, 108–9, 136
thesauri, 236
Thompson, Carl, 12
Thoreau, Henry David, 184
Thrasher, John Sidney, 9
Ticknor, George, 80–1
Times, 18, 124
Topham, Jonathan, 10, 15–16, 17
Torquay, 166–9
Traité de mécanique céleste (Laplace), 11–12
Translating Women (Flotow), 105
translation
  and abridgement, 131–43, 147
  and competition, 7, 9, 38, 161, 173–9, 179–84, 185, 187–91, 195–7, 209–13, 224–5
  copyright law, 196–7
  footnotes, 55–7
  illusion of transparency, 48
  literal, 57–9
  metaphor in, 108, 179, 181, 209
  and modernisation, 130
  and norms, 5, 24, 36–7, 237–8
  and pressure of time, 89, 170, 205–6, 223
  rates of payment for, 89–90, 205, 224, 232n
  register in, 210–11
  role of women in, 11–13, 105, 121–2, 125–6, 240–1
  of scientific writing, 4–5; *see also* Humboldt, Alexander von: English translations of major works
  as a social practice, 10
  sociological turn in, 10
  visibility in, 48–9, 113–14, 125, 162–3, 174, 187, 202–3, 207–8, 240; *see also* footnotes
voice, 56
  of writing style, 35–8, 39
translation process, 92
translation theory, 37, 129–30
translational habitus, 24
translational intertextuality, 160
translators
  and accountability, 64, 72
  for Bohn libraries, 121–2
  critics, 237–8
  information needs, 235–7
  role of, 2, 233–5, 235
  stylistic choices, 24, 85, 175–82, 184, 208–13
  see also Black, John; Bohn, Henry; Dallas, William Sweetland; Otté, Elise C.; Paul, Benjamin Horatio; Prichard, Augustin; Ross, Thomasina; Sabine, Elizabeth; Williams, Helen Maria
travel writing, 3, 12–13, 46, 50–1, 96, 151
Travels and Researches of Alexander von Humboldt (MacGillivray), 111–13
Tschudi, Johann Jakob von, 126–7

Ulloa, Antonio de, 62–3
Unsex'd Females, The (Polwhele), 76
Usteri, Paul, 146

Valle, Ellen, 132–3
Varnhagen von Ense, Karl, 22, 156, 229, 236
Venuti, Lawrence, 48
Verbreitung der Wärme auf der Oberfläche der Erde, Die (Dove), 162, 198
Vestiges of the Natural History of Creation (Chambers), 218–19
Views from the Lower Rhine (Forster), 163
Views of Nature (Humboldt, trans. Otté), 7, 151
  Bohn as publisher of, 151, 169–73
  footnotes, 173, 174
  illustrations, *152*
  letter from Humboldt to Bohn, *171*
  presentation and paratext, 173–4
  religion, 182–4

Views of Nature (cont.)
  style, 175–82, 184
  title page, 164–5, 172
  see also Ansichten der Natur
    (Humboldt); *Aspects of Nature*
    (Humboldt, trans. Sabine)
Vindication of the Rights of Woman
  (Wollstonecraft), 69
visibility in translation, 48–9, 113–14,
  125, 162–3, 174, 187, 202–3,
  207–8, 240; *see also* footnotes
volcanoes, 157
Voyage Round the World, A (Forster),
  34, 55
*Vues des Cordillères et monumens
  des peuples indigènes de
  l'Amérique* (Humboldt) *see
  Researches Concerning the
  Institutions and Monuments of the
  Ancient Inhabitants of America*
  (Humboldt, trans. Williams)

Weltbewußtsein [World-Consciousness]
  (Ette), 3
Werner, Petra, 195
Wesleyan-Methodist Magazine,
  216
Westminster Review, 192, 218
Williams, Helen Maria, 7, 75
  critics, 76–8
  engraving, 77
  financial difficulties, 87–8

  relationship between Humboldt and,
    88–9, 90–1
  relationship between Longman and,
    75, 89–90, 110
  salon, 79–82
  as translator, 83–6
  as writer, 82–3
  *see also Personal Narrative of Travels
    to the Equinoctial Regions of the
    New Continent* (Humboldt, trans.
    Williams); *Researches Concerning
    the Institutions and Monuments of
    the Ancient Inhabitants of America*
    (Humboldt, trans. Williams)
Wilson, James, 78
Wimmer, Gottlob August, 146
Withers, Charles, 11, 151, 239
witnessing, 136–7, 208
Wollstonecraft, Mary, 69
Wolzogen, Caroline von, 80
women
  as audience, 68–9
  in science, 10–11, 234–5
women translators, 11–13, 105, 121–2,
  125–6, 240–1
Wrangell, Ferdinand von, 161
Wright, Thomas, 122
writing *see* scientific writing
writing style *see* style
Wulf, Andrea, 18

Yeo, Richard, 23

EU representative:
Easy Access System Europe
Mustamäe tee 50, 10621 Tallinn, Estonia
Gpsr.requests@easproject.com

www.ingramcontent.com/pod-product-compliance
Lightning Source LLC
Chambersburg PA
CBHW071813300426
44116CB00009B/1294